An Introduction to
Behavioural Ecology

An Introduction to Behavioural Ecology

J.R. Krebs FRS

Royal Society Research Professor
at the Edward Grey Institute of Field Ornithology
Department of Zoology
University of Oxford
and Fellow of Pembroke College

N.B. Davies

Reader in Zoology
at the University of Cambridge
and Fellow of Pembroke College

Drawings by Jan Parr

THIRD EDITION

Blackwell
Science

© 1981, 1987, 1993 by
Blackwell Science Ltd
Editorial Offices:
Osney Mead, Oxford OX2 0EL
25 John Street, London WC1N 2BL
23 Ainslie Place, Edinburgh EH3 6AJ
350 Main Street, Malden
  MA 02148 5018, USA
54 University Street, Carlton
  Victoria 3053, Australia
10, rue Casimir Delavigne
  75006 Paris, France

Other Editorial Offices:
Blackwell Wissenschafts-Verlag GmbH
Kurfürstendamm 57
10707 Berlin, Germany

Blackwell Science KK
MG Kodenmacho Building
7-10 Kodenmacho Nihombashi
Chuo-ku, Tokyo 104, Japan

First published 1981
Reprinted 1982 (twice), 1983, 1985
Second edition 1987
Reprinted 1989, 1990, 1991 (twice)
Third edition 1993
Reprinted 1993, 1994 (twice), 1995,
1996, 1998

German translation 1984
Japanese translation 1984
Italian translation 1987
Hungarian translation 1988

Set by Setrite Typesetters, Hong Kong,
printed and bound in Great Britain
at The Alden Press, Oxford

The Blackwell Science logo is a
trade mark of Blackwell Science Ltd,
registered at the United Kingdom
Trade Marks Registry

DISTRIBUTORS

  Marston Book Services Ltd
  PO Box 269
  Abingdon, Oxon OX14 4YN
  (*Orders*: Tel: 01235 465500
          Fax: 01235 465555)

USA
  Blackwell Science, Inc.
  Commerce Place
  350 Main Street
  Malden, MA 02148 5018
  (*Orders*: Tel: 800 759 6102
          781 388 8250
        Fax: 718 388 8255)

Canada
  Login Brothers Book Company
  324 Saulteaux Crescent
  Winnipeg, Manitoba R3J 3T2
  (*Orders*: Tel: 204 224-4068)

Australia
  Blackwell Science Pty Ltd
  54 University Street
  Carlton, Victoria 3053
  (*Orders*: Tel: 3 9347 0300
          Fax: 3 9349 5001)

A catalogue record for this title
is available from the British Library

ISBN 0-632-03546-3

Library of Congress
Cataloging-in-Publication Data

Krebs, J.R. (John R.)
  An introduction to behavioural
  ecology/J.R. Krebs, N.B. Davies:
  drawings by Jan Parr. – 3rd 3d.
      p.   cm.
    Includes bibliographical references
    (p.   ) and indexes.
    ISBN 0-632-03546-3
    1. Animal behaviour.
    2. Behaviour evolution.
    3. Animal ecology.
  I. Davies, N.B. (Nicholas B.),
  1952– II Title.
  QL751.K815 1993   92–39740
  591.51 – dc20   CIP

Contents

Acknowledgements, xii

Introduction, 1

1 **Natural Selection, Ecology and Behaviour, 4**
  Questions about behaviour, 4
    Reproductive behaviour in lions, 5
  Natural selection, 8
  Genes and behaviour, 10
    (a) Use of genetic mutants, 10
    (b) Artificial selection experiments, 11
    (c) Studying populations with genetic differences, 11
  Selfish individuals or group advantage?, 14
    Theoretical objections, 15
    Empirical studies, 16
  Behaviour, ecology and evolution, 21
  Summary, 22
  Further reading, 22
  Topics for discussion, 23

2 **Testing Hypotheses in Behavioural Ecology, 24**
  The comparative approach, 25
    Social organization in weaver birds, 25
    Social organization in African ungulates, 28
  Adaptation or story-telling?, 29
    Alternative hypotheses, 29
    Cause and effect, 29
    Confounding variables, 30
    Alternative adaptive peaks or non-adaptive differences, 30
  Primate social organization, 31
    Home range size, 35
    Sexual dimorphism in body weight, 37
    Sexual dimorphism in tooth size, 38
    Testis size and breeding system, 40
  The comparative approach reviewed, 41
  Experimental studies of adaptation, 42
    Optimality models, 44
  Gathering food, 44
    Crows and whelks, 44
  Summary, 45
  Further reading, 47
  Topics for discussion, 47

3 **Economic Decisions and the Individual, 48**
  The economics of carrying a load, 48
    Starlings, 48
    Bees, 53

The economics of prey choice, 59
Sampling and information, 62
The risk of starvation, 63
Variability in searching or handling time, 64
Environmental variability, body reserves and food-storing, 65
Feeding and danger: a trade-off, 66
Nutrient constraints: herbivores and plants, 70
Optimality models and behaviour: an overview, 73
Summary, 74
Further reading, 76
Topics for discussion, 76

4   **Predators versus Prey: Evolutionary Arms Races, 77**
Predators versus cryptic prey, 79
  Testing functional hypotheses about adaptation, 79
  Does even crude crypsis confer an advantage?, 84
The advantage and evolution of warning coloration, 86
  Evolution of warning coloration, 88
The trade-off between conspicuousness and crypsis, 90
Predator–prey arms races, 92
Brood parasites and their hosts, 93
  Cuckoos versus hosts, 93
  Evolutionary equilibrium or continuing arms race?, 96
    (a) Continuing arms race, 96
    (b) Evolutionary equilibrium, 98
  Conclusion, 99
Summary, 100
Further reading, 100
Topics for discussion, 101

5   **Competing for Resources, 102**
Competition by exploitation: the ideal free distribution, 102
Competition by resource defence: the despotic distribution, 105
The ideal free distribution with unequal competitors, 106
The economics of resource defence, 110
    (a) Economic defendability, 110
    (b) Optimal territory size, 113
    (c) Graphical models of optimal territory size: a caveat, 114
  Shared resource defence, 114
  Interspecific territoriality, 117
Summary, 118
Further reading, 118
Topics for discussion, 119

6   **Living in Groups, 120**
Living in groups and avoiding predation, 120
  Increased vigilance, 120
  Dilution and cover, 123
  Group defence, 126
  Costs of being in a group, 126
Living in groups and getting food, 126

Finding good sites, 126
Catching difficult prey, 130
Harvesting renewing food, 130
Costs associated with feeding, 131
Weighing up costs and benefits — optimal group size, 133
Comparative studies, 134
Coloniality in cliff swallows: a case study, 135
Time budgets, 138
Are groups of optimal size stable?, 143
Individual differences in a group, 144
Evolution of group living: schooling in guppies, 144
Summary, 146
Further reading, 146
Topics for discussion, 146

7   **Fighting and Assessment, 147**
The war of attrition, or 'the waiting game', 147
(a) The problem, 147
(b) The theory, 148
(c) Testing the theory, 150
The evolution of conventional fighting, 150
Hawks and Doves, 151
Hawk, Dove and Bourgeois, 154
Simple models and reality, 155
Examples of animal contests, 156
(a) Serious fights, 156
(b) Respect for ownership, 157
(c) The influence of resource value, 159
(d) Contests of strength, 160
Sequential assessment, 164
Contests with differences in both resource value and fighting ability, 164
Theory: the asymmetric war of attrition, 165
Data: contests between male spiders, 167
(a) Measuring resource value, V, 167
(b) Measuring the costs of a fight, K, 168
(c) Results of staged encounters, 168
Badges of status, 171
Summary, 173
Further reading, 173
Topics for discussion, 174

8   **Sexual Conflict and Sexual Selection, 175**
Males and females, 175
Females as a scarce resource, 176
The sex ratio, 177
(a) Local mate competition, 180
(b) Local resource competition or enhancement, 181
(c) Maternal condition, 181
(d) Population sex ratio, 182
Sexual selection, 183

Ardent males, 183
Reluctant females, 186
(a) Non-genetic benefits: good resources and parental ability, 187
(b) Genetic benefits, 188
Elaborate ornaments: Fisher's hypothesis and the handicap
hypothesis, 190
(a) Examples of female preference for elaborate male displays, 190
(b) Fisher's hypothesis, 192
(c) The handicap hypothesis, 194
(d) Evidence for the Fisher and handicap hypotheses, 196
Male investment, 197
Sexual conflict, 202
(a) Mating decisions, 202
(b) Parental investment, 204
(c) Infanticide, 204
(d) Multiple matings, 204
The significance of courtship, 205
Summary, 206
Further reading, 206
Topics for discussion, 207

9 **Parental Care and Mating Systems, 208**
Proximate constraints on parental care, 209
Birds, 209
Mammals, 210
Fish, 210
Hypothesis 1. Paternity certainty, 211
Hypothesis 2. Order of gamete release, 211
Hypothesis 3. Association, 212
Ancestral origin of uniparental care, 212
An ESS model of parental investment, 213
Mating systems with no male parental care, 214
(a) Grey-sided voles, *Clethrionomys rufocanus*, 215
(b) Blue-headed wrasse, *Thalassoma bifasciatum*, 216
Variation in mating systems, 216
A comparative survey of mammalian mating systems, 217
(a) Daily female movements predictable, 219
(b) Daily female movements not predictable, 219
Leks and choruses, 220
Mating systems with male parental care, 225
Monogamy, 225
Extra-pair matings and intra-specific brood parasitism, 226
Polygyny, 230
(a) No cost of polygyny to females, 231
(b) Cost of polygyny to females, 232
Sexual conflict and polygamy, 234
The pied flycatcher, *Ficedula hypoleuca*, 235
The dunnock, *Prunella modularis*, 237
Female desertion and sex role reversal, 238
Ecology and dispersal, 239
Consequences of sex differences in dispersal, 241

Conclusion, 241
Summary, 242
Further reading, 243
Topics for discussion, 243

10 Alternative Breeding Strategies, 244
Hypotheses for the occurrence of alternative strategies within a
    species, 244
  Changing environment, 245
  Making the best of a bad job, 246
  Alternative strategies in evolutionary equilibrium, 249
    (a) Salmon: hooknoses and jacks, 249
    (b) Figwasps: fighters and dispersers, 252
    (c) Nesting strategies in female digger wasps: diggers and
       enterers, 254
Problems of measuring costs and benefits of alternative strategies, 256
Sex change as an alternative strategy, 260
  Changing from female to male, 260
  Changing from male to female, 262
  Sex change versus sneaking, 263
Summary, 263
Further reading, 264
Topics for discussion, 264

11 On Selfishness and Altruism, 265
Kin selection, 265
  Examples of altruism between relatives, 269
    (a) Co-operation and alarm calls in ground squirrels and prairie
       dogs, 269
    (b) Wife sharing in the Tasmanian native hen, 272
  How do individuals recognize kin?, 275
  Conclusions about kin selection, 277
Mutualism, 278
Manipulation, 279
Reciprocity, 280
  The Prisoner's dilemma, 280
    Predictions of the model, 283
  Examples of reciprocity, 285
    (a) Spawning in the black hamlet fish, 285
    (b) Regurgitation of blood by vampire bats, 285
    (c) Alliances in primates, 287
Summary, 289
Further reading, 290
Topics for discussion, 290

**12 Co-operation and Helping in Birds, Mammals
and Fish, 291**
Genetic predispositions and ecological constraints, 291
An example of helping in birds — the Florida scrub jay, 293
    (a) Breeders benefit from the presence of helpers, 294
    (b) Habitat saturation is an ecological constraint, 295

(c) Males benefit by inheriting a breeding space, 296
The scrub jay pattern in other species, 299
  Do helpers really help? Experimental evidence, 302
  Experimental evidence for breeding constraints, 302
Helpers are not always relatives, 304
  (a) The dwarf mongoose: unrelated helpers and
      pseudopregnancies, 304
  (b) Anemone fish, 305
  (c) Pied kingfisher: primary and secondary helpers, 306
An alternative hypothesis for the evolution of helping, 308
Conflict in breeding groups, 310
  (a) Groove-billed anis, 311
  (b) Acorn woodpeckers, 312
Division of labour and specialized helpers, 314
Conclusions, 315
Summary, 316
Further reading, 316
Topics for discussion, 317

13 Altruism in the Social Insects, 318
The social insects, 318
  The problem, 318
  The definition of 'social insect', 318
The life cycle and natural history of a social insect, 320
How eusociality evolved: two pathways, 322
  Hypothesis 1. Staying at home to help, 322
    (a) Ecological constraints, 322
    (b) Genetic predispositions, 324
  Hypothesis 2. Sharing a nest, 325
    (a) Ecological constraints, 326
    (b) Genetic predispositions, 326
Haplodiploidy and altruism, 328
  Conflict between workers and queen, 332
  Conflict over the sex ratio, 332
  Tests of worker−queen conflict, 334
Haplodiploidy and the origin of eusociality, 338
Parental manipulation or daughter advantage, 343
The importance of demography, 344
Comparison of vertebrates and insects, 344
Summary, 347
Further reading, 347
Topics for discussion, 348

14 The Design of Signals: Ecology and Evolution, 349
Ecological constraints and communication, 350
  Communication in ants, 351
  Bird and primate calls, 352
Reactors and the design of signals, 358
  How signals originate, 358
  How signals are modified during evolution: ritualization, 361
  Hypotheses for the evolution of signal design, 363

(a) Reduction of ambiguity, 363
(b) Manipulation, 365
(c) Honesty, 367
Evidence from present-day signals, 367
An example of bluff or manipulative signalling: mantis shrimps, 367
Honest signalling: Thompson's gazelles, 368
Receiver psychology: Tungara frogs and swordfish, 368
Variability of signals and information, 369
Signalling, manipulation, and the animal mind, 371
Summary, 373
Further reading, 374
Topics for discussion, 374

15 Conclusion, 375
How plausible are our main premises?, 375
Selfish genes, 375
Group selection, 376
Optimality models and ESSs, 378
Causal and functional explanations, 382
A final comment, 384
Summary, 386
Further reading, 386

References, 387

Author index, 407

Subject index, 412

Colour plates face p. 212

Acknowledgements

1st Edition

This book is based on lectures given by us at Cambridge and Oxford Universities and we thank our students for providing a stimulating and critical audience.

We thank especially Tim Birkhead for reading the first draft of the whole manuscript and the following for comments on particular chapters: Anthony Arak, Patrick Bateson, Jane Brockmann, Tim Clutton-Brock, William Foster, Peter de Groot, Paul Harvey and Geoff Parker.

For sending manuscripts and allowing us to quote unpublished work we thank: Jeffrey Baylis, Lew Oring, Richard Wrangham, Robert Hinde, Dan Rubenstein, Peter de Groot, Uli Reyer, Bob Metcalf, Ron Ydenberg, Ric Charnov, Haven Wiley, Clive Catchpole and Malte Andersson.

Finally we thank Robert Campbell for his encouragement and enthusiasm during the preparation of this book.

2nd Edition

We thank Tim Clutton-Brock, William Foster, Paul Harvey, Nadav Nur and Alan Grafen for their valuable comments. Finally we thank Sara Trevitt and Robert Campbell of Blackwell Scientific Publications for their help in the preparation of the manuscript.

3rd Edition

We thank Dale Clayton, Tim Clutton-Brock, William Foster, Anne Houde and Anne Magurran for advice, and Susan Sternberg of Blackwell Scientific Publications for her gentle persuasion.

Introduction

This brief introduction describes the organization and contents of our book. The book is about the survival value of behaviour. We call this subject 'behavioural ecology' because the way in which behaviour contributes to survival and reproduction depends on ecology. If, for example, we want to answer the question 'How does living in a group contribute to an individual's survival?', we have to start thinking in terms of the animal's ecology; the kind of food it eats, its enemies, its nesting requirements and so on. These ecological pressures will determine whether grouping is favoured or penalized by selection. Behavioural ecology is not only concerned with the animal's struggle to survive by exploiting resources and avoiding predators, but also with how behaviour contributes to reproductive success. Much of the book is therefore about competition between individuals for the chance to reproduce and pass on their genes to future generations.

The book emphasizes the theoretical background to each subject discussed, but we prefer to illustrate the theory with examples after a very brief general introduction, rather than developing long, abstract, theoretical arguments. Although none of the ideas we have used are difficult to understand we have placed some of the more complicated arguments and details in boxes which can be ignored if the reader is in a hurry.

Chapter 1 is a general introduction to the book, in which we distinguish between different kinds of questions that one can ask about behaviour. In particular we emphasize the difference between questions about survival value or function and those concerned with causal mechanisms. We show that natural selection should favour individuals who are best able to propagate their genes to future generations.

In Chapter 2 we discuss how to test hypotheses for the adaptive advantage of behaviour. One method is comparison among species. The rationale here is that differences between species in behaviour can be correlated with differences in their ecology. From these correlations, inferences can be drawn about the adaptive significance of behavioural traits. We illustrate this approach with reference to social organization in weaver birds, antelope and primates. The second method is to perform experiments, for example to change behaviour and measure the consequences this has for the individual's chances of survival and for its reproductive success.

In Chapter 3 we focus on the individual. Animals are viewed

as making 'decisions' between alternative courses of action and the decisions can be analysed in terms of their costs and benefits. A powerful tool in this approach is optimality theory, which allows us to test hypotheses about the importance of various costs and benefits by predicting their effects on the animal's decision rules. By considering the basic decisions underlying behaviour patterns, we show how the same models can be used to understand what at first sight seem very different problems, such as feeding and searching for mates. Chapter 4 looks at decisions over evolutionary time, and how these change during arms races between predators and prey, and brood parasites and hosts.

In the next three chapters we consider how individuals should behave when they have to compete with others for scarce resources such as food, territories or mates. We discuss how competitors should be distributed in relation to resource distribution and abundance (Chapter 5) and the costs and benefits of living in groups (Chapter 6). In Chapter 7 we introduce the idea of game theory as a technique for analysing how individuals behave in contests for resources.

Chapters 8, 9 and 10 are concerned with sexual reproduction. A consideration of the basic differences between males and females leads to the idea that members of one sex (usually male) may compete for access to members of the other (Chapter 8). This is the theory of sexual selection. The differences between male and female also suggest that the interests of the two sexes during reproduction often differ (the theory of sexual conflict). Chapter 9 discusses how these battles within and between the sexes are influenced by ecology. Here we rely heavily on the comparative approach, correlating differences between species in sexual strategies with differences in ecology. From differences between species we turn to differences between individuals (Chapter 10). We introduce the idea that different individuals within a species sometimes adopt different sexual strategies. These differences may be related to age or size, or they may simply be equally profitable, alternative, ways of achieving the same end.

In Chapter 11 we examine how altruistic behaviour can evolve. We then illustrate the theoretical arguments with reference to 'helpers'; individuals that help others to rear young instead of producing their own. Chapter 12 deals with birds, mammals and fish, and Chapter 13 is devoted entirely to the social insects, where helping reaches its most sophisticated level of development. Many of the earlier chapters refer to communication as a behavioural mechanism of competition for resources and social interaction. In Chapter 14 we tie together these threads in a general discussion of animal signals. We follow the pattern set in earlier chapters by considering both ecological constraints and intraspecific selection pressures. In the final chapter we reassess the view that the survival value of behaviour can be understood within a neo-Darwinian framework using methods such as optimality models and game theory.

Finally a word about the style of presentation. We generally use convenient and informal shorthand rather than traditional formal scientific style. A phrase such as 'Offspring are selected to demand more food than the parent wants to give' is short for 'During the course of evolution selection acting on genetic differences in the begging behaviour of offspring will have favoured an increase in the intensity of begging. This increase will have been favoured to the extent where the level of begging by any individual offspring exceeds the optimum level for the parent'.

Some readers may wonder whether our informal shorthand, together with catchy descriptive labels for various behaviour patterns such as 'manipulation' and 'sneaker', are a sign of sloppy thinking. There is no doubt that loose terminology can indicate imprecise thinking and half-formulated ideas. But it is equally easy to conceal woolly arguments behind an obfuscating screen of scientific jargon. We have used a simple direct style in order to make our arguments clear and not because behavioural ecology is woolly subject. This point is nowhere better illustrated than by George Orwell in his brilliant essay 'Politics and the English Language' (1946). He translates the following well-known verse from *Ecclesiastes* into modern English: 'I returned and saw under the sun, that the race is not always to the swift, nor the battle to the strong, neither yet bread to the wise, nor yet riches to men of understanding, nor yet favour to men of skill; but time and chance happeneth to them all.'

And now the translation. 'Objective consideration of contemporary phenomena compels the conclusion that success or failure in competitive activities exhibits no tendency to be commensurate with innate capacity, but that a considerable element of the unpredictable must invariably be taken into account.'

This translation is not only tired and ugly, lacking the fresh, vivid imagery of the biblical passage, but it replaces precise illustrations with woolly generalization. While we cannot hope to emulate the clarity and brilliance of the writer of *Ecclesiastes*, or indeed of George Orwell, we hope we have avoided the worst excesses of the Orwellian parody and presented our ideas in simple but precise language.

Chapter 1. Natural Selection, Ecology and Behaviour

Questions about behaviour

In this book we will explore the relationships between animal behaviour, ecology and evolution. We shall describe how animals behave under particular ecological conditions and then ask 'Why has this behaviour evolved?' For example, we shall attempt to understand why some animals are solitary while others go around in groups and why most individuals court before they copulate. Why do some birds have songs consisting of pure whistles while others produce buzzes and trills? We shall also ask some precise, quantitative questions such as why do great tits lay clutches of 8 eggs and why does the male dungfly copulate for on average 41 min?

Niko Tinbergen, one of the founders of ethology, emphasized that there are several different ways of answering the question 'Why?' in biology. These have come to be known as Tinbergen's four questions (Tinbergen 1963). For example, if we asked why starlings, *Sturnus vulgaris*, sing in the spring, we could answer as follows.

1 In terms of *survival value or function*. Starlings sing to attract mates for breeding.
2 In terms of *causation*. Because increasing daylength triggers off changes in hormone levels in the body, or because of the way air flows through the syrinx and sets up membrane vibrations. These are answers about the internal and external factors which cause starlings to sing.
3 In terms of *development*. Starlings sing because they have learned the songs from their parents and neighbours.
4 In terms of *evolutionary history*. This answer would be about how song had evolved in starlings from their avian ancestors. The most primitive living birds make very simple sounds, so it is reasonable to assume that the complex songs of starlings and other song birds have evolved from simpler ancestral calls.

It is important to distinguish these various kinds of answer or otherwise time can be wasted in sterile debate. If someone said that swallows migrate south in the autumn because they are searching for richer food supplies while someone else said they migrated because of decreasing daylength, it would be pointless to argue about who was correct. Both answers may be right: the first is in terms of survival value or function and the second is in terms of causation. Factors influencing survival value are some times called 'ultimate' while causal factors are referred to as 'proximate'. It is these two answers that are the most frequently muddled up and so to make the distinction clear we will discuss an example in detail.

REPRODUCTIVE BEHAVIOUR IN LIONS

In the Serengeti National Park, Tanzania, lions (*Panthera leo*) live in prides consisting of between 3 and 12 adult females, from 1 to 6 adult males and several cubs. The group defends a territory in which it hunts for prey, especially gazelle and zebra. Within a pride all the females are related; they are sisters, mothers and daughters, cousins, and so on. All were born and reared in the pride and all stay there to breed. Females reproduce from the age of 4 to 18 years and so enjoy a long reproductive life.

For the males, life is very different. When they are 3 years old, young related males (sometimes brothers) leave their natal pride. After a couple of years as nomads they attempt to take over another pride from old and weak males. After a successful take-over they stay in the pride for 2 to 3 years before they, in turn, are driven out by new males. A male's reproductive life is therefore short.

The lion pride thus consists of a permanent group of closely related females and a smaller group of separately interrelated males present for a shorter time. We will consider three interesting observations about reproductive behaviour in a pride (Bertram 1975).

1 Lions may breed throughout the year but although different prides may breed at different times, within a pride all the females tend to come into oestrus at about the same time. The mechanism, or causal explanation, may be the influence of an individual's pheromones on the oestrus cycles of other females in the pride. A similar phenomenon occurs in schools, where girls living in the same dormitory may also synchronize their menstrual cycles, perhaps due to the effect of pheromones (McClintock 1971).

The function of oestrus synchrony in lionesses is that different litters in the pride are born at the same time and cubs born synchronously survive better. This is because there is communal suckling and with all the females lactating together a cub may suckle from another female if its mother is out hunting (Fig. 1.1). In addition, with synchronous births there is a greater chance that a young male will have a companion when it reaches the age at which it leaves the pride. With a companion a male is more likely to achieve a successful take-over of another pride (Bygott *et al.* 1979).

Female lions show synchronous oestrus . . .

Fig. 1.1 Top: when a new male takes over a lion pride, he kills the young cubs fathered by the previous males. Bottom: a female suckles her sister's cub alongside her own.

... and frequent
copulation

2   A lioness comes into heat every month or so when she is not pregnant. She is on heat for 2 to 4 days during which time she copulates once every 15 min throughout the day and night. Despite this phenomenal rate of copulation the birth rate is low. Even for those cubs that are born, only 20 per cent will survive to adulthood. It can be calculated that there are 3000 copulations for each offspring that attains the adult stage.

   The causal explanation for why lion matings are so unsuccessful is not the failure of the male to ejaculate but rather the high probability of ovulation failure by the female or a high rate of abortion. But why are females designed in this apparently inefficient way?

One hypothesis is that it may be advantageous to the female to be receptive even at times when conception is unlikely, because this means that each copulation is devalued. For a male there is only a 1:3000 chance that a given copulation will produce a surviving cub and so it is not worth fighting with other males in the pride over a single mating opportunity. Given that males may also kill cubs that are not their own (see below), it may pay a female to mate with all the males in the pride to increase paternity uncertainty. Ideally a female may give each male a sufficient chance of being the father of her cubs that it does not pay him to kill them!

Males kill cubs

3 When a new male, or group of males, takes over a pride they sometimes kill the cubs already present (Fig. 1.1). The causal explanation for this behaviour may be the unfamiliar odour of the cubs which induces the male to destroy them. A similar effect, known as the Bruce Effect, occurs in rodents where the presence of a strange male prevents the implantation of a fertilized egg or induces abortion.

The advantage of the infanticide for the male that takes over the pride is that killing the cubs fathered by a previous male brings the female into reproductive condition again much quicker and so hastens the day that he can father his own offspring. If the cubs were left intact then the female would not come into oestrus again for 25 months. By killing the cubs she becomes ready for mating after only 9 months. Remember that a male's reproductive life in the pride is short, so any individual that practises infanticide when he takes over a pride will father more of his own offspring and therefore the tendency to commit infanticide will spread by natural selection.

Causal and functional explanations of lion behaviour

The take-over of a pride by a new coalition of adult males also contributes to the reproductive synchrony of the females; because all the dependent offspring are either killed or evicted during the take-over, the females will all tend to come into oestrus again at about the same time (Packer & Pusey 1983a). Interestingly, the heightened sexual activity of the females is most marked during the first few months after a take-over. The females play an active role in soliciting copulations from several males and this appears to elicit competition between different male coalitions for the control of the pride, with the result that larger coalitions eventually become resident. This is of adaptive advantage to the female because she needs protection from male harassment of her cubs for over 2 years in order to rear her cubs successfully (3.5 months gestation plus 1.5 to 2 years with dependent young) and only large male coalitions are likely to remain in the pride for more than 2 years. High sexual activity in females may therefore incite male–male competition and so result in the best protectors taking over the pride (Packer & Pusey 1983b).

The differences between the causal and functional explanations of these three aspects of reproductive behaviour in the lions are summarized in Table 1.1.

Natural selection

Throughout this book we will be focusing on functional questions about behaviour. Our aim is to try and understand how an animal's behaviour is adapted to the environment in which it lives. When we discuss adaptations we are referring to changes brought about during evolution by the process of natural selection. For Charles Darwin, adaptation was an obvious fact. It was obvious to him, that eyes were well designed for vision, legs for running, wings for flying and so on. What he attempted to explain was how

Table 1.1 Summary of causal and functional explanations for three aspects of reproductive behaviour in lions. From Bertram (1975) and Packer and Pusey (1983a,b)

Observation	Causal explanations	Functional explanations
1 Females are synchronous in oestrus	Chemical cues? Take-overs by males	Better cub survival Young males survive better and have greater reproductive success when they leave pride if in a group
2 High rate of copulation	Female infertility Time of ovulation concealed	Each copulation of less value to a male Increased paternity uncertainty may protect cubs from males Elicits competition between male coalitions, so females get best protectors taking over pride
3 Young die when new males take over pride	Abortion (?chemical) Take-over males kill or evict young	Females come into oestrus quicker Male removes older cubs which would compete with his young

adaptation could have arisen without a creator. His theory of natural selection, published in the *Origin of Species* in 1859, can be summarized as follows.

1 Individuals within a species differ in their morphology, physiology and behaviour (*variation*).

Heritable variation with competition for survival and reproduction

2 Some of this variation is *heritable*; on average offspring tend to resemble their parents more than other individuals in the population.

3 Organisms have a huge capacity for increase in numbers; they produce far more offspring than give rise to breeding individuals. This capacity is not realized because the number of individuals within a population tends to remain more or less constant over time. Therefore there must be *competition* between individuals for scarce resources such as food, mates and places to live.

4 As a result of this competition, some variants will leave more offspring than others. These will inherit the characteristics of their parents and so evolutionary change will take place by *natural selection*.

5 As a consequence of natural selection organisms will come to be *adapted* to their environment. The individuals that are selected will be those best able to find food and mates, avoid predators and so on.

When Darwin formulated his idea he had no knowledge of the mechanism of heredity. The modern statement of the theory of natural selection is in terms of genes. Although selection acts on differences in survival and reproductive success between individual organisms, or phenotypes, what changes during evolution is the relative frequency of genes.

We can restate Darwin's theory in modern genetic terms as follows.

1 All organisms have genes which code for protein synthesis. These proteins regulate the development of the nervous system, muscles and structure of the individual and so determine its behaviour.

Selection causes changes in gene frequency

2 Within a population many genes are present in two or more alternative forms, or alleles, which code for slightly different forms of the same protein. These will cause differences in development and so there will be variation within a population.

3 There will be competition between the alleles of a gene for a particular site (*locus*) on the chromosomes.

4 Any allele that can make more surviving copies of itself than its alternative will eventually replace the alternative form in the population. Natural selection is the differential survival of alternative alleles.

The individual can be regarded as a temporary vehicle or

survival machine by which genes survive and are replicated (Dawkins 1976). Because selection of genes is mediated through phenotypes, the most successful genes will be those which promote most effectively an individual's survival and reproductive success (and that of relatives, see later). As a result we would therefore expect individuals to behave so as to promote gene survival.

Before we discuss how thinking about genes can help us to understand the evolution of behaviour, we should examine the evidence that gene differences can cause differences in behaviour.

Genes and behaviour

Behavioural ecology is concerned with the evolution of adaptive behaviour in relation to ecological circumstances. Natural selection can only work on genetic differences and so for behaviour to evolve (a) there must be, or must have been in the past, behavioural alternatives in the population, (b) the differences must be, or must have been, heritable; in other words a proportion of the variation must be genetic in origin, and (c) some behavioural alternatives must confer greater reproductive success than others.

Three main methods have been used to study the ways in which genes influence behaviour.

(a) Use of genetic mutants

Seymour Benzer (1973) used mutagens (radiation or chemicals) to produce genetic mutations that change behaviour in the fruit fly, *Drosophila*. In one mutant, known colourfully as 'stuck', the male fails to disengage from the female after the normal 20-min period of copulation. Mutation of another gene produces 'coitus interruptus' males which disengage after only 10 min and fail to produce any offspring. Benzer was able to trace the cause of these mutations and show that they resulted from abnormalities in the sensory receptors, nervous system or muscles of the flies. Mutants which exhibit deficiencies in learning ability have also been isolated (Dudai & Quinn 1980). Normal *Drosophila* learn to selectively avoid an odour which is associated with an electric shock. Mutant 'dunce' flies do not learn to avoid the shock though they show normal behaviour in other respects and can learn visual tasks. Dunce flies are produced by an abnormality of a complex gene, at least part of which codes for the enzyme cyclic AMP phosphodiesterase, which breaks down the intracellular second messenger cAMP. Dunce flies have an abnormally high level of cAMP and, furthermore, normal flies display poor

Behavioural differences may have a genetic basis

Learning mutants in Drosophila

learning after being fed on phosphodiesterase inhibitors. This suggests that the enzyme is necessary for associative learning. Other learning mutants, such as 'amnesiac', which learns normally but forgets very rapidly, have also been isolated. In all cases the mutations perturb second messenger systems (Dudai 1989).

(b) Artificial selection experiments

Selected lines are produced by choosing as parents in each generation those individuals which show the most extreme values of some behaviour character. For example, Aubrey Manning (1961) was able to select for two different mating speeds in the fruit fly, *Drosophila melanogaster*, by selectively breeding from fast and slow maters. Such selection experiments nearly always work, showing that much of the continuous phenotypic variation seen in populations has some genetic basis. Another example is provided by male field crickets, *Gryllus integer*, which either call to attract females or silently intercept females attracted to the callers. Cade (1981) was able to select for males which called a lot and those which called only rarely, thus showing that variation in calling duration had a genetic component.

Genetic differences in cricket calling . . .

(c) Studying populations with genetic differences

Geographically distinct populations of a species often have a different morphology and behaviour, reflecting adaptations to differing ecological conditions. Stevan Arnold (1981) studied the garter snake, *Thamnophis elegans*, in the south-west United States. Inland populations are very aquatic and commonly feed underwater on frogs, fish and leeches. Coastal populations are terrestrial foragers and mainly eat slugs. In laboratory choice experiments it was found that wild-caught inland snakes refused to eat slugs, though the coastal ones readily accepted them. Tests with naive newborn snakes showed that 73 per cent of the coastal individuals attacked and ate slugs while only 35 per cent of the inland snakes did so. Young snakes are incubated inside their mother and so one possibility is that the mother's diet could influence the young directly. Arnold arranged matings between inland and coastal individuals and found that the offspring tended to show an intermediate incidence of slug eating. They did not tend to resemble their mother as opposed to their father and so the influence of maternal diet can be ruled out. These results suggest that differences in food preference are correlated with genetic differences, and that garter snakes in different areas have been selected to respond to different prey types.

. . . and food preferences of snakes

Most species of warblers are summer visitors to Europe. If individuals are kept in a cage, they show a period of 'restlessness' in the autumn which corresponds with the time at which they would migrate south to Africa for the winter. Quantitative comparisons of the nocturnal restlessness of caged experimental birds showed that the duration of the restlessness correlated well with the distance which the individuals migrated.

Peter Berthold (Berthold *et al.* 1990a,b) has carried out a series of large-scale experiments on blackcaps (*Sylvia atricapilla*) to investigate the genetic basis of both the duration of migratory restlessness and the direction of migratory orientation, measured in cages. Blackcaps are ideal for such a study because different populations show differing degrees of migratory behaviour: in southern Germany, for example, all birds migrate, in southern France only a part of the population migrates, and in the Cape Verde Islands, the population is entirely sedentary. To what extent do these inter-population differences reflect genetic differences? Berthold has investigated this question by two kinds of experiment: cross-breeding of birds from different populations, and selection experiments. Both kinds of study involved breeding blackcaps in captivity and studying the migratory behaviour of offspring in cages equipped with electronic perches to record migratory restlessness.

Selection for migratory behaviour in warblers

Hybrids between parents from southern Germany (migratory) and the Cape Verde Islands (resident) had intermediate behaviour. About 40 per cent of the offspring showed migratory restlessness and the remainder did not. Furthermore, the preferred average directional heading shown by the migratory hybrid offspring was indistinguishable from that of the German parents. These results show that the difference between the two populations has a genetic basis. They suggest that more than one gene is involved (otherwise the offspring would all resemble the dominant parent) and that the effect of several genes is likely to have a threshold effect (otherwise all offspring would be intermediate between the two parents).

Berthold's selection experiments also confirm a genetic basis to differences in migratory behaviour. Among 267 hand-raised blackcaps from a population in the Rhone Valley of southern France, three-quarters were migratory and one-quarter resident, when tested in the laboratory. By selectively breeding from migratory and non-migratory parents, Berthold was able to produce strains of blackcaps that were either 100 per cent migratory (in three generations) or 100 per cent residents (in six generations) (Fig. 1.2). Not only does this experiment confirm a genetic basis to differences in migratory behaviour, but it also shows how an

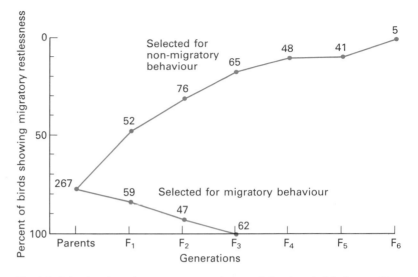

Fig. 1.2 Selection for migratory or non-migratory behaviour in blackcaps. The numbers indicate how many birds were hand-raised in each generation. Starting from a parental stock in which about 75 per cent of birds migrate, the selection experiment produced a population of non-migrators in six generations, and another line of migrators in three generations. In each generation, half the pairs were first-year birds and half were mixed first-year × older bird pairs. (Note that the vertical axis of the graph has a reversed scale.)

ecologically important trait such as migration may respond rapidly to selection.

Behaviour differences caused by gene differences

These various examples show that *genetic* differences between individuals can lead to differences in *behaviour* (mating behaviour, learning, singing, foraging and migration). Three points need to be emphasized. First, when we talk about 'genes for' a particular structure or behaviour, we do not imply that one gene alone codes for the trait. Genes work in concert and it is likely that many genes together will influence an individual's mating preference, foraging habits and migration patterns. However, a *difference* in behaviour between two individuals may be due to a *difference* in one gene. A useful analogy is the baking of a cake. A difference in one word of a recipe may mean that the taste of the whole cake is different, but this does not mean that the one word is responsible for the entire cake (Dawkins 1978). Whenever we talk about 'genes for' certain traits this is shorthand for gene differences bringing about differences in behaviour.

Second, genes will often influence behaviour in simple ways. Gene differences can result in behavioural differences because genes code for enzymes which influence the development of the sensory, nervous and muscle systems of the animal which in turn affect its behaviour. For example, a gene could influence a moth's preference for a particular background by coding for certain visual pigments in the eye. Third, just because it can be shown that genes influence behaviour this does not imply that genes alone produce the behaviour, or even that the behaviour can be usefully divided up into genetic and environmental components. The way in which behaviour develops is the result of a complex interaction between genes and environment. Imagine, for example, that a behavioural ecologist comes across the nest of a long tailed tit (*Aegithalos caudatus*). He or she would be immediately impressed by how well the nest was adapted to the bird's way of life, the wonderful crypsis of the nest, its strength and warmth. There are three different ways in which this adaptation could develop in an individual (Bateson 1983). Individuals could all learn by trial and error how to build good nests. Alternatively they could copy another, more experienced bird. Finally, during evolution genes necessary for the expression of actions leading to the construction of good nests could have spread through the population by natural selection because individuals building the best nests would have left most young. Thus, all individuals may be able to build good nests without practice or observing others. However, even if nest building required learning for its proper development, genetic differences in learning ability may still be involved in its evolution.

Genetic differences and developmental processes

Selfish individuals or group advantage?

We now return to our theme of studying the adaptive significance of behaviour, how it contributes to an individual's chances of survival and its reproductive success. We interpreted the behaviour of the lions in relation to individual advantage, reflecting Darwin's emphasis on evolution as a struggle between individuals to outcompete others in the population. Many traits evolve because of their advantage to the individual even though they are disadvantageous to others in the population. For example, it's not to the species' advantage to have a cub killed when a new male takes over a lion pride. It's not to the lionesses' advantage either! However, she is smaller than the male and there is probably not much that she can do about it. Infanticide has evolved simply because of its advantage to the male that practises it.

Behaviour of advantage to individuals may be disadvantageous to the group

A few years ago, however, many people thought that animals behaved for the good of the group, or of the species. It was common to read (and sometimes still is) explanations like, 'lions rarely fight to the death because if they did so, this would endanger survival of the species' or, 'salmon migrate thousands of miles from the open ocean into a small stream where they spawn and die, killing themselves with exhaustion to ensure survival of the species'.

Because 'group thinking' is so easy to adopt, it is worth going into a little detail to examine why it is the wrong way to frame evolutionary arguments.

The main proponent of the idea that animals behave for the good of the group is V.C. Wynne-Edwards (1962, 1986). He suggested that if a population over-exploited its food resources it would go extinct and so adaptations have evolved to ensure that each group or species controls its rate of consumption. Wynne-Edwards proposed that individuals restrict their birth rate to prevent over-population, by producing fewer young, not breeding every year, delaying the onset of breeding, and so on. This is an attractive idea because it is what humans ought to do to control their own populations. However there are two reasons for thinking that it is unlikely to work for animal populations.

THEORETICAL OBJECTIONS

Imagine a species of bird in which each pair lays 2 eggs and there is no over-exploitation of the food resources. Suppose the tendency to lay 2 eggs is inherited. Now consider a mutant which lays 6 eggs. Since the population is not over-exploiting its food supplies, there will be plenty of food for the young to survive and the 6-egg genotype will become more common very rapidly.

Will the 6-egg type be replaced by birds that lay 7 eggs? The answer is yes, as long as individuals laying more eggs produce more surviving young. Eventually a point will be reached where the brood is so large that the parents cannot look after it as efficiently as a smaller one. The clutch size we would expect to see in nature will be the one that results in the most surviving young because natural selection will favour individuals that do the best. A system of voluntary birth control for the good of the group will not evolve because it is unstable; there is nothing to stop individuals behaving in their own selfish interests.

Group selection

Wynne-Edwards realized this and so proposed the idea of 'group selection' to explain the evolution of behaviour that was for the good of the group. He suggested that groups consisting of selfish individuals died out because they over-exploited their food resources. Groups that had individuals who restricted their birth rate did not over-exploit their resources and so survived. By a process of differential survival of groups, behaviour evolved that was for the good of the group.

In theory this might just work, but groups must be selected during evolution; some groups must die out faster than others. In practice, however, groups do not go extinct fast enough for group selection to be an important force in evolution. Individuals will nearly always die more often than groups and so individual selection will be more powerful. In addition, for group selection to work populations must be isolated. Otherwise there would be nothing to stop the migration of selfish individuals into a population of individuals all practising reproductive restraint. Once they had arrived their genotype would soon spread. In nature, groups are rarely isolated sufficiently to prevent such immigration. So group selection is usually going to be a weak force and probably rarely very important (Williams 1966; Maynard Smith 1976a).

EMPIRICAL STUDIES

Apart from these theoretical objections, there is good field evidence that individuals do not restrict their birth rate for the good of the group but in fact reproduce as fast as they can. A good example is David Lack's long-term study of the great tit (*Parus major*) in Wytham Woods, near Oxford, England (Perrins 1965; Lack 1966).

Optimal clutch size in birds

In this population the great tits nest in boxes and lay a single clutch of eggs in the spring. All the adults and young are marked individually with small numbered metal rings round their legs. The eggs of each pair are counted, the young are weighed and their survival after they leave the nest is measured by retrapping ringed birds. This intensive field study involves several people working full-time on the great tits throughout the year, and it has been going on for 40 years! Most pairs lay 8 to 9 eggs (Fig.

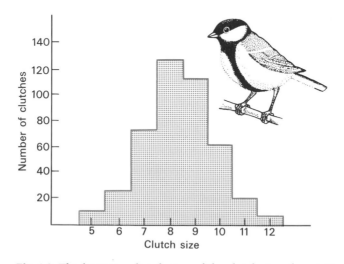

Fig. 1.3 The frequency distribution of the clutch size of great tits in Wytham Woods. Most pairs lay 8–9 eggs. From Perrins (1965).

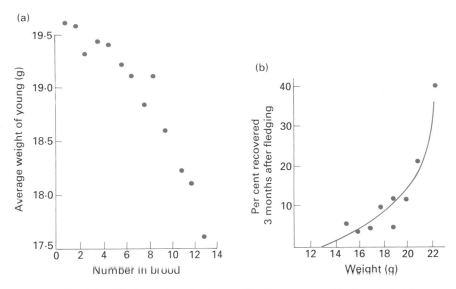

Fig. 1.4 (a) In larger broods of great tits the young weigh less at fledging because the parents cannot feed them so efficiently. (b) The weight of a nestling at fledging determines its chances of survival. Heavier chicks survive better. From Perrins (1965).

1.3). The limit is not set by an incubation constraint because when more eggs are added the pair can still incubate them successfully. However, the parents cannot feed larger broods so well. Chicks in larger broods get fed less often, are given smaller caterpillars and consequently weigh less when they leave the nest (Fig. 1.4a). It is not surprising that feeding the young produces a limit for the parents because they have to be out searching for food from dawn to dusk and may deliver over 1000 items per day to the nest at the peak of nestling growth.

The significance of nestling weight is that heavier chicks survive better (Fig. 1.4b). An over-ambitious parent will leave fewer surviving young because it cannot feed its nestlings adequately. By creating broods of different sizes experimentally it can be demonstrated that there is an optimum brood size, one that maximizes the number of surviving young from a selfish individual's point of view (Fig. 1.5). The commonest observed clutch size (Fig. 1.3) is close to the predicted optimum but slightly lower. Why is this? A possible reason is that the optimum in Fig. 1.5 is the one which maximizes the number of surviving young *per brood* whereas, at least in stable populations, we would expect natural selection to design animals to maximize their *lifetime* reproductive output. Figure 1.6 shows how the mortality costs of raising larger broods can be incorporated into our argument so as to predict the brood size which maximizes overall lifetime

Survival versus reproduction

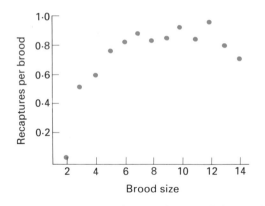

Fig. 1.5 Experimental manipulations of the number of young in a nest show that the optimal brood size for a pair of great tits is between 8 and 12 eggs. This is the brood size which maximizes the number of surviving young. From Perrins (1979).

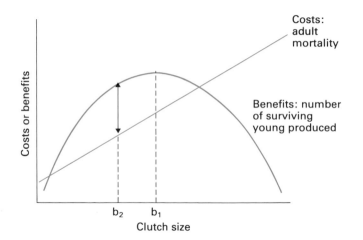

Fig. 1.6 The influence of adult mortality on the optimal clutch size. The number of young produced versus clutch size follows a curve, as in Fig. 1.5, with b_1 being the clutch size which maximizes the number of young produced per brood. Increased clutch size, however, has the cost of increased adult mortality, shown here for simplicity as a straight line. The clutch size which maximizes lifetime reproductive success is b_2, where the distance between the benefit and cost curves is a maximum. This is less than the clutch size b_1, which maximizes reproductive success per brood. From Charnov and Krebs (1974).

reproductive success. In general, the clutch size which maximizes lifetime breeding success will be slightly less than that which maximizes success per breeding attempt. Box 1.1 gives a more general model for the optimal trade-off between current and future reproductive effort.

Box 1.1 *The optimal trade-off between survival and reproductive effort. From Pianka and Parker (1975) and Bell (1980).*

The more effort an individual puts into reproduction, the lower its chances of survival and so the lower its expectation of future reproductive success. Reproductive costs include allocation of resources to reproduction which would otherwise have been spent on own growth and survival and the increased risks entailed in reproduction, such as exposure to predators. The optimal life history depends on the shape of the curve relating profits in terms of present offspring to costs in terms of future offspring.

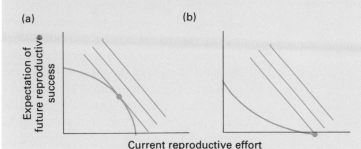

The families of straight lines represent fitness isoclines, i.e. equal lifetime production of offspring. In a stable population, present and future offspring will be of equal value and these lines will have slopes of −1. In an expanding population, current offspring are worth more than future offspring (current offspring gain a greater contribution to the gene pool) and the slopes are steeper. In a declining population future offspring are worth more and slopes will be less than −1.

The point of intersection of the curves relating the trade-off between current and future reproductive success, with the fitness isocline furthest from the origin gives the optimal reproductive tactic (indicated by solid dot). When the trade-off curve is convex (a), fitness is maximized by allocating part of the resources to current reproduction and part to survival (i.e. iteroparity, or repeated breeding). When the curve is concave (b), it is best to allocate all resources to current reproduction even at the expense of own survival (semelparity, or 'big bang' suicidal reproduction). If maximal future reproductive success is greater than maximal current reproductive success in case (b), then the optimal tactic is to not breed and save all resources for the future.

As yet there is no evidence for increased mortality costs from raising larger broods in the great tit. When Pettifor *et al.* (1988) manipulated brood size by adding or removing 3–4 young soon after hatching, parents raising enlarged broods survived just as well to the next season as those which raised their own natural brood size or a decreased brood. Whether this means that there is no cost of reproduction, or that the cost is hard to measure, needs further work. In blue tits (*Parus caeruleus*) similar brood manipulations did reveal significant survival costs in some years but not in others, so longer-term studies are needed for a proper evaluation (Nur 1988).

Individuals may have different optima

Our conclusion is that reproductive rate is close to that which maximizes individual success. Thus clutch size is optimal from the selfish individual's point of view. Of course, the exact clutch size may vary a little from year to year and during the season depending on food supplies, so individuals do show some variation. However the variations are in relation to their own selfish optima, not for the good of the group. A good example of this is provided by Goran Högstedt's study (1980) of magpies, *Pica pica*, breeding in southern Sweden. Observed clutch sizes varied from 5 to 8 depending on feeding conditions in different territories. Our hypothesis would be that some females laid only 5 eggs because this was the maximum number of young they could raise efficiently on their particular territories. Högstedt manipulated clutch sizes experimentally and found that pairs that had produced large clutches did best with large broods, while those which had laid small clutches did best with smaller broods (Fig. 1.7). Variation

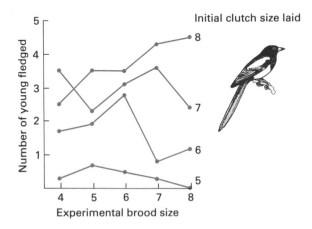

Fig. 1.7 Experiments on clutch size in magpies. Pairs which had initially laid 5, 6, 7 or 8 eggs were given experimentally reduced or enlarged broods. Pairs which had naturally laid large clutches did better with large broods and vice versa. From Högstedt (1980).

Table 1.2 A comparison of the clutch size of European passerine birds grouped into two ecological categories. Hole nesting species have larger clutches. From Lack (1968)

Nest type	Predation	Average clutch size	Average length of nestling period (days)
Holes	Low	6.9	17
Open	High	5.1	13

in clutch size occurred because there was a range of territory quality and each pair raised a brood size appropriate for its own particular territory. Experiments have shown similar individual optimization of clutch size in great tits (Pettifor *et al.* 1988) and collared flycatchers (Gustafsson & Sutherland 1988).

One of the main themes of this book is that different life history strategies will be favoured in different ecological circumstances. For example, continuing our discussion of clutch size, if the European passerine bird species are grouped into two ecological categories, namely those that nest in holes and those that build open nests, it is found that the hole nesters lay larger clutches (Table 1.2). The same relationship occurs in ducks where open nesting species again have smaller clutches than hole nesters (Lack 1968). In holes the young are relatively safe from predators but in the open there is a premium for getting the young out of the vulnerable nest as soon as possible. The same quantity of food could be used to rear a small brood quickly or a larger brood more slowly. In open nesting species the higher risks of predation have apparently selected for a smaller clutch size and rapid nestling growth.

Hole-nesting species lay larger clutches

Behaviour, ecology and evolution

We are now in a position to summarize the main principles which underlie the arguments in this book. First, during evolution natural selection will favour individuals who adopt life history strategies which maximize their gene contribution to future generations. Second, the way in which adult survival and reproductive effort are best traded-off in a life history will depend on ecology, the physical environment an individual lives in, its competitors, food, predators, and so on. Third, because an individual's success at surviving and reproducing depends critically on its behaviour, selection will tend to design animals as efficient foragers, efficient predator avoiders, efficient copulators, efficient parents, and so on. What will be 'optimal' will depend on the behavioural alternatives available which will depend on various constraints imposed by phylogeny, physiology and ecology.

Behavioural ecology, therefore, is a meeting point for behaviour, ecology and evolution. We can think of ecology as setting the stage on which animals must perform their behaviour, and evolution as a process which selects individuals whose behaviour results in greatest success in the struggle to contribute genes to the population's gene pool. The aims of the subject are to understand why different species behave in different ways and why, within a species, there may also be individual differences in behaviour. To tackle these problems we have to learn about a species' ecology and also about how individuals in a population compete for scarce resources, such as food, mates and territories.

Summary

Behavioural ecology is concerned with functional questions about behaviour, namely how a particular behaviour pattern contributes to an animal's chances of survival and its reproductive success. Experiments show that differences in behaviour can result from differences in genes. Natural selection will favour genes which best promote an individual's chances of passing those genes on to future generations. Individuals are expected to behave in their own selfish interests and not for the good of the species or group. Ecological conditions will determine which behaviour patterns are favoured during evolution.

Further reading

The books by G.C. Williams (1966) and Richard Dawkins (1976, 1982) are excellent discussions of behaviour and evolution. Williams emphasizes the evolution of individual selfishness as opposed to behaviour for the good of the group. Dawkins champions the view that we should think in terms of genes rather than individuals in order to understand the evolution of behaviour.

Partridge (1983) and Bateson (1983) discuss how genes influence behaviour. Southwood (1981) and Lessells (1991) show how different life histories are favoured under different ecological conditions.

In this chapter we have discussed optimal clutch size from the parent's point of view. In fact there will be conflicts of interest between the male and female (see Chapter 9), between parents and offspring (Trivers 1974), and between members of the brood. Mock (1984, 1985) shows how chicks within a brood may kill their siblings when competing for food brought by the parents.

Montgomerie and Weatherhead (1988) discuss how the trade-off between current and future reproductive success influences nest defence by parent birds.

Topics for discussion

1 How would you test the causal and functional explanations for lion behaviour in Table 1.1?
2 Under what conditions would the trade-off between present and future reproduction look like (a) or (b) in Box 1.1?
3 What do you understand by the term 'optimal clutch size'?

Chapter 2. Testing Hypotheses in Behavioural Ecology

A rigorous scientific approach to the function of behaviour will involve four stages: observations, hypotheses, predictions and tests. The first two, observation and hypotheses, often go hand in hand. It may take many years getting to know a particular species before it is possible to ask good questions about its behaviour and ecology. Niko Tinbergen's work (1953) on the herring gull, *Larus argentatus*, was the result of over 20 years' painstaking observations of the bird's behavioural repertoire and the environment in which it lives. Having observed some aspect of the behaviour of an animal that we do not understand, how should we proceed?

Let us assume, for example, that we want to discover why our animal lives in a group as opposed to on its own. We may get a strong hint about the function of this simply from observation. If the animal only lived in a group in the breeding season we might suspect that it gained some advantage in terms of increased mating efficiency, for example, whereas if it only lived in a group in winter we may suspect some advantage concerned with improved feeding efficiency or avoiding predation. We can test our ideas in three main ways.

Three methods of hypothesis testing

1   *Comparison between individuals within a species.* Individuals in groups may have greater success at feeding, mating or avoiding predators than solitary individuals. Furthermore, success may vary with group size. The problem is, however, that there may be confounding variables: solitary individuals may be poorer competitors and this, rather than their solitary existence *per se*, may explain their lower success, or individuals in groups may live in better quality habitats, and so on.

2   *Experiments.* It is often better, therefore, to perform an experiment. With an experiment we can vary one factor at a time; for example we could change group size and see how this influenced success under a particular set of conditions. Niko Tinbergen pioneered the method of elegant field experimentation to answer functional questions. For example, to test the hypothesis that spacing out of gull nests functioned to reduce predation he put out experimental plots of eggs with different spacing patterns and found that those with a clumped distribution suffered greater predation than those that were spaced out as in nature (Tinbergen *et al.* 1967).

3   *Comparison among species.* Different species have evolved in relation to different ecological conditions and so comparison among species may help us to understand how differences in

feeding ecology or predation pressure, for example, influence the tendency to live in groups or to be solitary. Using the comparative method is rather like looking at the result of experiments done by natural selection over evolutionary time. The results of these 'experiments' are the designs of the various species' behaviour which we now obeve. For example, group living may occur in species which experience particular feeding conditions.

In this chapter we will discuss in detail these last two methods for investigating adaptation.

The comparative approach

The idea of comparison lies at the heart of most hypotheses about adaptation. It is the comparative study of different species which gives us a feel for the range of strategies that animals adopt in nature. When we ask functional questions about the behaviour of particular species we are usually asking why it is different from other species. Why does species A live in groups compared with species B which is solitary? Why do males of species B mate monogamously compared with males of species A which are polygynous, and so on? A powerful method for studying adaptation is to compare groups of related species and attempt to find out exactly how differences in their behaviour reflect differences in ecology. We will first describe two examples which pioneered the comparative approach and inspired workers to use the method with other animal groups. Then we will point out some of the methodological difficulties in formulating and testing hypotheses based on comparison. Finally we will describe some recent examples of the comparative method which have attempted to overcome these problems.

Correlating differences in behaviour with differences in ecology

SOCIAL ORGANIZATION IN WEAVER BIRDS

The first person to attempt a systematic analysis of this kind was John Crook (1964) who studied about 90 species of weaver birds (Ploceinae). These are small finches which live throughout Africa and Asia, and although they all look rather alike there are some striking differences in their social organization. Some are solitary, some go around in large flocks. Some build cryptic nests in large defended territories while others cluster their nests together in colonies. Some are monogamous, with male and female forming a permanent pair bond; others are polygamous, the males mating with several females and contributing little to care of the offspring. How can we explain the evolution of this great diversity in behaviour?

Weaver bird social
behaviour and mating
systems are correlated
with diet

Crook's approach was to search for correlations between these aspects of social organization and the species' ecology. The ecological variables he considered were the type of food, its distribution and abundance, predators and nest sites. His analysis showed that the weaver birds fell into two broad categories.

1 Species living in the forest tended to be insectivorous, solitary feeders, defend large territories and build cryptic solitary nests. They were monogamous and males and females had similar plumage.

2 Species living in the savannah tended to eat seeds, feed in flocks and nest colonially in bulky conspicuous nests. They were polygamous and there was sexual dimorphism in plumage, the males being brightly coloured and the females rather dull (Fig. 2.1).

Why is the behaviour and morphology of the weaver birds linked to their ecology in such a striking way? Crook invoked predation and food as the main selective pressures that have influenced the evolution of social organization. His argument was as follows.

1 In the forest, insect food is dispersed. Therefore it is best for the birds to feed solitarily and defend their scattered food resources as a territory. Because the food is difficult to find, both parents have to feed the young and therefore stay together as a pair throughout the breeding season. With the male and female visiting the nest, both must be dull coloured to avoid attracting predators. Cryptic nests spaced out from those of neighbours decrease their vulnerability to predation.

2 In the savannah, seeds are patchy in distribution and locally superabundant. It is more efficient to find patches of seeds by being in a group because groups are able to cover a wider area in their search. Furthermore the patches contain so much food that there is little competition within the flock while the birds are feeding.

In open country the birds cannot hide their nests and so they seek safety in protected sites, such as spiny acacia trees. Nests are bulky to provide thermal insulation against the heat of the sun. Because good breeding sites are few and scattered, many birds nest together in the same tree. Within a colony, males compete for nest sites and those that defend the best sites attract several females while males in the poorer parts of the colony fail to breed. In addition, because food is abundant, the female can feed the young by herself and so the male is emancipated from parental care and can spend most of his time trying to attact more females. This has favoured brighter plumage coloration in males and the evolution of polygamy.

Predation and food
dispersion are key
selective forces

Fig. 2.1 Top: some species of weaver birds, like this one (*Malimbus scutatus*) are insectivorous, building cryptic solitary nests in the forest and feeding alone in large territories. Bottom: other species like this one (*Ploceus cucullatus*) feed on seeds out in the open savannah. They build conspicuous nests in colonies and feed in flocks. The males are often brightly coloured.

Supporting evidence for this interpretation comes from species with intermediate ecology. The grassland seed-eaters have patchy food supplies so group living is favoured for efficient food finding. However, in grassland the nests are vulnerable, so predation favours spacing out. The result is a compromise; these species have an intermediate social organization, nesting in loose colonies and feeding in flocks.

These results show clearly how food and predation may be important in determining social organization. They also reveal how several different traits such as nests, feeding behaviour, plumage colour and mating system can all be considered together as a result of the same ecological variables. Crook's work with the weaver birds inspired several people to use the comparative method to study social organization in other groups. David Lack (1968) extended the argument to include all bird species and Peter Jarman (1974) used the same approach for the African ungulates.

SOCIAL ORGANIZATION IN AFRICAN UNGULATES

Jarman (1974) considered 74 species of ungulates; all eat plant material but differences in the precise type of food eaten are correlated with differences in movements, mating systems and antipredator behaviour. The species were grouped into five ecological categories (Table 2.1). Just as in the weaver birds, several adaptations seem to go together.

Body size, diet and social organization

The major correlate of diet and social organization is body size. Small species have a higher metabolic requirement per unit weight and need to select high quality patches of food such as berries and shoots. These tend to occur in the forest and are scattered in distribution, so the small species are forced to live a solitary existence. The best way to avoid predators in the forest is to hide. Because the females are dispersed, the males must also

Table 2.1 The social organization of African ungulates in relation to their ecology. From Jarman (1974)

	Exemplary groups	Body weight (kg)	Habitat	Diet	Group size	Reproductive unit	Antipredator behaviour
Grade I	Dikdik Duiker	3–60	Forest	Selective browsing; fruit, buds	1 or 2	Pair	Hide
Grade II	Reedbuck Gerenuk	20–80	Brush, riverine grassland	Selective browsing or grazing	2 to 12	Male with harem	Hide, flee
Grade III	Impala Gazelle Kob	20–250	Riverine woodland, dry grassland	Graze or browse	2 to 100	Males territorial in breeding season	Flee, hide in herd
Grade IV	Wildebeest Hartebeest	90–270	Grassland	Graze	Up to 150 (thousands on migration)	Defence of females within herd	Hide in herd, flee
Grade V	Eland Buffalo	300–900	Grassland	Graze unselectively	Up to 1000	Male dominance hierarchy in herd	Mass defence against predators

be dispersed and the commonest mating system is for a pair to occupy a territory together.

At the other extreme, the largest species eat poor quality food in bulk and graze less selectively on the plains. It is not economical to defend such food supplies and these species wander in herds, following the rains and fresh grazing. In these large herds there is potential for the strongest males to monopolize several females by defence of a harem or a dominance hierarchy of mating rights. When predators come along these species cannot hide on the open plains and so either flee or rely on safety in numbers in the herd. Ungulates of intermediate size show aspects of ecology and social organization in between these two extremes (Table 2.1).

Adaptation or story-telling?

This comparative approach to adaptation is persuasive, but there are problems (Clutton-Brock & Harvey 1979; Gould & Lewontin 1979). Many of the following problems are not unique to comparative studies and it is worth bearing them in mind throughout the book.

Problems in interpreting
comparative data

ALTERNATIVE HYPOTHESES

The explanations for the differences in social organization are plausible, but alternative hypotheses have not been considered in a rigorous manner.

The ecological variables, such as predator pressure and patchy environment, have also been used in rather a vague way. Indeed, in the weaver birds we invoked a patchy food distribution as responsible for the evolution of flocking while in the ungulates we said that high quality patchy foods favour a solitary existence. There is the obvious danger here of explaining things too easily without any rigorous quantification of the ecological factors concerned.

CAUSE AND EFFECT

Consider the observation that weaver birds with a diet of seeds go about in flocks. Our explanation was that seed eating selects for flocking because this is the best way to find a patchy food supply. However we could equally well have suggested that predation selects for flocking and, as a consequence, the birds are forced to select locally abundant food so all the flock can get enough to eat. In this case a diet of seeds is a consequence, or effect, of flocking, not a cause. Maybe predation also selects for flocking in the forest insectivores but because their diet is incompatible with flocking they have to forage singly.

CONFOUNDING VARIABLES

With the comparative approach there are often confounding variables. For example, we observe the giraffe with its long neck feeding at the tops of trees and the buffalo with its short neck feeding on the ground. We then say that a long neck is an adaptation for feeding high up in the trees. But a long neck could equally well aid predator detection. How can we control for such confounding variables and decide which selective pressure has favoured the trait? Perhaps it is both?

Controlling for differences in body size

A particularly important confounding variable in comparative studies is body size. Jarman controlled for this in his analysis of the ungulates by dividing the species up into categories of different body weight (Table 2.1). Most biological traits do not increase in a 1 : 1 relationship with body size; their relation to body size is said to be allometric (Gould 1966). For example, the brain mass of different bird species increases at about the two-thirds power of body weight. In this case, before we can examine the ecological correlates of brain size we first have to remove the effects of body size. This can be done by calculating the appropriate line of best fit when brain mass is plotted against body weight and then measuring deviations from the line to see whether the size of the trait is greater or less than expected from body weight.

ALTERNATIVE ADAPTIVE PEAKS
OR NON-ADAPTIVE DIFFERENCES

It is tempting when comparing between species to assume that differences are always adaptive but some differences may simply be alternative solutions to the same ecological pressures. An ecologist from Mars who visited the Earth would observe that in the United States people drive their cars on the right hand side of the road while in Britain they drive on the left. He would then perhaps make lots of measurements in an attempt to find ecological correlates to explain the adaptive significance of the difference. In fact driving on the right and driving on the left may just be equally good alternatives for preventing accidents (Dawkins 1980).

Some differences between animals may be like this. Sheep use horns for fighting and deer use antlers. Horns are derived from skin while antlers are derived from bone (Modell 1969). The

Fig. 2.2 The horns of the sheep (left) and antlers of the deer (right) are both used in fighting. Horns are derived from skin and antlers from bone.

Some differences between species may reflect different solutions to the same problem

differences between horns and antlers need not necessarily reflect ecological differences; it may simply be a case of evolution working with different raw materials to produce the same functional end (Fig. 2.2). The problem with non-adaptive explanations is that they are hypotheses of the last resort. Further scientific enquiry is stifled. Maybe there is an adaptive explanation for the difference but we just haven't discovered it yet. For example, antlers are dropped and then renewed each year whereas horns are not. Perhaps this difference is related to the extent of seasonal variation in mating competition and food supply?

These criticisms are important, but they certainly do not mean that the comparative method is a failure. On the contrary, the approach is impressive in the way it brings together such a wide diversity of behavioural and morphological traits within the same ecological framework. Crook's study of the weaver birds and Jarman's work on the antelopes have served as models for ecological work on other groups of species. However the most recent comparative studies have attempted to control for these various problems, and we will now discuss another example, bearing the criticisms in mind, to illustrate how changes in methodology have made comparison between species a more rigorous exercise.

Primate social organization

Early knowledge of primate behaviour came mainly from studies in zoos. In 1932 Lord Zuckerman suggested that primates tend to be social animals because they have continuous sexual activity. It should be clear from the last chapter that this is a causal explanation and leaves the functional significance of sociality

Fig. 2.3 Four photographs illustrating the variety of primate social organization. (a) Table 2.2, Grade I: A solitary insectivorous prosimian, the lesser bushbaby (*Galago senegalensis*) (photo by Caroline Harcourt). (b) Table 2.2, Grade II: A monogamous pair of folivorous, arboreal, black gibbons (*Hylobates concolor*). The male is on the left. © Ron Tilson/BPS. (c) Table 2.2, Grade IV: Part of a troop of folivorous, arboreal, dusky langurs (*Presbytes obscurus*). © Ron Tilson/BPS. (d) Table 2.2, Grade V: A troop of savannah-dwelling olive baboons (*Papio anubis*). There are two subordinate males in the foreground and the dominant males are in the background near the females and young. Photo by Irven De Vore (Anthro-Photo).

unexplained. In the 1950s, the first studies of primates in the field (e.g. Carpenter 1954) revealed that sexual activity was not in fact continuous. It also became clear that different species have very different social organizations (Fig. 2.3). Tiny tarsiers and lemurs hunt solitarily in the tree tops for insects at night. Some monkeys go around in small groups in the trees by day, feeding on leaves or fruit. Others are terrestrial and live in large troops. Among the apes, the orang-utan is solitary, the gibbon lives in pairs and small family groups while the chimpanzee may live in bands of up to 50.

How can we explain the evolution of this bewildering array of social organization? It soon became apparent that ecological factors were important. For example, De Vore (1965) noticed that, compared with other species of primates, anubis baboons live in large groups, the males are large and they have big teeth. He suggested these may all be adaptations to predator defence in a terrestrial environment. By 1966, there were sufficient field data for John Crook and Stephen Gartlan to apply the first comparative approach to a large number of primates.

Like the weaver bird and antelope work, they categorized the species into several groups based on ecology and behaviour. Table 2.2 shows that at one extreme, insectivorous primates are nocturnal, forest animals which are solitary; then there is a variety of fruit and leaf eating species which are diurnal and live in small to large groups; finally, at the other extreme, are the vegetarian browsers of the open country which live in large groups and

Primate group size, mating systems and diet

Table 2.2 Crook and Gartlan's (1966) division of the primates into five 'adaptive grades'

	Exemplary species	Habitat	Diet	Activity	Group size	Reproductive unit	Sexual dimorphism
Grade I	*Galago Lepilemur Microcebus*	Forest	Insects	Nocturnal	Solitary	Pairs	Slight
Grade II	*Indiri Lemur Hylobates*	Forest	Fruit or leaves	Crepuscular or diurnal	Very small groups	Single male with family	Slight
Grade III	*Colobus Saimiri Gorilla*	Forest or forest edge	Fruit or fruit and leaves	Diurnal	Small groups	Multi-male groups	Slight to fairly marked
Grade IV	*Macaca Cercopithecus aethiops Pan*	Forest edge to tree savannah	Vegetation /omnivore	Diurnal	Medium to large groups	Multi-male groups	Marked
Grade V	*Erythrocebus patas Papio hamadryas Theropithecus gelada*	Grassland or arid savannah	Vegetation /omnivore	Diurnal	Medium to large groups	One-male groups	Marked

show intensive competition between males for females and marked sexual dimorphism.

Food and predation seem to be the main selective pressures

Once again, food and predation were suggested as the main selective pressures responsible for this link between social organization and habitat. Insects are dispersed and difficult to find, so just like the insectivorous weaver birds, these primates are solitary. In open country, predation favours grouping for safety and food is locally abundant so this allows many individuals to congregate at a food source; like the open-country weaver birds, these primates live in groups. In a large group, males compete with each other for mating rights and hence large male body size has been selected for.

Crook and Gartlan's approach was to categorize the primates into a small number of discrete groups. This raises two main problems. First of all, variation in features such as home range size and group size is continuous and so division into hard and fast groups is a bit arbitrary. Because the groups are subjectively defined, it is difficult for subsequent workers to categorize new species in the scheme. Second, different aspects of social organization such as breeding system and group size do not necessarily vary together in the same way. For example, two species of primate could have the same breeding system but live in different sized groups.

Tim Clutton-Brock and Paul Harvey tried to avoid these problems, firstly by measuring the various aspects of social behaviour and morphology on a continuous scale. Secondly, they used multivariate statistics to tease out the effects of several different ecological variables on the same traits, and to analyse the influence of ecological factors on each aspect of social organization independently. Their third improvement in approach was a careful consideration about which taxonomic level should be used for analysis, e.g. species, genus, subfamily or family.

Comparative analyses should be based on independent evolutionary events . . .

This last problem is one about the independence of data points. Imagine plotting all the species of primates on a graph in order to investigate the relationship between body weight and some interesting variable such as home range size, brain size or mating system (e.g. females per male in a breeding group). On our graph we would find that within a genus all the species will be clumped together in a cluster of points. For example, all six species of gibbons are of similar body weight, all are monogamous, arboreal and eat fruit. Our problem is whether we should treat these as six independent points or just one point in any statistical analysis. If we treated them as six independent points our analysis may be biased because it would reflect phylogeny, rather than ecology; all six gibbons may be descended from a single ancestor which was monogamous, arboreal and ate fruit. Because species within a genus tend to have similar characteristics due to phylogenetic constraints, analysis of species data will be statistically biased by those genera containing large numbers of species.

A number of different solutions to the problem of taxonomic independence have been proposed. One conservative option is to base comparative analyses on higher taxonomic levels such as genera, or even families, taking one average value for, say, all species within a genus or genera within a family. This method might succeed in eliminating problems of non-independence, but it is also likely to throw away useful data, for example where evolutionary divergence between species within a genus represents genuinely independent evolutionary events.

... which can be identified from a phylogenetic tree

Evolutionary biologists now agree that the ideal way to carry out a comparative analysis is to reconstruct a phylogenetic tree of the group under study and to use this tree to make comparisons between data points that can be clearly defined as independent evolutionary events (Harvey & Pagel 1991; Harvey & Purvis 1991; Box 2.1). In practice, such phylogenetic trees are not always available, in which case either they can be approximated (Box 2.1) or the comparative biologist can revert to the approximate, conservative approach of using higher taxonomic levels for statistical comparisons, as do the examples in the following sections.

We shall now consider some examples of the comparative approach to primate social organization and morphology, treating different genera as independent points for analysis, to illustrate how comparison has become a more rigorous and objective exercise.

HOME RANGE SIZE

Larger animals need to eat more food and so, in general, we would expect them to have larger home ranges. Therefore, if we want to examine the influence of an ecological variable, such as diet, on home range size, we have to control for body weight as a confounding variable. When home range size is plotted against the total weight of the group that inhabits it, as expected the larger the group weight the larger the home range (Fig. 2.4).

Specialist feeders have larger home ranges than do generalists

The influence of diet on home range size can be seen when the specialist feeders (insectivores, frugivores) are separated from the leaf eaters (folivores); the specialist feeders have larger home ranges for a given group weight. The probable explanation is that fruit and insects are more widely dispersed than leaves and so specialist feeders need a larger foraging area in which to find enough food.

Box 2.1 *The comparative method of independent comarisons.*

1 In carrying out a comparative analysis, it is essential to base the statistical analysis on independent evolutionary events. For example, Höglund (1989) found a strong association between lekking behaviour and sexual dimorphism in birds when he used species as the units of comparison, but the relationship disappeared when he did the correct comparison of independent points. Be wary of species as data points!
2 The ideal method is to construct a phylogenetic tree, as illustrated below:

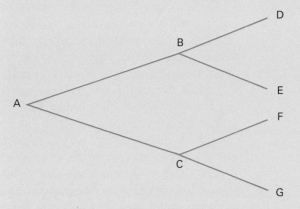

In this tree A gave rise to two descendants, B and C, each of which gave rise to two more, D, E, F and G.
3 If the phylogeny and the 'character states' (values of the relevant variables) are fully known, comparisons can be made between each ancestor–descendant pair as independent events (A and B, B and D, B and E, A and C, C and F, C and G). Thus one would have six comparisons of, say, sexual dimorphism and home range size.
4 Most often the ancestral character states are not known. The living species D, E, F and G can be studied directly, but the behaviour of A, B and C can only be guessed. In this case comparisons can be made between D and E and between F and G on the basis of measured values, and between B and C on the basis of an estimated value (e.g. half-way between the two descendant species). There is considerable debate about how best to estimate the ancestral values B and C (Harvey & Purvis 1991). In order to avoid such debates, some authors prefer just to analyse differences between pairs of extant species (Møller & Birkhead 1992).

5 An example of using the method of independent comparisons. De Voogd *et al.* (1992) (Fig. a) measured the volume of a brain nucleus (the higher vocal centre, HVC) involved in song learning in 45 species of passerine birds. They examined the relationship between the size of the brain nucleus relative to the rest of the brain and an estimate of song learning — the size of song repertoire. The phylogenetic tree of birds constructed from genetic divergence by Sibley and Monroe (1990) was used to apply the technique described in paragraph 3. A significant relationship between the size of song repertoire and the brain nucleus was observed. The results are expressed as differences ('contrasts') in the two measures between pairs of taxa that have been identified as independent divergences.

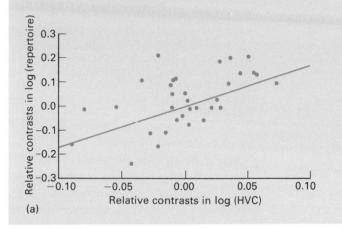

(a)

These general trends are confirmed by more detailed studies of particular species. The red colobus monkey (*Colobus badius*) is a specialist feeder, eating shoots, fruit and flowers. The food occurs in scattered clumps and this species wanders over a large home range of about 70 ha. The black and white colobus (*C. geureza*) is a generalist, eating leaves of all ages. Its food supply is dense and evenly distributed and its home range is only 15 ha (Clutton-Brock 1975).

SEXUAL DIMORPHISM IN BODY WEIGHT

In primates, males are often larger than females. Two hypotheses could explain this observation. Sexual dimorphism could enable males and females to exploit different food niches and thus avoid competition (Selander 1972). If this was true, then we might

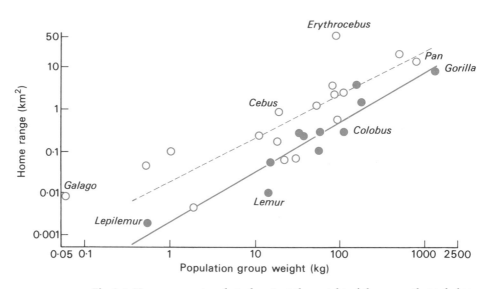

Fig. 2.4 Home range size plotted against the weight of the group that inhabits the home range for different genera of primates. The solid circles (●) are folivores, through which there is a solid regression line. The open circles (○) are specialist feeders (insectivores or frugivores) and the regression line through these points is dashed. Some of the genera are indicated by name. From Clutton-Brock and Harvey (1977).

Sexual dimorphism in body size results from sexual competition

predict that dimorphism would be greatest in monogamous species where male and female usually associate together and feed in the same areas. Alternatively it could have evolved through sexual selection, large body size in males being favoured because this increases success when competing for females (Darwin 1871). If sexual competition is important then we would predict that dimorphism should be greater in polygamous species, where large male size would be especially advantageous because a male could potentially monopolize several females.

The comparative data show no sign of the trend predicted by the niche separation hypothesis but do support the sexual competition hypothesis; the more females per male in the breeding group, the larger the male is in relation to the female (Fig. 2.5).

SEXUAL DIMORPHISM IN TOOTH SIZE

Males often have larger teeth than females. Again, two hypotheses can be suggested (Harvey *et al.* 1978). Large teeth may have evolved in males for defence of the group against predators. Alternatively males may have larger teeth for competition with other males over access to females. There is the problem here of body weight as a confounding variable; males are larger than

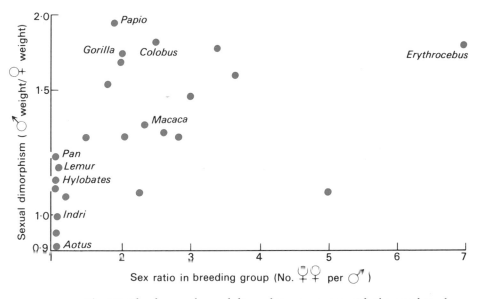

Fig. 2.5 The degree of sexual dimorphism increases with the number of females per male in the breeding group. Each point is a different genus, some of which are indicated by name. From Clutton-Brock and Harvey (1977).

females and so a difference between the sexes in tooth size could just reflect a difference in body size.

This can be controlled for by calculating the line of best fit when female tooth size is plotted against body weight. If the tooth size of a male is now plotted on the same graph, it can be seen whether its size is greater than expected for a female of the same body weight. The results show that in monogamous species male tooth size is as expected for a female of equivalent body weight. However it is larger than expected in harem-forming species. These data support the sexual competition hypothesis for the evolution of larger teeth in males. However we cannot exclude the predator defence hypothesis because maybe the harem-forming species arc the ones most vulnerable to predation.

The analysis can be taken a step further by considering species where several males live together in a group (multi-male troops). It is found that, within this type of social organization, the males of terrestrial species have larger teeth for their body size than arboreal species. Therefore even within the same mating system there is a difference in tooth size in different habitats. The terrestrial environment is usually thought to present greater risks of predation and so predation pressure may have been responsible for the evolution of larger teeth in terrestrial species.

Our conclusion is that both sexual competition and predation may have influenced the evolution of sexual dimorphism in tooth

Sexual dimorphism in tooth size is related to sexual competition and defence against predators

size. There is also the further possibility that differences in tooth size are important in reducing diet overlap between the sexes and so preventing competition for food. This example shows that, even with careful analysis, it may be difficult to tease out the effect of several variables on the evolution of a trait.

TESTIS SIZE AND BREEDING SYSTEM

The heaviest primates, the gorilla (*Gorilla gorilla*) and orangutan (*Pongo pygmaeus*) have breeding systems that involve one male monopolizing mating with several females, and have testes that weigh 30 g and 35 g respectively (average weight of both testes). The smaller chimpanzee (*Pan troglodytes*), by contrast, has a breeding system where several males copulate with each oestrus female and this species has testes weighing 120 g! It seems likely that the marked differences in testes weights are related to differences in breeding system. In single-male breeding systems (gorilla and orang-utan) each male need ejaculate only enough sperm to ensure fertilization. In multi-male systems (chimpanzee), however, a male's sperm has to compete with sperm from other males. Selection should, therefore, favour increased sperm production and hence larger testes.

Large testes occur in multi-male groups

Harcourt *et al.* (1981) tested this hypothesis by comparing 20 genera of primates, varying in body size from the 320-g marmoset (*Callithrix*) to the 170-kg gorilla. Figure 2.6 shows that, as ex-

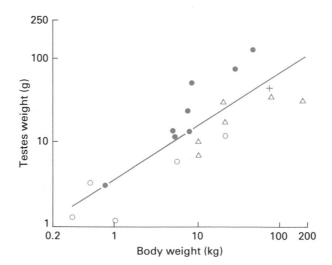

Fig. 2.6 Log combined testes weight (g) versus log body weight (kg) for different primate genera. ●, Multi-male breeding system; ○, monogamous; △, single-male; +, *Homo*. From Harcourt *et al.* (1981).

pected, testes weight increases with body weight. For a given body weight, however, it is clear that genera with multi-male breeding systems have heavier testes than genera with single-male or monogamous breeding systems. The data points for the former group lie above the line, and those for the latter lie below ('single-male' indicates that there is only one breeding male although, as in the gorilla, there may be more than one male in the social group; 'monogamous' indicates that there is just one male and one female in a group). These data therefore support the sperm competition hypothesis.

The comparative approach reviewed

The statistical approach we have described for the primates is certainly a major improvement on the first applications of the comparative method. To summarize, the main improvements are as follows.
1   Different aspects of social organization are treated independently and as continuous variables.
2   Confounding variables are dealt with in a rigorous manner.
3   Care is taken to choose the most appropriate taxonomic level for analysis.
4   The data are used wherever possible to discriminate between alternative hypotheses such as predation or sexual competition.

The comparative method can be used for testing hypotheses not amenable to experimentation

The end result of many of the analyses is a plausible interpretation which may be treated as a hypothesis for further testing. In conclusion, the comparative approach is very useful for looking at broad trends in evolution and the general relationship between social organization and ecology. It generates hypotheses which can be used as predictions for other groups of animals. It can also be used to test hypotheses which are not amenable to experimentation, such as the effect of polygamy on sexual dimorphism. Furthermore, it is impressive in the way it shows how diet, predation, social behaviour and body size, for example, can all be interrelated.

However, we need a different approach to understand in detail why animals adopt particular strategies in relation to their ecology. Can we actually measure patch structure and predation risk and then come up with precise predictions as to how an animal will behave? Can we explain why a monkey goes round in a group of 20 rather than one of 16 or 25, why its home range is 10 ha rather than 8 or 12 ha, and why it spends 1 hour in a patch of fruit? Indeed we can attempt to answer precise questions like these. No one has yet attempted this for anything as complicated as primate social behaviour. However a start has been made for simpler kinds of behaviour using optimality theory and an experimental approach.

Experimental studies of adaptation

We now turn to a different, and complementary, way of looking at how selection moulds behaviour. Instead of broad scale comparisons between species, the emphasis will be on the behaviour of individuals of the same species and analysing their behaviour in terms of *costs* and *benefits*.

The idea of trying to measure costs and benefits grew out of Niko Tinbergen's experimental approach to studying the survival value of behaviour. For example, Tinbergen observed that in a colony of black-headed gulls (*Larus ridibundus*) nesting on sand dunes in north-western England, incubating parents always pick up the broken eggshell after a chick has hatched and carry it away from the nest (Fig. 2.7). Although carrying the shell takes only a few minutes each year it is crucial for the survival of the young. The eggs and young of the black-headed gull are well camouflaged against the grass, sand and twigs around the nest. The inside of the broken shell, however, is white and highly conspicuous. Tinbergen carried out an experiment to test the hypothesis that the conspicuous white broken shell reduces the camouflage of the nest. He painted hens' eggs to resemble cryptic gull eggs and laid them out at regular intervals in the gull colony. Next to some he placed a broken shell. The results confirmed his prediction that the cryptic eggs were much more likely to be discovered and eaten by predators such as crows if they were close to a broken shell. So it is easy to visualize why the parent benefits by removing the conspicuous empty shell soon after the chick has hatched: the camouflage of the brood is preserved and the likelihood of the parent perpetuating its genes is increased.

But there is more to the story than this. The parent does not remove the eggshell immediately; it stays with the newly hatched chick for an hour or more and then goes off with the shell. In order to explain the delay in removing the shell we have to introduce the idea of a trade-off between costs and benefits. If the parent flies off with the shell at once, it has to leave the newly hatched chick unattended (the second parent is away at the feeding grounds fuelling up for its next stint at the nest). Tinbergen observed that the new chick, with its plumage still wet and matted, is easily swallowed and therefore makes a tempting meal for a cannibalistic neighbouring adult. However, when the chick's

Costs and benefits tested experimentally: eggshell removal in gulls

Fig. 2.7 A black-headed gull removing an eggshell from its nest. Photo by N. Tinbergen.

down has dried out and become fluffy it is much harder for a gull to swallow, and is therefore less vulnerable to attacks from neighbours. The parent's delay before removing the shell therefore probably reflects a balance between the benefits of maintaining the camouflage of the brood and the costs associated with leaving a newly hatched chick at its most vulnerable moment.

A different balance of selective pressures in oystercatchers

When the balance between costs and benefits is changed, the length of the parent's delay might also be expected to change. This is borne out by observations of the oystercatcher, another ground nesting bird with camouflaged eggs and young. The oystercatcher (*Haematopus ostralegus*) is a solitary nester and cannibalism by neighbours is therefore not a risk associated with leaving the newly hatched chicks. The parents benefit by restoring camouflage of the nest as soon as possible after hatching and as expected the parent removes broken eggshells more or less as soon as a chick has hatched and before its down is dry.

OPTIMALITY MODELS

Tinbergen's study of eggshell removal illustrates how experimental studies of costs and benefits can be used to unravel behavioural adaptations, but it has an important limitation. The hypothesis about the trade-off between camouflage and chick vulnerability made only a qualitative prediction. The idea would be consistent with observations of gulls removing eggshells 1, 2, 3 or perhaps even 4 hours after the chick has hatched, so that it is hard to test whether the hypothesis is right or wrong. One way of trying to make a hypothesis more easily testable is to try to

Quantitative models of costs and benefits

generate quantitative predictions. If one could predict that the parent gull should remove its eggshell after 73.5 min then one would have produced a very testable model indeed. This is an approach which has been developed by using *optimality models* to study adaptations. An optimality model seeks to predict which particular trade-off between costs and benefits will give the maximum net benefit to the individual.

Thinking back to the gulls, if one could measure exactly how much the survival of the brood is reduced by the conspicuous broken eggshell next to the nest and exactly how the risk of cannibalism by neighbours changes with time since the chick hatched, one could start to calculate the optimum time for the parent to delay removal of the shell. In this case the optimum might well be defined as the time that maximizes total reproductive success for the season. But the currency of an optimality model does not have to be survival or production of young. The overall success of an individual at passing on its genes may depend on finding enough food, choosing a good place to nest, attracting many mates, and so on. In solving any of these problems an animal makes decisions, and the decisions can be analysed in terms of an optimal trade-off between appropriate costs and benefits. For a foraging animal, for example, currencies might be energy and time.

Gathering food

CROWS AND WHELKS

On the west coast of Canada, as in many coastal areas, crows feed on shellfish. They hunt for whelks at low tide, and having found one they carry it to a nearby rock, hover and drop it from the air to smash the shell on the rock and expose the meat inside. Reto Zach (1979) observed the behaviour of north-western crows in detail and he noted that they take only the largest whelks and on average drop the shell from a height of about 5 m. Zach carried out experiments in which he dropped whelks of different sizes from various heights. This, together with data on the energetic costs of flying and searching, gave him the information to carry out calculations of the costs and benefits associated with foraging. The benefit obtained by the crow and the cost paid could both be measured in calories, and Zach's calculations revealed that only the largest whelks (which contain the most calories and break open most readily) give enough energy for the crow to make a net profit while foraging. As predicted from these calculations, the crows ignored all but the very largest whelks even when different sizes were laid out in a dish on the beach.

Usually the crow has to drop each whelk twice or more in order to break it open. Since ascending flight is very costly, Zach thought that the crow might have chosen the dropping height which would minimize the total expenditure of energy in upward flight. If each drop is made from close to the ground, a very large number of drops is required to break open the shell, while at greater and greater heights the shell becomes more and more likely to break open on the first drop (Fig. 2.8a). The experiment of dropping shells from different heights allowed Zach to calculate the total vertical flight needed to break an average shell from different dropping heights (Fig. 2.8b). The crows fly to an average height of 5.2 m and at this height the total vertical flight per whelk is close to its minimum. However, the crow would have to undertake almost the same total upward flight even if each drop was made from a height somewhat greater than 5.2 m (this is indicated by the very shallow U-shaped curve of Fig. 2.8b) because slightly fewer drops would be needed. Zach suggests that there may be an additional penalty for dropping from too great a height: the whelk may bounce away and be lost from view or may break into so many fragments that the pieces are too small to retrieve.

The story of crows and whelks shows how calculations of costs and benefits can be used to produce a quantitative prediction. The crow seems to be programmed to choose a dropping height that comes close to minimizing the total vertical flight per whelk. Other currencies, such as maximizing net rate of energy gain, predict much greater drop heights (Plowright *et al.* 1989).

Crows minimize the upward flight distance needed to break a whelk

Summary

Throughout this book we will be using both the comparative method and the experimental analysis of costs and benefits to individuals to test hypotheses about behavioural adaptations. The

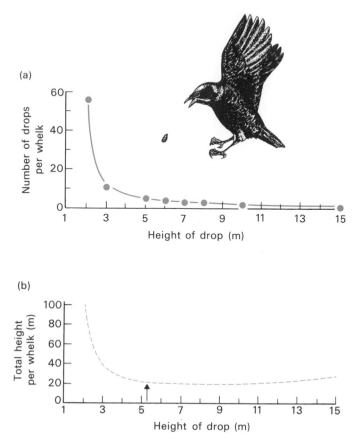

(a)

Number of drops per whelk

Height of drop (m)

(b)

Total height per whelk (m)

Height of drop (m)

Fig. 2.8 Results obtained by dropping whelk shells from different heights. (a) Fewer drops are needed to break the shell if it is dropped from a greater height. (b) The total upward flight needed to break a shell (no. of drops × height of each drop) is close to its minimum at the height most commonly used by the crows (shown by arrow). From Zach (1979).

comparative method involves comparing different species to see whether differences in behaviour are correlated with differences in ecology. In weaver birds, antelopes and primates the main factors determining the evolution of social behaviour are the distribution and abundance of food, predators and mates. These have been shown to influence group size, home range size, mating behaviour and sexual dimorphism. Two of the main problems in comparative studies are those of confounding variables and the choice of taxonomic level for comparison.

The experimental approach involves a detailed study of the costs and benefits of a behaviour pattern to an individual of a particular species. Behaviour can be viewed as having costs and benefits and animals should be designed by natural selection to maximize net benefit. Ultimately the net benefit must be measured in terms of gene contribution to future generations. This will depend on various shorter-term goals such as foraging efficiency, mating efficiency and efficiency of avoiding predators. Optimality models can be used to predict which particular trade-off between costs and benefits gives maximum net benefit.

Further reading

Harvey and Pagel (1991) provide a good discussion of some of the problems in using the comparative approach. David Lack's (1968) book applies the comparative method to the breeding biology of birds. Mark Ridley (1983) discusses the methodology of the comparative approach, particularly the problem of what to use as independent data points. Niko Tinbergen's book on gulls (1953) and his book *Curious Naturalists* (1974) give a marvellous sense of the excitement of observing animals in the field and show the power of performing simple field experiments to help us understand why animals behave the way they do.

Topics for discussion

1 What are the relative merits of the comparative method and the experimental approach for studying the adaptive significance of behaviour?
2 How can we decide what are the independent units of observation in comparative studies?
3 How could Jarman's (1974) analysis of ungulate social organization be improved by applying Harvey and Pagel's (1991) methods?

Chapter 3. Economic Decisions and the Individual

In this chapter we will describe in more detail how the idea of economic analysis of costs and benefits can be used to understand behaviour. Most of our examples will refer to foraging, but the same principles apply to other aspects of behaviour.

The economics of carrying a load

STARLINGS

Starlings feed their young mainly on leatherjackets (*Tipula* larvae) and other soil invertebrates. A busy parent at the height of the breeding season makes up to 400 round trips from its nest to feeding sites every day, ferrying a load of food to its insatiable nestlings on each trip (Fig. 3.1). In this section we are going to focus our miscroscope of economic analysis on one aspect of the parent starling's behaviour and ask: How many leatherjackets should the parent bring home on each trip? This may seem like

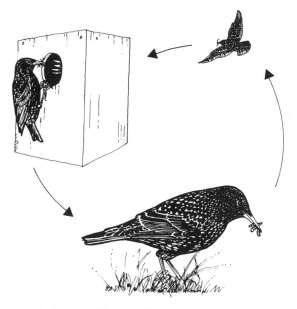

Fig. 3.1 Starlings fly from their nest to a feeding site, search for a beak-full of leatherjackets by probing in the grass, then take them home to the nestlings. The question examined in the first part of this chapter is how many items the parent should bring on each trip in order to maximize the rate of delivery of food to the nestlings.

an inconsequential question, but the size of load has a critical effect on the parent's overall rate of delivering food to the nest, which in turn determines whether or not the chicks survive to become healthy fledglings. As we saw in Chapter 1, the reproductive success of small birds is often limited by their ability to feed their young. There is therefore strong selective pressure on the parents to perform as effective food deliverers.

Optimal load size in starlings: diminishing returns

It is not hard to imagine why it would be a poor strategy to bring one leatherjacket on each trip, given that the parent is in fact capable of carrying several at a time: the amount of travelling back and forth for the parent would be unnecessarily high. It would be as if you did your week's shopping at the supermarket by making a separate trip for each individual item instead of collecting a whole bag full of groceries on one trip. Should the starling, therefore, bring back the biggest load it can carry on every trip? The answer is no, because of a crucial fact about starling foraging behaviour. A starling hunts for food by probing with its bill in a special way. It thrusts its closed bill into the turf of a meadow and then spreads its mandibles to part the vegetation and expose the leatherjackets just below the soil surface. The bird is very efficient at probing, but when it has a load of leatherjackets already in its bill it becomes less efficient. For this reason it is not necessarily the best thing for the starling to fully load its beak before flying home to unload and start again. A similar kind of problem faces many animals that carry a load back to their nest or some other central place such as a cache of food.

The starling's problem of load size can be summarized as a graph (Fig. 3.2a). The graph shows time along the horizontal axis and load (measured in leatherjackets) on the vertical axis. Consider a starling at the nest about to embark on a round trip. It has to fly to (and eventually from) the feeding site; the times of these two trips are added together and plotted on the graph as 'travelling time'. When it arrives at the part of the meadow where the leatherjackets are abundant it starts to load up with food. The first couple of leatherjackets are found quickly and easily, but because of the encumbrance of the prey in its beak, the bird takes longer and longer to find each successive prey. The result is a 'loading curve' (or 'gain curve' as it is sometimes called) that rises steeply at first but then flattens off. This is a curve of diminishing returns — the longer the starling has been foraging, the less likely it is to find another leatherjacket in the next few moments — and the starling's problem, as summarized in the graph, is when to give up on this curve. If it gives up too early it spends a lot of time travelling for a small load; if it struggles on too long it spends time in ineffective search which could be

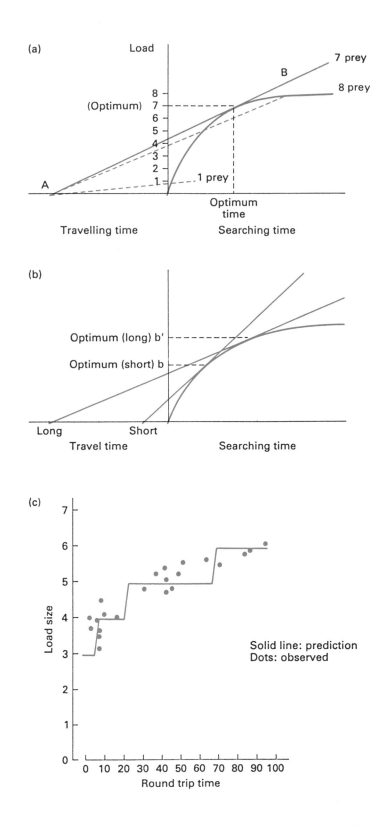

(a)

Load

8
(Optimum) 7
6
5
4
3
2
1

B
7 prey
8 prey

1 prey

A

Optimum
time

Travelling time   Searching time

(b)

Optimum (long) b'
Optimum (short) b

Long   Short
Travel time   Searching time

(c)

Load size

Solid line: prediction
Dots: observed

Round trip time

better spent by going home to dump its load and starting again at the beginning of the loading curve. Somewhere in between these extremes is the starling's 'best' option. A reasonable hypothesis (but at the moment it is no more than this), is that for the starling 'best' means 'providing the maximum net rate of delivery of food to the chicks'. Any starling that is slightly better at producing chicks than its rivals will be at a selective advantage, so selection should in the long run favour behaviour that maximizes chick production.

The theoretical optimum can be shown graphically

The best load can be found by drawing the line AB in Fig. 3.2a. The slope of this line is (load/[travel-time + foraging-time]) or in other words rate of delivery of food; this can be seen by the fact that it forms the hypotenuse of a right angled triangle with a base measured in 'time' and a vertical corresponding to 'load'. Now, travel time and the loading curve are constraints — fixed properties of the environment (or more precisely of the interaction between the starling and its environment), so any line such as AB representing the starling's rate of delivery of food has to start at the beginning of the round trip travel time and touch the loading curve somewhere. Any other line you could draw with these constraints will have a shallower slope (that is, give a lower rate of delivery) than the line AB. A couple of examples are shown in Fig. 3.2a.

The model predicts smaller loads with shorter travel times

Figure 3.2b develops the argument a little further. Suppose that the starling now switches to feeding at a closer site with a short travel time, how should its load per trip change? Using the same method as before, we can now draw two lines (Fig. 3.2b):

Fig. 3.2 *Opposite.* (a) The starling's problem of load size. The horizontal axis shows 'time' and the vertical axis shows 'load'. The curve represents the cumulative number of leatherjackets found as a function of time spent searching. The line AB represents the starling's maximum rate of delivery of food to the nestlings. This rate is achieved by taking a load of 7 leatherjackets on each trip. Two other lines, corresponding to loads bigger (8) and smaller (1) than 7 are shown to make the point that these loads result in lower rates of delivery (shallower slopes). Note that although the cumulative load is shown here as a smooth curve, in reality it is a stepped line since each food item is a discrete package. (b) When the round trip travel time is increased from short to long the load size that maximizes delivery rate increases from b to b'. (c) When starlings were trained to collect mealworms from a feeder, they brought bigger loads from greater distances. Each dot is the mean of a large number of observations of loads brought from a particular distance. The predicted line goes up in steps because the bird is predicted to change its load size in steps of one worm (of course the mean loads do not have to be integers). The prediction shown here is one based on the model of Fig. 3.2b, but it also includes the refinement of taking into account the energetic costs to the parent of foraging and to the chicks of begging. From Kacelnik (1984).

when the travel time is shorter, the load that maximizes rate is smaller. One way to think of this is to imagine the starling at its moment of decision to go home. If it goes it loses the opportunity to continue foraging, if it stays it loses the opportunity to go home and start afresh. When it is far away the expected returns from going home are relatively low, since there is a long way to fly before the next chance to forage. It therefore pays to persist a little longer on the present trip, until current gains drop to a slightly lower level.

The predicted and observed relationships between load size and travel time are similar

Alex Kacelnik (1984) tested this prediction of the model of load size in the following way. He trained parent starlings in the field to collect mealworms for their young from a wooden tray onto which he could drop mealworms through a long piece of plastic pipe. Rather than letting the birds generate their own loading curve by diminishing search efficiency, Kacelnik generated the curve for them by dropping mealworms at successively longer and longer intervals. The trained bird would simply wait on the wooden tray for the next worm to arrive, until eventually it flew home with a beak-load for its chicks. The beauty of this experimental method is that Kacelnik knew the shape of the loading curve precisely and was hence able to present exactly the same loading curve at randomly varying distances (ranging from 8 to 600 m) from the nest on different days. The results were striking (Fig. 3.2c): not only did the load size increase with distance from the feeder to the nest, but there was a close quantitative correspondence between the observed load sizes and those predicted by the model of maximizing delivery rate.

Optimality models include assumptions about currencies and constraints

Let us briefly summarize what the results of the starling study show. We started off by considering load size from the point of view of costs and benefits. We formulated a specific hypothesis about how costs and benefits might influence load size in the form of a model (Fig. 3.2a), and then used the model to generate a quantitative prediction (Fig. 3.2b). In making the model we did three important things. First we expressed a general conviction that starlings are designed by natural selection to be good at their job of parenting. This is not something that we aimed to test, but it is our general background assumption to justify thinking in terms of maximizing pay-off in relation to costs and benefits. Second, we made a guess about the *currency* of costs and benefits; we suggested that for a parent starling the crucial feature of doing a good job is maximizing net rate of delivery of food to the nestlings (rather than, say, waking them up as little as possible which is what human parents of young babies might favour!). Third, we specified certain *constraints* on the starling's behaviour. Some of these constraints are to do with features of the environ-

(a) (b)

Fig. 3.3 (a) The model in Fig. 3.2 can be used to predict the duration of copula in dungflies (Box 3.1) and (b) to make predictions about many other examples of animals exploiting patchily distributed resources, such as these goldfinches (*Carduelis carduelis*) feeding on teazel heads.

ment (the time required to travel, the shape of the loading curve) and in the experiment of Kacelnik, they were clearly identifiable. Another important assumption about constraints is that the starling is assumed to 'know', or at least to behave as if it knows, the travel time and the shape of the loading curve. When we worked out the optimum load size we assumed that these were known. The experimental results supported the predictions of the model, and in so doing they supported the hypotheses about the currency and the constraints that were used to construct the model. Kacelnik did actually compare the predictions of models based on several different currencies, and he found for example, that one based on energetic efficiency (energy gained/energy spent) as opposed to rate gave a rather poor fit to the data.

The importance of comparing alternative models

The model of load size (usually called the 'marginal value theorem' (Charnov 1976)) is applicable to lots of situations in which an animal exploits a resource that occurs in discrete sites or patches, and within a patch it experiences diminishing returns (see Box 3.1 and Fig. 3.3).

BEES

How much nectar should a bee carry home?

A similar problem to that of the starling is faced by a worker honey bee as it flies from flower to flower filling its honey crop with nectar to take back to the hive. Bees, like the starlings, often return to the hive with less than the maximum load they could carry and their behaviour can be explained by a model

In bees, diminishing
returns arise from the
cost of carrying nectar

similar to that used for the starling. There is, however, an important difference: the bee experiences a curve of diminishing returns neither because the nectar in its crop makes it less able to suck more flowers nor because of resource depression (Box 3.1) but because the weight of nectar in the crop adds an appreciable energetic cost to flight. The more the bee loads up its crop the more of its load it will burn up as fuel before it gets home. As a consequence, while the gross quantity of nectar harvested increases at a constant rate, the *net* yield of energy for the hive increases at a diminishing rate as the crop fills (producing, in effect, a loading curve like that of the starling).

Paul Schmid-Hempel (Schmid-Hempel *et al.* 1985) tested whether these diminishing net returns influence the bee's decision about when to go home and empty its crop. He trained bees to fly from the hive to a cluster of artificial flowers, each containing 0.6 mg of nectar. By varying the amount of flight the bee had to do between each flower in the cluster, he could alter the total cost of carrying the crop load and therefore the extent to which the bee experienced diminishing net returns. If, for example, the bee could collect a load of 10 flower's-worth of nectar while flying for a total of 5 s, it would experience little decrease in returns as it loaded up, while a bee collecting the same load by flying for a total of 50 s would suffer sharply diminishing returns. As predicted, the bees went home with smaller loads when they were forced to fly a greater distance between flowers (Fig. 3.4a). Figure 3.4a also shows two predicted lines based on maximizing two different currencies. One is based on the currency used for the starlings, net rate of energy delivery, while the other is based on a currency that did not work for the starlings, energetic efficiency. In contrast to the starlings, the second currency but not the first accounts for the bees' behaviour.

Bees maximize efficiency
and not rate of energy
gain

Box 3.1 *The marginal value theorem and reproductive decisions.*

The model of load carrying for starlings is applicable to many other situations in which animals experience diminishing returns within a patch. It has been used to predict how much time an animal foraging for itself (as opposed to carrying loads) will spend in each site before moving on (Cowie 1977). Diminishing returns in each patch (generally referred to as 'resource depression') might arise, for example, simply because of depletion, or because prey in the patch take evasive action and become harder to catch, or because the predator becomes less likely to search new areas in the patch (it crosses its own path more) as time goes by, or because the predator starts with the easy prey and then goes on to hunt for those that are more difficult to catch or are less rewarding. An example of the last of these is when bumble-bees or other nectar-feeders visit the biggest and most reward-ing flowers on an inflorescence first, and then go on to the smaller flowers which hold less nectar (Hodges & Wolf 1981).

Reproductive decisions can be analysed as though they are foraging decisions with the same model. An example is Geoff Parker's (1978) analysis of how male dungflies search for mates. Males compete with one another for the chance to mate with females arriving at cowpats to lay their eggs. Often one male will succeed in kicking another off a female during copulation and take her over. When two males mate with the same female the second one is the individual whose sperm fertilizes most of the eggs. Parker (1978) showed this by the clever technique of irradiating males with cobalt-60, which sterilizes them but does not alter sperm activity (the sperm can still fertilize an egg but the egg does not develop). If a normal male is allowed to mate after a sterile one about 80 per cent of the eggs hatch, whereas if the sterile male mates second only 20 per cent of them hatch. The conclusion from these 'sperm competition' experiments is clear: the second male's sperm fertilizes about 80 per cent of the eggs. It is not surprising, therefore, that after a male has copulated he sits on top of the female and guards her until the eggs are laid, only relinquishing his position to a rival male after a severe struggle.

When a second male takes over (or when a male en-counters a virgin) how long should he spend copulating? Parker carried out sperm competition experiments in which he interrupted the second male's copulation after different times; this showed that the longer the second male mates the more eggs he fertilizes, but the returns for extra copu-lation time diminish rapidly (see Fig. on p. 56). There is a cost associated with a long copulation: the male misses the chance to go and search for a new female. After the male has copulated for long enough to fertilize about 80 per cent of the eggs, the returns for further copulating are rather small and the male might do better by searching elsewhere for a new mate.

The analogue of travel time in the starling model is the time the male dungfly must spend guarding the present

Continued on p. 56

Box 3.1 *Continued*

female until she has laid her eggs plus the time he spends
searching for a new female. This total is 156 min on average.
As shown below, this estimate of travel time can be used to
predict with reasonable accuracy how long the male spends
copulating with a female.

Some other reproductive decisions that can be analysed
in a similar way are discussed by Charnov and Skinner
(1984).

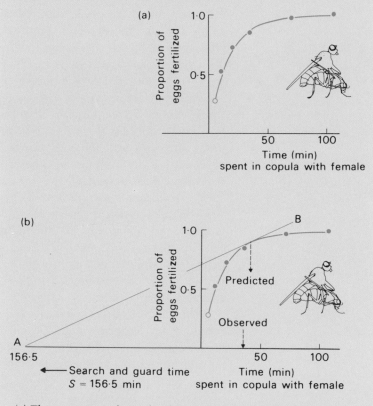

(a) The proportion of eggs fertilized by a male dungfly (*Scatophaga
stercoraria*) as a function of copulation time: results from sperm
competition experiments. (b) The optimal copulation time (that which
maximizes the proportion of eggs fertilized per minute), given the
shape of the fertilization curve and the fact that it takes 156 min to
search for and guard a female, is 41 min. The optimal time is found by
drawing the line AB. From Parker (1978).

Why should there be this difference between bees and starlings?
A simple example shows why the 'starling currency' is normally
a sensible one to consider. Compare a starling that forages for

(a)

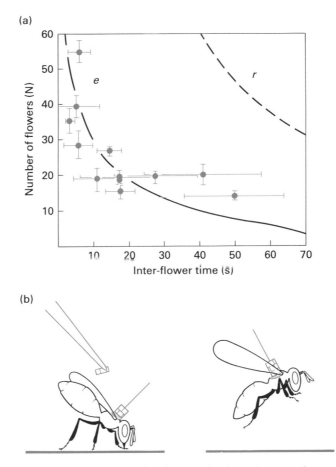

(b)

Fig. 3.4 (a) The relationship between load size (expressed as number of flowers visited) carried home by worker bees and flight time between flowers in a patch. Each dot is the mean of an individual bee and the two lines are predictions based on maximizing efficiency (*e*) and maximizing rate (*r*). From Schmid-Hempel *et al.* (1985). (b) By placing tiny weights on the bee's back while it is foraging Schmid-Hempel was able to study the bee's rule of thumb for departure from a patch to go home to the hive with a load of nectar. The weights, in the form of brass nuts, are placed on a fine rod that is permanently glued to the bee's back. They can be added or removed to simulate loading and unloading. From Schmid-Hempel (1986).

1 hour, spending 1 kJ and gaining 9 kJ with one that spends 10 and gains 90. Both have the same efficiency (9) but the former has 8 kJ to spend on its chicks and the latter has 80. In other words net rate ((gain − cost)/time) tells us how much the animal has left at the end of the day to spend on reproduction or survival while efficiency does not. On the other hand efficiency may be a sensible currency when the crucial variable for the animal is not

Life expectancy of bees
depends on work load

just the amount of gain, but also the amount spent. If, for example you have to drive from A to B on a fixed amount of fuel, efficiency might be very important indeed. It turns out that bees may be in this position. The equivalent for a foraging bee to having a fixed amount of fuel would be if it had a more or less fixed total lifetime capacity for expenditure of energy. Wolf and Schmid-Hempel tested this idea by manipulating the rate of energy expenditure of individuals, either by varying the time they were allowed to forage each day (Schmid-Hempel & Wolf 1988) or by fixing different sized permanent weights onto their backs (Wolf & Schmid-Hempel 1989). Both experiments showed that the bees that worked hardest survived for a shorter time than controls. For example, when workers carried a permanent weight of >20 mg, their survival was reduced from 10.8 to 7.5 days. These experiments lend some support to the hypothesis that workers, by maximizing efficiency, might extend their lifespan and thus contribute more nectar overall to the colony than they would by maximizing net rate.

The contrast between bees and starlings serves to underline the point that one of the aims of economic cost−benefit analyses is to compare alternative currencies and to try to understand why a particular currency is appropriate in each case. In each study one of the major advantages of the quantitative analysis was that it allowed us to see when there was a *discrepancy* between observed and predicted results. Without this potential for discrepancy it would have been impossible, for example, to tell whether bees were maximizing rate or efficiency, or nothing at all.

The bee example also illustrates another important point. We have been thinking of animals as well-designed problem solvers making decisions that maximize an appropriate currency, but of course we do not believe that bees and other animals calculate their solutions in the same way as the behavioural ecologist. Instead the animals are programmed to follow rules of thumb which give more or less the right answer. The bees, for example, might use a rule that involved a threshold body weight ('if weight greater than x then go home'). Schmid-Hempel (1986) investigated this by adding tiny (7 mg) weights to the bee's back while it was foraging (Fig. 3.4b). He found that when he added five weights at intervals during a foraging bout the bees went home with a smaller load, as predicted if they were using a threshold weight rule. However, another experiment showed that the rule is not this simple. Instead of adding five weights gradually, Schmid-Hempel added five weights at the start of a foraging bout and then took them off gradually as the bee filled its crop. These bees

Adding weight to the
bee's back causes it to fly
home with a smaller load

also went home with smaller loads than unmanipulated bees (or than controls where the weights were placed on the bee's back for a brief moment). The most reasonable interpretation of these results is that the bee in some way integrates the total weight it has carried since arriving at the foraging site.

The economics of prey choice

The same kind of economic approach that we have used for bees and starlings can also be used to account for the kinds of prey items that predators decide to eat.

When shore crabs are given a choice of different sized mussels they prefer the size which gives them the highest rate of energy return (Fig. 3.5). Very large mussels take so long for the crab to

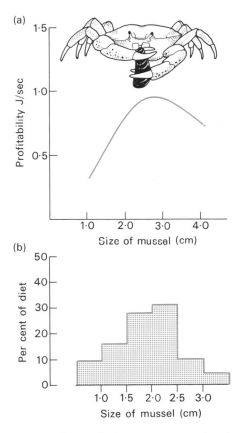

Fig. 3.5 Shore crabs (*Carcinus maenas*) prefer to eat the size of mussel which gives the highest rate of energy return. (a) The curve shows the calorie yield per second of time used by the crab in breaking open the shell and (b) the histogram shows the sizes eaten by crabs when offered a choice of equal numbers of each size in an aquarium. From Elner and Hughes (1978).

crack open in its chelae that they are less profitable in terms of energy yield per unit breaking time (E/h) than the preferred, intermediate sized, shells. Very small mussels are easy to crack open, but contain so little flesh that they are hardly worth the trouble. However the story cannot be as simple as this, because the crabs eat a range of sizes centred around the most profitable ones. Why should they sometimes eat smaller and larger mussels? One possible hypothesis to explain why several sizes are eaten is that the time taken to search for the most profitable sizes influences the choice. If it takes a long time to find a profitable mussel, the crab might be able to obtain a higher overall rate of energy intake by eating some of the less profitable sizes.

Optimal prey choice depends on energy values, handling time ...

In order to calculate exactly how many different sizes should be eaten we need to develop a more precise argument based on handling time, searching time, and the energy values of the various prey (Box 3.2). The equations in Box 3.2 show the following for the simple example of a predator faced with a choice of two sizes of prey. First, when the more profitable type (higher E/h) is very abundant the predator should specialize on this alone. This is intuitively obvious: if something giving a high rate of return is readily available, an efficient predator should not bother with less profitable items. Secondly the availability of the less profitable prey should have no effect on the decision to specialize on the better prey. This also makes sense: if good prey are encountered sufficiently often to make it worthwhile to ignore the bad ones, it is never worth taking time out to handle bad prey regardless of how common they are. The third conclusion from Box 3.2 is that as the availability of the good prey increases, there should be a sudden change from no preference (the predator eats both types when encountered) to complete preference (the predator eats only the good prey and always ignores the bad ones).

... and search time

A test of the optimal diet model

An experiment which tested these predictions is illustrated in Fig. 3.6. The predators were small birds (great tits) and the prey were large and small pieces of mealworm. In order to control precisely the predator's encounter rate with the large and small worms the experiment involved the unusual step of making the prey move past the predator rather than vice versa (Fig. 3.6a). The big worms in the experiment were twice as large as the small ones ($E_1/E_2 = 2$) and h_1 and h_2 could be accurately measured as the time needed for the bird to pick up a worm and eat it. During the experiment the bird's encounter rate with large worms was varied so as to cross the predicted threshold from non-selective to selective foraging (equation 3.2) in Box 3.2. The results were qualitatively but not quantitatively as predicted, the main difference between observed and expected results being that the switch

Box 3.2 *A model of choice between big and small prey.*

Consider a predator which encounters two prey types, big prey$_1$ with energy value E_1 and handling time h_1, and small prey$_2$ with energy value E_2 and handling time h_2. The profitability of each prey (energy gain per unit handling time) is E/h. Imagine that the big prey are more profitable, so

$$\frac{E_1}{h_1} > \frac{E_2}{h_2}$$

How should the predator choose prey so as to maximize its overall rate of gain? Let us assume that the predator has encountered a prey — should it eat it or ignore it?

(a) If it encounters prey$_1$, it should obviously always eat it. Therefore choice of the more profitable prey$_1$ does not depend on the abundance of prey$_2$.

(b) If it encounters prey$_2$, it should eat it provided that

Gain from eating > gain from rejection and searching for a more profitable prey$_1$

i.e. if

$$\frac{E_2}{h_2} > \frac{E_1}{S_1 + h_1} \tag{3.1}$$

where S_1 is the search time for prey$_1$.

Re-arranging, the predator should eat prey$_2$ if

$$S_1 > \frac{E_1 h_2}{E_2} - h_1 \tag{3.2}$$

Thus the choice of the less profitable prey, prey$_2$, does depend on the abundance of the more profitable prey, prey$_1$.

This model makes three predictions. First, the predator should either just eat prey$_1$ (specialize) or eat both prey$_1$ and prey$_2$ (generalize). Second, the decision to specialize depends on S_1, not S_2. Third, the switch from specializing on prey$_1$ to eating both prey should be sudden and should occur when S_1 increases such that equation (3.2) is true. Only when the two sides of the equation are exactly equal will it make no difference to the predator whether it eats one or both types of prey.

was not a step but a gradual change (Fig. 3.6b). When big worms were abundant the birds, as predicted, were selective even if small worms were extremely common.

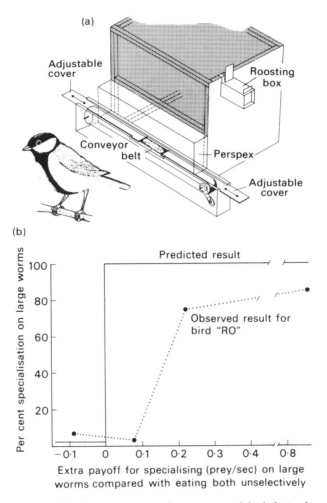

Fig. 3.6 (a) The apparatus used to test a model of choice between big and small worms in great tits (*Parus major*). The bird sits in a cage by a long conveyor belt on which the worms pass by. The worms are visible for half a second as they pass a gap in the cover over the top of the belt and the bird makes its choice in this brief period. If it picks up a worm it misses the opportunity to choose ones that go by while it is eating. (b) An example of the results obtained. As the rate of encounter with large worms increases the birds become more selective. The x-axis of the graph is the extra benefit obtained from selective predation. As shown in Box 3.2, the benefit becomes positive at a critical value of S_1, the search for worms. The bird becomes more selective about the predicted point, but in contrast with the model's prediction this change is not a step function. From Krebs *et al.* (1977).

Sampling and information

The discussion so far has referred to animals that know their environment. In Fig. 3.2a, the model assumes that the animal knows the quality of each patch and the travel time before it

Optimal sampling by woodpeckers

starts. Sometimes this may be a reasonable assumption, but at other times it may be more realistic to assume that the animal learns as it goes along. Steve Lima (1984) studied this kind of problem in downy woodpeckers. He trained woodpeckers in the field to hunt for seeds hidden in holes drilled in hanging logs. Each log had 24 holes and in each experiment some logs were quite empty and others had seeds hidden in some or all of the holes. The woodpeckers could not tell in advance which were the empty logs so they had to use information gathered at the start of foraging on each log to decide whether or not it was likely to be empty and therefore should be abandoned. When the logs contained 0 or 24 seeds the task was easy: looking in a single hole in theory gave sufficient information to decide and the woodpeckers in fact took an average of 1.7 looks. The task was more complicated when the two kinds of log contained 0 and 6 or 0 and 12 seeds: finding one empty hole is not longer enough to reject a log, but there must be some point at which the information gained from seeing a succession of empty holes makes it worthwhile giving up. Lima calculated how many empty holes the woodpeckers ought to check before giving up on a log in order to maximize their rate of food intake. The calculated values were 6 and 3 while the observed means were 6.3 and 3.5; thus, the woodpeckers use information gleaned while foraging in a way that comes close to maximizing their overall rate of intake.

The risk of starvation

Two kinds of currency for foraging animals — rate of food intake (starlings, great tits) and efficiency (bees) — have come up so far. Another currency that may be important for foraging animals is the risk of starvation. This is especially likely to be important when the animal lives in an environment that is unpredictable; the exact amount of food the animal will obtain is uncertain.

Choosing a variable foraging option may minimize the risk of starvation

For example, imagine you are offered the choice of two daily food rations: one is fixed at 10 sausages per day, the other is uncertain; on half the days you get 5 sausages and on the other half, 20 sausages. Although the *average* of the second diet is higher than that of the first, it is a riskier option since there is no way of telling whether you will get 5 or 20 on any particular day. Which is the better option? The answer depends on the benefit (or 'utility' in economic jargon) of eating different numbers of sausages per day. If a diet of 10 is enough to survive on while 5 is not, then nothing is to be gained by choosing the risky option. If, on the other hand, 10 is not quite enough to survive on, the only viable option may be to take the risk and hope for 20 sausages.

This option offers a 50 per cent chance of survival while the certain option offers no chance.

In short, animals should be sensitive not only to the mean rate of return from a particular foraging option but also the variability. Whether or not animals prefer high variability should depend on the relationship between the animal's needs (usually called its *state*) and the expected rewards. If energy requirements are less than the average expected reward, it pays to choose the less variable option (*risk-averse behaviour*) whilst, if requirements are above average, it usually pays to choose the more variable option (*risk-prone behaviour*).

'Risk-sensitive' behaviour

This idea has been tested in an experiment by Caraco *et al.* (1990). They offered yellow-eyed juncos (*Junco phaeonotus*) (small birds) in an aviary a sequence of choices between two feeding options: one variable and one with a fixed pay-off. For example, the variable option in one treatment was either 0 or 6 seeds with a probability of 0.5 each, whilst the corresponding fixed option was always 3 seeds. The experiment was carried out at two temperatures: 1°C and 19°C. At the low temperature the rewards from the fixed option were inadequate to meet daily energy needs, whilst at 19°C they were sufficient. As predicted by the theoretical argument, the birds switched from risk-averse behaviour at 19°C to risk-prone behaviour at 1°C. An equivalent result was obtained by Cartar and Dill (1990) in a study of bumblebee foraging. They augmented or depleted the energy reserves of the nest, and found that workers switched to risk-prone behaviour when the reserves were low. In this case the reserves of the colony as a whole were treated as equivalent to the reserves of an individual.

Cold juncos are risk prone

These experiments suggest that foragers are able to respond to variability in the amount of reward obtained, and that preference depends on state. However, they do not investigate the question of whether preference changes with time. Two examples of how time of day could be important are suggested by Houston and McNamara (1982, 1985). First, if the animal starts off the day risk-prone, but has good luck in its first few choices, it might be expected to become risk-averse later on. Second, as dusk approaches, for a diurnal forager the long period of enforced overnight fasting might favour a switch to risk-prone behaviour to increase the likelihood of overnight survival. These hypotheses still remain to be tested.

Variability in searching or handling time

The theoretical discussion and experimental studies described above refer only to variability in amount of food obtained (e.g. 0

or 6 seeds in Caraco *et al.*'s experiment). It has been implicitly assumed that there is no difference between options in search or handling time. In fact many experiments have shown that animals usually strongly prefer foraging options with variability in time. Most of these studies have been done by psychologists with rats or pigeons in a 'Skinner box' in which rewards are obtained by pressing a lever/pecking a key. If pecking one of the keys results in a standard reward after a fixed time, say 5 s, and pecking the other key produces rewards after a variable time but with a mean of 5 s, pigeons prefer the variable option (e.g. Mazur 1984).

Discounting the future may explain preference for variable delays

The theoretical interpretation of this is probably different from that for risk-prone and risk-averse behaviour with respect to amount of food. The animals behave as if they are 'future discounting', that is, placing a high value on rewards obtained soon rather than on rewards obtained after a long delay. If the weighting given to immediate rewards is sufficiently high, the animals will prefer the variable time option because the occasional very short time delays far outweigh the occasional long ones in terms of perceived benefit.

Why should animals discount the future in this way? Humans do it 'because you never know what might happen between now and then'. For foragers under natural selection a similar logic might well apply. Rewards in the future are less valuable than rewards here and now because interruptions, death by predator attack, or other changes may mean that future rewards are never collected.

Environmental variability, body reserves and food-storing

Small birds carry optimal, not maximal reserves

Small birds in winter often experience large daily fluctuations in body mass: the 20-g great tit, for example, typically loses 10−15 per cent of its body mass overnight in winter and regains the mass during the following day (Owen 1954). The daily gain and overnight loss is almost entirely made up of fat, which acts as fuel for overnight survival: thus each day in winter a small bird faces an uphill struggle to build up sufficient reserves for surviving the next night. Given this observation, should we expect small birds to carry as much fat as possible at all times, as an insurance against starvation? Both empirical observation and optimality models suggest that in fact birds usually carry less than the maximum reserves. In winter, birds are usually heaviest on the coldest/harshest days, suggesting that on other days they are carrying fewer reserves than the maximum. Furthermore, if one examines the trajectory of weight gain through the day one finds that birds increase their weight rapidly in the afternoon (Owen

1954; Bednekoff 1992), implying that earlier in the day they do not carry as much fat as they could. Lima (1986) and McNamara and Houston (1990) explained these observations by hypothesizing that the reserves carried by a bird reflect an optimal trade-off between costs and benefits. The benefit of carrying extra reserves is reduced risk of overnight starvation, whilst the cost is increased danger of death from predation. The danger might arise simply because heavier birds are less agile at escaping or, more subtly, because birds with more reserves spend more time foraging rather than hiding from predators. This hypothesis predicts that the optimal level of reserves will increase (i.e. birds will be heavier) when the energy cost of overnight survival is higher, or more unpredictable, or when the danger of predation is lower (Fig. 3.7a).

Many animals store food reserves in the environment as well as on their bodies (VanderWall 1990). Among the best studied species are members of the corvid family such as Clark's nut-cracker (*Nucifraga columbiana*) and the European jay (*Garrulus glandarius*) and members of the tit family such as the black-capped chickadee (*Parus atricapillus*) and marsh tit (*Parus palustris*). Nutcrackers and jays store seeds from pine and oak trees respectively in the autumn and retrieve them the following spring or summer to feed themselves and their young. The smaller tit species, on the other hand, seem to store and retrieve their food on a much shorter cycle of hours to days (Stephens & Krebs 1986; Brodin 1992). In both corvids and tits, food storing is an adaptation to survive in a variable environment. The long-term storers take advantage of the autumn abundance of seeds and utilize the food at other times of year when seeds would otherwise be scarce. The short-term storers use storing, like fat reserves, as an insurance against overnight starvation in an unpredictable environment (McNamara *et al.* 1990) (Fig. 3.7b).

Small birds may carry more reserves in a variable environment

Some birds store food instead of fat

Feeding and danger: a trade-off

If you watch a squirrel eating chocolate chip cookies in the park, as Steve Lima and colleagues did (Lima *et al.* 1985) you will notice that the squirrel generally comes to your picnic table, grabs a cookie and retreats to a tree to eat it. If you put out small fragments of cookie the squirrel will often make repeated sorties to the table and take each morsel back to the tree to eat it. This is obviously not a very efficient way to eat food: if maximizing net rate of energy intake or efficiency was the only important factor for a squirrel it would simply sit on the table and eat pieces of cookie until it was full. One interpretation of the

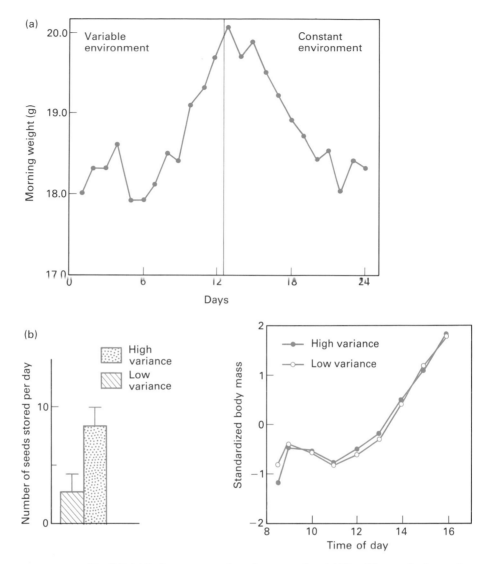

Fig. 3.7 (a) Body reserves and environmental variability. The graph shows the body mass of a captive great tit (one of eight in the experiment) which was transferred from a constant to a variable environment for 12 days before returning to the constant environment. Variability in this experiment was produced by randomly altering the length of the night-time period of no foraging. From Bednekoff (1992). (b) Food-storing and variability. In this experiment, captive marsh tits (one example is shown) stored more food (left), but did not put on more body reserves (right), in a more variable environment. These results suggest that food-storing, like fat storage, is a method of coping with environmental variability: whilst great tits, which do not store food, cope with environmental variability by putting on extra fat reserves, marsh tits store extra food in the environment. The right-hand graph also shows the daily weight trajectory of a marsh tit. In the afternoon, the bird transfers food from its hoards to its body, so reserves rise steeply towards the end of the day. From Hurly (1992).

squirrel's behaviour is that it is balancing the demands of feeding and safety from predators. It could feed at maximum rate and run a good chance of being killed by a cat by staying on the table or it could be completely safe from cats but die of starvation in the trees. Neither of these is the best solution to maximize survival, so the squirrel does a mixture of the two. Lima *et al.* argued that the squirrel should be more prone to seek safety in the trees while feeding when this involves a smaller sacrifice in terms of feeding rate. Consistent with this they found that when the feeding table was close to the trees the squirrels were more likely to take each item to cover. Big pieces of cookie were more likely to be taken to cover than small ones; they take a long time to eat and are therefore more dangerous to handle out in the open and when handling time is long the relative cost of travelling back and forth is reduced.

Hungry sticklebacks accept danger of predation to obtain high intake rates

The balance between the benefits of feeding and of avoiding danger is also influenced by an animal's hunger. On a very cold day in winter normally shy birds become quite tame at the garden bird table, presumably because their increased need for food overrides the danger of coming into the open. Manfred Milinski and Rolf Heller (1978, 1979) studied a similar problem with sticklebacks (*Gasterosteus aculeatus*). They placed hungry fish in a small tank and offered them a simultaneous choice of different densities of water fleas, a favourite food. When the fish were very hungry they went for the highest density of prey where the potential feeding rate was high, but when they were less hungry the fish preferred lower densities of prey. Milinski and Heller hypothesized that when the fish feeds in a high density area it has to concentrate hard to pick out water fleas from the swarm darting around in its field of vision, so it is less able to keep watch for predators, as shown by Milinski (1984b). A very hungry fish runs a relatively high chance of dying from starvation and so is willing to sacrifice vigilance in order to reduce its food deficit quickly. When the stickleback is not so hungry it places a higher premium on vigilance than on feeding quickly, so it prefers the low density of prey. The balance of costs and benefits shifts from feeding to vigilance as the stickleback becomes less hungry.

Consistent with this hypothesis Milinski and Heller found that predation risk influences choice of feeding rate. When they flew a model kingfisher (*Alecedo atthis*) (a predator on sticklebacks) over a tank containing hungry fish they found that the sticklebacks preferred to attack low rather than high prey densities (Fig. 3.8). This is to be expected if the hungry fish, in spite of its high chance of starvation, places a very high premium on vigilance when a predator is in the vicinity.

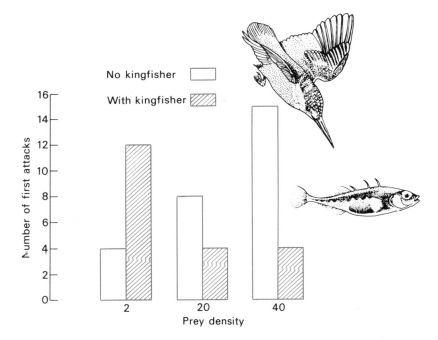

Fig. 3.8 Hungry sticklebacks normally prefer to attack high density areas of prey but after a model kingfisher was flown over the tank they preferred to attack low density areas. From Milinski and Heller (1978).

An important difference between Milinski and Heller's analysis of foraging and those described earlier is that the cost–benefit calculations include the animal's hunger state. An optimization model in which the animal's state changes as a result of its behaviour (the fish becomes less hungry as a result of feeding) is referred to as a dynamic, as opposed to static, model. In fact the traditional view that an animal's internal state controls its behaviour can be turned on its head and the animal can be seen as using its behavioural repertoire to control the internal state in an optimal way. The influence of the kingfisher on the stickleback is to alter the optimal allocation of time to feeding and vigilance, so that the fish decreases its hunger at a slower rate.

The idea of a dynamic feedback between foraging, body reserves and danger of predation has been used by Jim Gilliam (1982) to predict how an individual should shift between habitats as it grows up. His analysis applies to the bluegill sunfish (*Lepomis macrochirus*). In experimental ponds in Michigan, Earl Werner found that these fish could obtain a higher rate of food intake by foraging on benthic invertebrates such as chironomid larvae than they could by foraging either on the plankton or near the emergent vegetation at the edge of the pond. As might be expected the fish spend most of their time (more than 75 per cent) foraging on the benthos. However, when predators in the form of largemouth bass (*Micropterus salmoides*) were added to the pond, a significant change in habitat use by the sunfish was seen. The bass could eat only the smallest sunfish (the others were too big) and these fish now spent more than half their foraging time in the reeds feeding on plankton where they were relatively safe, even though as a result their food intake was reduced by about one-third and their seasonal growth rate by 27 per cent. The bigger sunfish continued to forage with equanimity on the benthos (Werner *et al.* 1983). The little fish thus face a trade-off: is it better to stay in the relative safety of the reeds and grow slowly, prolonging the period of vulnerability to predators, or is it better to gamble on rapid growth to a safe size in the benthos? Gilliam was able to show that the best thing (to maximize its total chance of survival) for the fish to do is to stay in the safety of the reeds until a certain size is reached and then to go for the benthic prey. This accords with observation: the young fish in the presence of predators tend to feed in safe places, and as they get bigger they shift to the better feeding areas.

Bluegill sunfish start off in safe habitats and move to dangerous places as they grow bigger

Nutrient constraints: herbivores and plants

The examples we have described so far have illustrated the variety of currencies that might be important for foraging animals, but we have said little about the range of constraints that might be important. The diet of herbivores is a good example to illustrate the importance of constraints.

As a general rule nutrient quality of food is more important to herbivores than to carnivores and insectivores. This is because plants often lack essential dietary components and only by careful selection of plant species can a herbivore obtain a balanced intake. For example, the diet of moose (*Alces alces*) on the shores of Lake Superior in Michigan is strongly influenced by sodium requirements. The moose feed in two habitats: forest, where they browse on deciduous leaves, and small lakes, where they crop plants growing under water. The aquatic plants are rich in sodium but relatively poor in energy, while terrestrial plants have little sodium but a high energy content. The moose need both energy and sodium to survive and therefore have to eat a mixed diet, but to predict the exact mixture involves making an optimality model.

Since the diet of the moose contains two components we can plot it as a point on a graph, the axes of which are intake of terrestrial and aquatic plants (Fig. 3.9a). If, for example, the moose

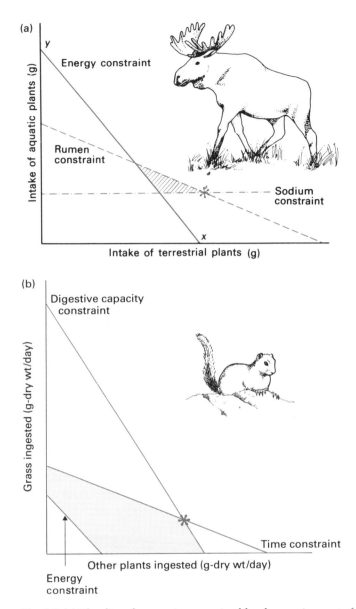

Fig. 3.9 (a) The diet of moose is constrained by the requirements for sodium and energy: the daily requirements are shown as the dot-dash and solid lines respectively and the moose has to eat a mixture of plants which lies in the space above these two lines. The third constraint is the size of the moose's rumen (broken line). Aquatic plants are bulkier than terrestrial ones so fewer grams of these can be fitted into the rumen. The moose's diet was found to lie at the point inside the triangle that maximizes daily energy intake (indicated by a star). From Belovsky (1978). (b) The Columbian ground squirrel (*Spermophilus columbianus*) also maximizes daily energy intake subject to constraints. In this case digestive capacity (equivalent to the moose's rumen constraint), time available and minimum energy requirements define the feasible set of diets. From Belovsky (1986a).

ate largely terrestrial plants with just an occasional aquatic, its diet would be represented by a point at the lower right-hand side of the graph. Now as we have already said, the daily diet has to contain a certain minimum amount of sodium. This is represented in the graph as a constraint line: the horizontal dot-dash line shows the minimum intake of aquatic plants needed to satisfy sodium requirements. But this is not the only constraint on the animal's diet. It also needs a certain amount of energy per day. This could be obtained by eating a pure diet of y grams of aquatic plants of x grams or terrestrial plants or a mixture of the two, as shown by the solid line in Fig. 3.9a. This line shows the mixture of plants that would provide just enough energy to survive the day. Finally, the diet is constrained by the size of the moose's rumen. The moose has a specially modified stomach, the rumen, in which food is slowly fermented by micro-organisms prior to digestion. The size of the rumen sets an upper limit on the amount of food that can be processed at any one time, and therefore limits the total daily intake. The broken line on Fig. 3.9a shows the maximum amount of food that could be eaten per day with different combinations of terrestrial and aquatic plants.

The total effect of these constraints can now be assessed. Only the diets inside the small shaded triangle on Fig. 3.9a satisfy all three constraints. The diet has to be above the sodium line, above the energy line, and below the rumen line. But where inside the triangle is the optimal diet? This depends on the goal or optimality criterion. If the moose is designed, for example, to maximize its daily sodium intake, the diet should include as much aquatic material as possible and lie in the top left corner of the triangle. If the moose was designed to minimize the time spent in water each day, its diet might be near the bottom right of the triangle. Gary Belovsky (1978) carried out a detailed study of the diet and found that the mixture of plants eaten was at the point within the triangle which would be predicted if the moose maximizes its daily energy intake subject to the constraints of sodium need and rumen size. This point is at the right-hand corner of the triangle (marked with a star). To see why, think of lines parallel to the energy constraint line (energy isoclines) at different distances from the origin. Any point along one of these lines is a point of equal energy intake per day, while lines further from the origin represent higher energy intake. The highest energy isocline that the moose could achieve is at the bottom right-hand corner of the triangle of feasible diets.

Belovsky (1986a,b) has extended his conclusions from the moose study to attempt to identify the currencies and constraints that determine herbivore diets in general. He concluded that herbivores maximize energy intake and that digestive capacity is often the key constraint. In other words, herbivores tend to select plants with the highest density of energy per unit of bulk. However, Hobbs (1990) has criticized Belovsky's conclusions. Belovsky assumes that the daily capacity of the digestive tract in herbivores is measured by the bulk (total of dry matter and water) of plants that can be ingested. Hobbs points to evidence that only the dry matter of plants limits daily intake, the water being relatively rapidly absorbed. Hobbs also shows that quantitative predictions of Belovsky's models are rather sensitive to the exact value assumed for the digestive constraint, so the uncertainty about what exactly limits daily intake is all the more of a problem.

The need for more physiological information

What should one conclude from this debate? First, it illustrates the point that all optimality models are dependent on assumptions about physiological processes. Sometimes, as in the herbivore model, the optimality analysis helps to identify areas of physiology where more detailed knowledge would help in predicting behaviour. Second, the sensitivity of Belovsky's model to its assumptions illustrates a general advantage of models that make quantitative predictions: you know exactly where you stand. Third, Belovsky's conclusion that herbivores maximize daily energy gain must remain a tentative hypothesis for the moment.

Optimality models and behaviour: an overview

In this chapter we have seen how optimality models can be used to analyse decisions about foraging and mating. This approach is an extension of the idea of interpreting behaviour in terms of costs and benefits that we introduced in the last chapter. Let us now try to summarize some of the advantages and limitations of optimality modelling. Three main advantages that are illustrated by this chapter are the following.

Optimality models: testability, explicit assumptions, generality

1 Optimality models often make testable, quantitative predictions so that it is relatively easy to tell whether the hypotheses that are represented in the model are right or wrong. For example, the honeybee workers were shown not to be maximizing net rate of energy delivery to the hive, but were maximizing efficiency in their foraging. The hypotheses that were tested in the bee study, and in all optimality studies, were hypotheses about the *currency* (net rate or efficiency) and about the *constraints* on the animal's performance (energy costs, handling times and so on). The currency is a hypothesis about the costs and benefits impinging on the animal, for example for bees energetic costs and benefits seem to be much more important than, say, predation and other dangers. The constraints are hypotheses about the mechanisms of

behaviour and the physiological limitations of the animal, whether it is able to recognize differences in nectar concentration, how fast it can fly and so on.

2 A second advantage is that the assumptions underlying the currency and constraint hypotheses are made explicit. In the model used to analyse load size of starlings, for example, we had to make explicit assumptions about the loading curve, about the fact that the bird could encounter only one patch at a time, about the time taken to fly to the nest, and so on. By making these things explicit in the model one is forced to think clearly about the problem.

3 Finally, optimality models emphasize the generality of simple decisions facing animals. The starling model applied equally to dungflies, and we could have illustrated it with reference to many other animals and other decisions.

Now for a difficulty with the optimality approach: deciding what to do when the model fails to predict what the animal does. Take the dungflies as an example: the model predicts reasonably well, but not exactly, the duration of copula. What should be done about the discrepancy? Should we ignore it, assuming it is within the acceptable range of error, or should we try to analyse if further? Assume for the moment that we wanted to take the latter course. One possibility is that the currency of the model was incorrect; dungflies may trade-off feeding and mating, or danger and mating, rather than simply maximizing rate of fertilization. A second possibility is that the currency is correct but that the constraints have not been identified correctly; perhaps males run out of energy reserves while in copula. Finally, the whole idea of dungflies or other animals maximizing a currency may be incorrect. Animals may simply not be that well tuned by the process of natural selection or they may be lagging behind when some aspect of the environment changes. This kind of argument, although it is often put forward, should really be saved as a last resort, since it is not very fruitful in leading to new experiments or observations. There is, however, no straightforward recipe for distinguishing between the first two possibilities, although one useful step is, as we saw in the honeybee and moose examples, to compare different currencies (or equally, different constraints). Another important step is to analyse more thoroughly the mechanisms underlying behavioural decisions.

Summary

Behaviour can be viewed as having costs and benefits and animals should be designed by natural selection to maximize net benefit.

Table 3.1 A summary of the decisions, currencies and constraints discussed in this chapter

Animal	Decision	Currency	Some constraints	Test
Starling	Load size	Maximize net rate of gain	Travel time, loading curve, energetic costs	Load versus distance
Bee	Crop load	Maximize efficiency	Travel time, sucking time, energetic costs	Load versus flight time
Dungfly	Copulation time	Maximize fertilization rate	Travel time, guarding time, fertilization curve	Predict copula duration
Great tit	Size of worms	Maximize net rate of gain	Handling time, search time	Choice of large or small prey
Downy woodpecker	Patch time	Maximize net rate of gain	Travel time, recognition time	Number of holes inspected
Yellow-eyed junco	Where to feed	Minimize risk of starvation	Handling time, daily energy budget	Choice of variable or certain reward
Great tit/ marsh tit	Body reserves/ hoard size	Maximize survival	Energetic cost of carrying reserves	Body reserves/ hoard size in predictable and unpredictable environments
Squirrel	Where to eat	Maximize survival	Travel time, handling time	Vary size of food and distance
Stickleback	Where to feed	Minimize danger and starvation	Vigilance and foraging incompatible	Vary hunger and danger
Bluegill sunfish	Habitat choice	Maximize survival	Growth depends on food intake, danger related to size	Habitats used at different ages
Moose	Food choice	Maximize daily energy intake	Sodium need, digestion limit, energy limit	Per cent composition of diet

This idea can be used as a basis for formulating optimality models in which the criterion of maximum benefit, the constraints on the animal, and the currency for measuring benefit are specified. Different kinds of currency might be appropriate for measuring benefits and costs of different behaviours, for example, with feeding behaviour rate of intake might be a good currency and with male mating behaviour rate of fertilizing eggs seems reasonable (Table 3.1).

The emphasis of this approach is on quantitative testable predictions. Often the results of experiments deviate from the predictions of simple models; these deviations can be just as valuable as successful predictions in helping to understand how behaviour is designed.

Further reading

The paper by Maynard Smith (1978) discusses some of the pros and cons of optimality modelling. Three books which contain useful reviews of foraging economics are the monograph by Stephens and Krebs (1986) for a summary of mathematical models; the volume edited by Kamil *et al.* (1987) for empirical studies and two more general reviews by Russell Grey (who is very critical of the approach) and by Kacelnik and Cuthill; Krebs and Kacelnik (1991), Houston *et al.* (1988) and Hughes (1990) provide recent overviews.

Topics for discussion

1 Do animals have to be clever to forage optimally?
2 Are laboratory experiments on decision-making in simple environments useful for understanding behaviour in the field?
3 Is average net rate of intake a sensible currency for foraging animals?
4 How might one investigate the mechanisms by which animals discriminate between fixed and variable amounts of food.

Chapter 4. Predators versus Prey: Evolutionary Arms Races

In Chapter 3 we examined the ways in which predators search for and select their prey. We regarded various parameters (e.g. handling time) as more or less constant. Over the evolutionary time, however, these may vary. For example, during evolution we expect natural selection to increase the efficiency with which predators detect and capture prey. On the other hand, we would also expect selection to improve the prey's ability to avoid detection and to escape. The complex adaptations and counter-adaptations we see between predators and their prey are testament to their long coexistence and reflect the result of an arms race over evolutionary time (Table 4.1; Fig. 4.1). In this chapter we will consider three questions about predator–prey arms races.

Three questions about arms races

1 *Adaptation or story telling?* Do the proposed adaptations of one party make functional sense given the adaptations by the other party in the arms race? This is not a trivial question. One temptation when presenting functional explanations in biology is to invent clever stories which are difficult to test. For example, we might suggest that flamingos are pink so as to blend in with the setting sun, so that lions find them difficult to see in the evening when out hunting. You may think this is an unlikely idea! But for all we know, the proposal that brown moths are so coloured to blend in with the tree trunks on which they rest may be equally unlikely. We need to perform experiments to test our functional hypotheses.

Table 4.1 Examples of predator adaptations and counter-adaptations by prey

Predator activity	Predator adaptations	Counter-adaptations by prey
Searching for prey	Improved visual acuity	Crypsis
	Search image	Polymorphism
	Search limited area where prey abundant	Space out
Recognition of prey	Learning	Mimicry
Catching prey	Motor skills (speed, agility)	Escape flights, 'startle' response
	Weapons of offence	Weapons of defence
Handling prey	Subduing skills	Active defence, spines, tough integument
	Detoxification ability	Toxins

(a)

(b)

(c)

Fig. 4.1 Examples of adaptations by prey. (a) The syrphid fly, *Metasyrphus americanus* (Diptera) is harmless to predators but gains protection by mimicry of the wasp's yellow and black pattern on the abdomen. (b) When the Brazilian toad, *Physalaemus nattereri*, is attacked it rears up to display two large eyespots on its sides. Many predators show avoidance responses to patterns resembling eyes. (c) A cryptic lizard, which blends in beautifully with its background.

2 *How can an arms race begin*? Prey cannot suddenly evolve perfect counter-adaptations any more than a vertebrate could instantly evolve a perfect complex structure, such as an eye. Presumably at the start of an arms race even slight and very crude counter-adaptations could confer a selective advantage. Improvements in the predators then selected for improvement in the prey, resulting in better and better counter-adaptations. Our second question is, therefore, can even crude counter-adaptations by prey decrease predation pressure and thus serve as a starting point for an evolutionary arms race?

3 *How do arms races end?* Why don't predators become so efficient that the prey are driven to extinction? Alternatively, why don't prey evolve to be so good at escaping that the predators become extinct?

Predators versus cryptic prey

Underwing moths . . .

We begin by attempting to answer the first two questions for one example of an arms race, namely birds as predators on cryptic prey. Pietrewicz and Kamil (1981) and Sargent (1981) studied underwing moths (*Catocala* spp.) in the deciduous woods of North America. There are up to 40 species living in a particular locality and they are hunted extensively by birds, including blue jays and flycatchers.

TESTING FUNCTIONAL HYPOTHESES ABOUT ADAPTATION

. . . the forewings appear to be cryptic and the hindwings may startle predators

The forewings of the moths appear cryptic, looking very much like the bark of the trees on which the moths rest (Fig. 4.2). The hindwings, on the other hand, are often strikingly coloured yellow, orange, red or pink (Fig. 4.2). The moths rest with their forewings covering the hindwings but when they are disturbed the hindwings are suddenly exposed. Our hypotheses, therefore, are that the forewings are coloured so as to decrease detection by a predator and the hindwings may have a 'startle' effect on a predator that has succeeded in detecting the moth, causing the bird to stop momentarily and thus giving the moth time to escape.

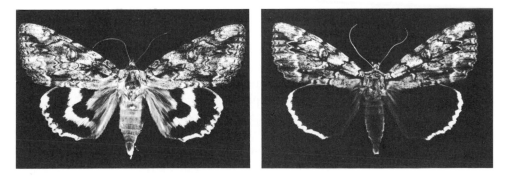

Fig. 4.2 Underwing moths, *Catocala* spp., have cryptic forewings and conspicuously coloured hind wings. Left: *Catocala neogama*; right: *C. retecta*. Photos by Ted Sargent.

Experimental test of
crypsis: jays hunt for
cryptic moths in a
slideshow

In support of the crypsis hypothesis for the forewings, different
species of underwing moths select different backgrounds which
match their own reflectance and thus maximize the cryptic effect.
Furthermore, they orient themselves in particular ways so that
the wing patterns merge in with the fissure patterns on the bark.
Pietrewicz and Kamil tested the importance of crypsis by giving a
slideshow to blue jays in an aviary (Fig. 4.3a). Slides were projected
on a screen and in some cases there was a moth present in the
picture while in other cases there was no moth. If there was a
moth, the jay was rewarded with a mealworm for pecking at the
slide, and the next slide was then shown after a short interval. If
there was no moth in the picture then the jay could peck a
second smaller 'advance' key to produce the next slide almost
immediately. If the jay made either of two errors, namely pecking at
the slide when no moth was present or pecking the advance key
when a moth was present, it was 'punished' in the form of a
delay to the next slide presentation.

This experimental procedure is ingenious for two reasons.
First, the predator is faced with a perceptual problem only; there
is no complication caused by other factors which might influence
predation such as prey taste, activity or escape efficiency. Second,
because the predator is stationary and the prey are, in effect,
moving past in front of it (in the form of a sequence of slides), it
is easy to control the frequency and order in which the predator
encounters prey. This would be more or less impossible if the
predator was moving about the cage searching for real moths. It
was found that the jay made many more mistakes if the moth
was presented on a cryptic background than if presented on a
conspicuous background (Fig. 4.3b). This provides direct support
for the hypothesis about crypsis.

In many species of underwing moths the forewings are polymorphic, that is there are different colour forms coexisting within
the same population. One hypothesis for this is that when a
predator discovers a moth it may form a 'search image' for that
particular colour pattern and concentrate on looking for another
which looks the same (Box 4.1). If all the population were of
exactly the same colour then all would be at risk, but if there
was a polymorphism then a predator which had a search image
for one morph may be more likely to miss the other morphs.
Pietrewicz and Kamil were able to test this idea by presenting
the jay with different sequences of slides. When, for example,
runs were presented of 'all morph A' or 'all morph B', the jay
quickly got its eye in and improved its success at pecking at the
moth as the trial proceeded. When, however, morphs A and B
were presented in random order, the jay did not improve its

'Searching image': jays
learn to see the cryptic
moths

(a)

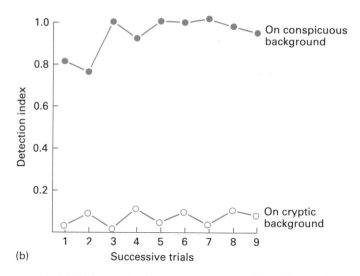

(b)

Fig. 4.3 (a) A blue jay in the testing apparatus. The slides are back projected on a screen in front of the bird. The advance key (see text) is to the right. A mealworm is delivered through the circular hole if the jay makes a correct response. Photo by Alan Kamil. (b) The jays were more likely to detect *Catocala* moths on a conspicuous background. A jay pecking indiscriminately at all slides gets a low score on the detection index. From Pietrewicz and Kamil (1981).

detection success with successive slides (Fig. 4.4). This shows that encountering polymorphic prey does indeed seem to prevent effective search image formation by the predator.

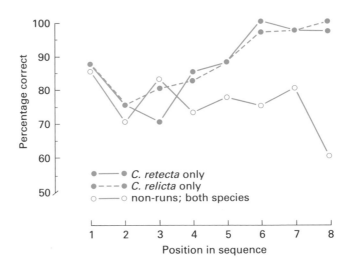

Fig. 4.4 The mean percentage of correct responses by jays when moths are presented in sequences of the same species (runs of either *Catocala retecta* or *Catocala relicta*) or in a sequence containing both species in random order. The jays improved their performance when runs of the same species were presented but not when the two species were presented in a mixed sequence. From Pietrewicz and Kamil (1981).

Box 4.1 *Search images.*

Luc Tinbergen (1960) studied the feeding behaviour of birds in Dutch pinewoods. He discovered that they did not eat certain insects when they first appeared in the spring but then suddenly started to include them in their diet. He suggested that the sudden change was due to an improvement in the birds' ability to see the cryptic insects, a process he called 'adopting a specific searching image'. There are, however, other hypotheses which could explain Tinbergen's observations. For example, the birds could have seen the insects all along but only decided to include them in their diet when their abundance increased sufficiently to make it profitable to search for them (Royama 1970). Alternatively, the birds may have been reluctant at first to eat novel prey or have improved in their ability to capture the prey.

Experiments by Marian Dawkins (1971) eliminated these alternative explanations and showed that predators can indeed undergo changes in their ability to see cryptic prey. Her predator was the convenient, if unspectacular, domestic chick and her prey were coloured rice grains. The clever design of the experiment was to keep the prey the same (therefore handling, acceptability constant) and to vary the background. Two examples are shown below. Chick (a) was presented with orange grains on a green background (solid line) and on an orange background (dashed line). Chick (b) had green grains on an orange background (solid line) and on a green background (dashed line). The two tests were run separately for each chick. In both cases the chicks found the prey quicker on a conspicuous background. On the cryptic background the chicks mainly pecked at background stones at first but after 3–4 min they eventually started to find the grain and by the end of the trial they were eating the cryptic prey at the same rate as when it was on the conspicuous background.

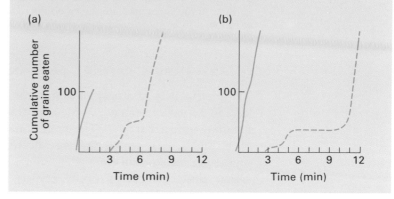

What about the bright-coloured hindwings? Debra Schlenoff (1985) tested the responses of blue jays to models which had variously patterned 'hindwings' concealed behind cardboard 'forewings'. The model moths were attached to a board and the jays were trained to remove them to get a food reward underneath. When the models were removed, the hindwings suddenly expanded from behind the forewings to mimic the reaction of the real moths. Jays which had been trained on models with grey hindwings showed a startle response when they were exposed to the brightly patterned hindwings typical of *Catocala*, whereas subjects trained on brightly patterned models did not startle to a novel grey hindwing. After repeated presentations the birds habituated to a particular *Catocala* pattern but a novel bright pattern elicited another startle responses. These results provide good evidence for the startle hypothesis, and the habituation effect suggests an adaptive advantage for the great diversity in hindwing patterns of different sympatric species of *Catocala*. Other experiments have shown that eyespots are particularly effective in causing a startle response in the predator (Fig. 4.5).

Experimental evidence for the startle effect

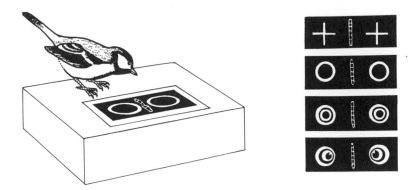

Fig. 4.5 A great tit about to pick up a mealworm (left). As it does so the light in the box is switched on to show a slide with an artificial 'wing display' on it. Some of the slides used are shown on the right; those towards the bottom were found to be most frightening to the bird. From Tinbergen (1974).

DOES EVEN CRUDE CRYPSIS CONFER AN ADVANTAGE?

The profitability of a prey to a predator is the value of the prey (e.g. in terms of energy) divided by the recognition time plus handling time (Chapter 3). The evolution of crypsis can be thought of as the way in which the prey decreases its profitability by increasing recognition time. Can even a small increase in recognition time, caused by crude crypsis, still bring a selective advantage? Erichsen *et al.* (1980) presented a great tit in a cage with a 'cafeteria' of prey moving past on a conveyor belt like the one in Fig. 3.6. As with the jays, the clever design of this experiment is that it enables the observer to control precisely the order in which the predator encounters prey and the rate of encounter. Three items came past the great tit on the belt.

1 *Inedible twigs.* In fact these were opaque pieces of drinking straw containing brown string. They had zero energy value, a handling time h_t (time for the great tit to pick up the item and reject it as inedible) and the encounter rate was λ_t (items per second).

2 *Large cryptic prey.* These were opaque pieces of drinking straw with a mealworm inside; energy value E_1, handling time h_1, (time to pick up and eat the prey) and encounter rate λ_1.

3 *Small conspicuous prey.* These were clear pieces of drinking straw with half a mealworm inside, clearly visible; energy value E_2, handling time h_2 (time to pick up and consume the prey), and encounter rate λ_2.

The large prey were worth more energy per unit handling time than the small prey $(E_1/h_1 > E_2/h_2)$. However, the problem of selecting large prey is that of recognizing them from the inedible

Great tits hunting for artificial cryptic prey

twigs; the opaque straw which came past on the belt had to be picked up and examined to see whether it contained a mealworm or just inedible string. The experimental design therefore mimics the problem faced by a predator searching for a profitable, but cryptic prey item.

If the predator was a *generalist*, picking up every item that came past, then in T_s seconds, its energy intake would be:

$$E = T_s \left(\lambda_1 E_1 + \lambda_2 E_2 \right)$$

in a total time

$$T = T_s + T_s \left(\lambda_1 h_1 + \lambda_2 h_2 + \lambda_t h_t \right).$$
(search time + handling time)

The rate of energy intake by a generalist would therefore be:

An optimal foraging model of crypsis . . .

$$\frac{E}{T} = \frac{\lambda_1 E_1 + \lambda_2 E_2}{1 + \lambda_1 h_1 + \lambda_2 h_2 + \lambda_t h_t} \qquad (4.1)$$

(Note that T_s has cancelled out in the division).

If the predator ignored the cryptic prey altogether (i.e. all the opaque straws) and just played safe and ate the conspicous small prey (i.e. picked up all the clear straws, all of which have a food reward inside), its rate of energy intake would be:

$$\frac{E}{T} = \frac{\lambda_2 E_2}{1 + \lambda_2 h_2} \qquad (4.2)$$

The predator should, therefore, specialize on the less profitable but conspicuous prey provided that the rate in equation 4.2 is greater than the rate in equation 4.1.

By similar reasoning, it will pay the predator to specialize on large, cryptic prey if:

$$\frac{\lambda_1 E_1}{1 + \lambda_1 h_1 + \lambda_t h_t} > \frac{\lambda_1 E_1 + \lambda_2 E_2}{1 + \lambda_1 h_1 + \lambda_2 h_2 + \lambda_t h_t} \qquad (4.3)$$

. . . predicts when cryptic prey should be ignored

The encounter rates were arranged so that in one treatment the birds were predicted to specialize on the small conspicuous prey and ignore the large cryptic prey (i.e. rate in equation 4.2 greater than rate in equation 4.1), and in another treatment they were predicted to specialize on the large cryptic prey (equation 4.3 holds). The tits behaved more or less as predicted (Fig. 4.6). Now for the important point: in the experiment h_t, the discrimination time for an inedible twig, was only 3 to 4 s. Therefore the tit, given time, could easily tell a twig from a large prey. Nevertheless, provided conspicuous prey were encountered sufficiently frequently, or provided many of the large items were twigs rather than real prey, it payed the tit to ignore the large prey altogether.

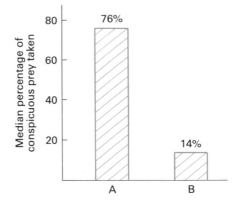

Fig. 4.6 Great tits foraging for artificial cryptic prey. The cost of recognizing cryptic profitable prey may cause the predator to specialize on other more conspicuous prey. In treatment A, twigs were four times as common as the large prey that resembled them; in B the large prey were four times as common as the twigs. The abundance of conspicuous small prey was constant in A and B. According to equations 4.1, 4.2 and 4.3, it pays the predator to specialize on conspicuous prey in treatment A, and on cryptic prey in B. From Erichsen *et al.* (1980).

Even slight crypsis may bring an advantage

The conclusion is that even slight crypsis, enough to impose only a second or two extra discrimination time, can still be of selective advantage to a prey. To gain an advantage the crypsis only has to be sufficient to make another prey item more profitable for the predator. The results of this experiment support the idea that crude crypsis can indeed provide a starting point for an evolutionary arms race.

The advantage and evolution of warning coloration

Why are some prey brightly-coloured?

Some prey are very brightly coloured, rather than cryptic. Fruit often becomes brightly coloured when ripe to increase the probability that it is eaten and so the seed is dispersed. This is an example of a prey which is in fact selected to be eaten by predators! On the other hand, many insects are also brightly coloured yet they are presumably selected to *avoid* predation (Fig. 4.7). Bright coloration is often associated with distastefulness, and A.R. Wallace suggested that 'some outward sign of distastefulness is necessary to indicate to its would-be destroyer that the prey is a disgusting morsel'. The hypothesis is, therefore, that gaudy colours are more easily learned by predators as representing noxious prey. Such 'warning coloration' is sometimes called aposematism.

Are brightly coloured prey really more easily recognized? Gittleman and Harvey (1980) tested this idea by presenting chicks with different coloured breadcrumbs. The chicks showed equal

Fig. 4.7 Cinnabar moth caterpillars (*Callimorpha jacobaenae*) are distasteful and warningly coloured with orange and black bands.

Warning colours help
predators to learn to
avoid unpalatable prey

. . .

preference for blue or green crumbs. In the experiment, the crumbs were made distasteful by dipping them in quinine sulphate and mustard powder. Four groups of chicks were used: (a) blue crumbs or (b) green crumbs on a blue background, and (c) blue crumbs or (d) green crumbs on a green background. Whatever the background colour, the chicks took more of the conspicuous prey early on in the experiment. However, overall the cryptic prey suffered the greatest predation (Fig. 4.8). This suggests that it does indeed pay a distasteful prey to be conspicuous. Chicks may learn to avoid conspicuous prey more readily simply because they are more easily recognized on account of their bright colour (Roper & Redston 1987), or the eating of a large number of prey in a short time may be a more powerful aversive experience than eating an even greater number of prey over a longer time period. Other experiments have shown that learning to avoid bright noxious prey can occur after just one trial and last for a long time. Miriam Rothschild had a pet starling which had just one experience of a brightly coloured distasteful caterpillar and it still refused to pick this species up a year later even though it had no further experience in the meantime. Sometimes predators will avoid brightly coloured and dangerous prey even though they have had no experience of them. For example, hand-reared 'naive' flycatchers (the great Kiskadee *Pitangus sulphuratus*) will avoid coral snakes (*Micrurus* spp.) which are coloured with red and yellow bands (Smith 1977).

There may be another advantage for the conspicuousness of warning coloration. Not only can predators learn to avoid con-

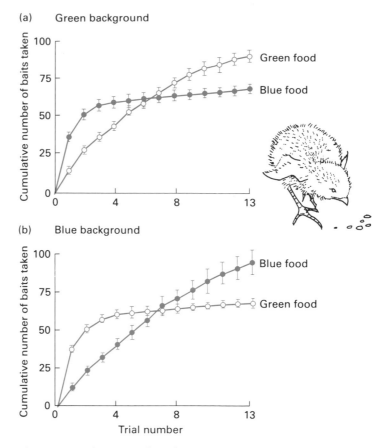

Fig. 4.8 Cumulative number of conspicuous and cryptic distasteful prey taken in successive trials by chicks. In (a) the green food is cryptic, in (b) the blue food is cryptic. In both experiments, the distasteful prey is eaten less when it is conspicuous. From Gittleman and Harvey (1980).

... and may reduce recognition errors

spicuous colours more easily, but having learned to avoid a prey type they are probably less likely to attack it by mistake if it is conspicuous (Guilford 1986). A bird, for example, often pecks at an item before it is certain that it is a prey. This behaviour often saves time and may be advantageous when prey are mobile. Given that predators may make hasty decisions, it may pay unpalatable prey to be conspicuous to reduce recognition errors.

EVOLUTION OF WARNING COLORATION

How did warning coloration evolve? One possibility is that conspicuous colours evolved first, followed by distastefulness. For example, some brightly coloured birds like kingfishers are distasteful (Cott 1940). Their colours may have been favoured for better mate attraction or territory defence and then, because they also increased conspicuousness to predators, this then favoured the evolution of distastefulness. The other possibility is that distastefulness came first. This may apply to those insects, such as caterpillars of the monarch butterfly *Danaus plexippus*, which feed on plants containing toxins and incorporate the toxins in their bodies as a defence against predation. It is plausible that here unpalatability evolved first followed by conspicuousness. In this case, then, bright coloration evolves specifically as a warning device.

Warning colours may have evolved through their effect on the survival of relatives in the same group

This last scenario poses an interesting problem. Imagine a population of unpalatable but cryptic larvae. A mutation arises in an adult which causes its larvae to be more conspicuous. These larvae would then surely be more obvious to predators and so more likely to perish. Although the predator may, as a result of its experience with the nasty taste, decide never to touch the brighter form again, because it is a rare mutant the predator is unlikely to encounter another one. Thus the mutation goes extinct during the sampling and never has the chance to spread. How, then, can warning colours ever evolve? R.A. Fisher (1930) was the first to propose a solution. He realized that distasteful, brightly coloured insects were often clumped in family groups (see Table 4.2 for an example). In this situation, because of the grouping, the predator does encounter others with the bright coloration, namely the siblings of the individual which perished in the sampling. Thus their lives are saved and some copies of the gene for conspicuous coloration get through to the next generation. This process is like that of kin selection (Chapter 11). Mathematical models have shown that with family grouping bright coloration can evolve in a distasteful species provided the brighter forms are not too conspicuous and provided the predator needs to sample fewer of them, compared to cryptic prey, in order to learn that they are distasteful (Harvey *et al.* 1982).

Brightly coloured insects often live in groups ...

Although Fisher's solution is ingenious, recent work has

Table 4.2 Brightly coloured species of caterpillars of British butterflies are more likely to be aggregated in family groups than cryptic species. From Harvey *et al.* (1983)

Dispersion	No. species of caterpillars	
	Aposematic	Cryptic
Large family groups	9	0
Solitary	11	44

challenged two of its assumptions. First, the assumption that the sampled individuals always perish may be wrong. Many brightly coloured insects have tough integuments which protect them against attacks by naive predators and they are released unharmed. Thus in some cases there may be a direct advantage to the individual in being conspicuous; so long as the bright colouring is more easily remembered, a bright distasteful caterpillar is better protected than a cryptic one in subsequent encounters with the same predator (Sillén-Tullberg 1985). Second, Fisher assumed that family grouping sets the stage for the evolution of warning coloration but a phylogenetic analysis (see Chapter 2) of butterflies suggests that warning coloration evolved *before* gregariousness (Sillén-Tullberg 1988). It seems possible, therefore, that warning coloration evolved because of the direct advantage it brought to individuals in decreasing their likelihood of attack, and clumping sometimes evolved afterwards through advantages such as the selfish herd effect (discussed in Chapter 6). Further work is needed to test the individual advantage of warning colours, but these recent studies suggest that family grouping is not critical for the evolution of bright colours. Note, as a further example, that many of the brightly coloured species in Table 4.2 are solitary.

... but grouping may have evolved after warning colours

The trade-off between conspicuousness and crypsis

The design of animals often represents a compromise between different selective pressures. As a defence against predators it may pay to be cryptic, but this may conflict with the advantage of being conspicuous for other activities such as territory defence or mate attraction. As an example of this trade-off, in many species of birds the males are brightly coloured in the breeding season but moult into duller female-like plumage after breeding.

In guppies, brighter males have a mating advantage ...

John Endler's (1980, 1983) work on the coloration of guppies (*Poecilia reticulata*) provides an illuminating experimental study of this trade-off. Endler studied several isolated populations of these little fish in the streams of Trinidad and Venezuela. Males are more colourful than females. Three types of colour can be distinguished. (a) Pigment colours (carotenoids — red, orange and yellow) which are obtained from the diet. If fish are fed carotenoid-free food then these colours fade within a few weeks (Kodric Brown 1989). (b) Structural colours (iridescent blue and bronze) produced by reflection of light from scales. (c) Black spots (melanin) which are partly under nervous control and can increase or decrease in size. Laboratory experiments showed that brighter colours brought a mating advantage. Females were particularly attracted to the orange spots (Houde 1988).

... but suffer
increased predation

To test whether there was counter-selection against bright colours due to predation, Endler sampled streams with different predator communities. He found that males living in streams with greater predation pressure were duller in colour, having both fewer colour spots per fish and also smaller spots (Fig. 4.9a). Not only did predation intensity influence guppy coloration but the type of predator was also important. In some streams the main predator was a prawn, which was red blind, and here the male guppies were significantly redder.

Finally, Endler showed how colours could change in response to changed selection pressure. In controlled laboratory populations, males kept with predators evolved to be duller, while those kept isolated from predators evolved to be brighter, with both larger and more spots (Fig. 4.9b). Similar effects could be shown under field conditions: when 200 dull males from a high predation stream were introduced into a new isolated stream in Trinidad, which had no guppies and no predators, the population evolved during a 2-year period into one with much more colourful males. Endler's experiments provide a very convincing example of how natural selection can change colour patterns in relation to a shifting balance between different selection pressures.

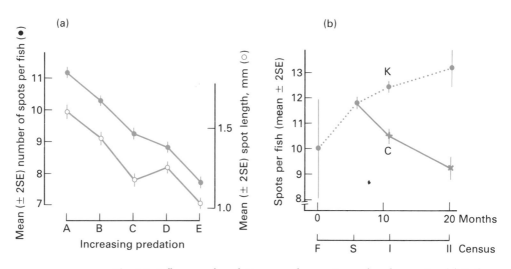

Fig. 4.9 Influence of predation on colour pattern of male guppies. (a) Both number of colour spots per fish and spot size are smaller in streams with greater predation. The main predators are other fish and prawns. Data from five streams in Venezuela with increasing levels of predation from A to E. From Endler (1983). (b) A selection experiment in the laboratory. F, foundation population of guppies kept with no predators. S, start of the experiment; predators added to population C but not to population K. Note the rapid change in population C after predation began. I and II are the dates of two censuses. From Endler (1980).

Predator–prey arms races

Why don't predators become so efficient that they drive their prey to extinction? Or why don't prey evolve such excellent counter-adaptations that the predators become extinct? In other words why do we see both parties in the arms race coexisting together? In many cases the complexity of their adaptations and counter-adaptations is such that they must have coexisted for a long time. We consider three hypotheses (Dawkins & Krebs 1979; Slatkin & Maynard Smith 1979).

Three hypotheses for prolonged coexistence between predator and prey

1 *Prudent predation.* Man has the capacity to be a prudent predator and avoid over-exploiting his food supplies to extinction. Should animal predators likewise be prudent? The problem with this idea is that it relies on group selection. In a population of prudent predators any individual that cheated and ate more than its 'fair share' would pass on more genes to future generations than prudent individuals (see Chapter 1). Prudent predation could, however, evolve where an individual had exclusive use of a resource (e.g. by defending a territory) and so was saving food for its own future use rather than for the good of the population.

2 *Group extinctions.* If group extinctions were common then the reason we see stable predator–prey systems in nature may be because all the unstable systems have indeed gone extinct.

3 *Prey are ahead in the arms race.* Predator–prey systems may be stable because the prey are always one step ahead in the arms race. One hypothesis for why this might be so can be described as the 'life–dinner principle'; rabbits run faster than foxes because the rabbit is running for its life while the fox is only running for its dinner. The cost of a mistake is clearly greater for the rabbit. As Dawkins and Krebs (1979) put it. 'A fox may reproduce after losing a race against a rabbit. No rabbit has ever reproduced after losing a race against a fox. Foxes who often fail to catch prey eventually starve to death, but they may get some reproduction in first'. Therefore, selection pressure will have been stronger on improving the ability of rabbits to escape than the ability of foxes to make successful captures.

In many cases the prey may also be ahead in the arms race because they have a shorter generation time than the predator and can therefore evolve at a faster rate. This will apply to examples such as weasels versus mice, and birds versus insects, but in some cases it is the predator which has the capacity to reproduce at a faster rate (aphids versus roses). We may also ask why prey do not become so efficient at escaping so as to drive their predators extinct. One hypothesis is that as predators become rare, because of increased prey efficiency, they exert little selection on the prey for further improvement.

Finally a predator may often be unlikely to drive a prey to extinction because if a prey species became rare, due for example to excessive predation, then the predator is likely to seek other prey species. Because adaptations for feeding on different prey are usually different, a predator is unlikely to be sufficiently specialized on any one species to drive it to extinction.

Brood parasites and their hosts

Some species of birds, fish and insects are brood parasites, laying all their eggs in the nests of other species. As with the case of predator versus prey, one party (the brood parasites) gain a benefit, namely their young are raised for free, while the other party (the hosts) suffer the cost of raising young which are of no genetic benefit to them. Clearly, we would expect selection to favour host defences against parasitism. This, in turn, should select for counter-adaptations by the parasite. For example, insects remove foreign young from their nests but some insect brood parasites are able to gain entrance by mimicking the chemical communication system of their hosts (Fig. 4.10). Recent experiments with birds have dissected the various stages of the evolutionary arms race between brood parasites and hosts.

CUCKOOS VERSUS HOSTS

An evolutionary arms race that can be studied experimentally

The cuckoo *Cuculus canorus* breeds throughout Europe and northern Asia. In Europe, there are about ten favourite host species but individual female cuckoos specialize on just one particular host and lay distinctive eggs which match, to varying degrees, the eggs of their respective hosts. For example, pipit-specialists lay brown, spotted eggs, wagtail-specialists lay white, spotted eggs, reed warbler-specialists lay greenish, spotted eggs and redstart-specialists lay pure blue eggs, all a reasonable match of the host eggs. It is not yet known how these various strains within the cuckoo species (known as gentes) are maintained. One possibility is that daughter cuckoos lay the same egg type as their mother and come to parasitize the same species of host that reared them, perhaps learning the host characteristics through imprinting.

The female cuckoo adopts a particular procedure when parasitizing a host nest. She usually finds nests by watching the hosts build. Then she waits until the hosts have begun their clutch and parasitizes the nest one afternoon during the hosts' laying period, laying just one egg per host nest. Before laying, she remains quietly on a perch nearby, sometimes for an hour or more. Then she suddenly glides down to the nest, removes one host egg and,

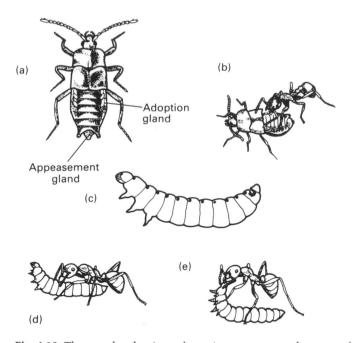

(a)

(b)

Adoption
gland

Appeasement
gland

(c)

(e)

(d)

Fig. 4.10 The rove beetle, *Atemeles*, gains entrance to the nests of its host, *Myrmica* ants, by mimicking the ant's chemical communication system. Secretion from the beetle's appeasement gland (a) suppresses the ant's aggressive behaviour towards intruders, and secretion from the adoption gland stimulates the ant to carry the beetle into its nest (b), where the beetle lays eggs. The beetle's developing larvae have rows of glands (c), the secretions from which stimulate the ant to regurgitate droplets of food to it (d, e). From Hölldobler (1971).

holding it in her bill, she then lays her own egg (Fig. 4.11a) and flies off, later swallowing the host egg whole. The total time spent at the host nest is less than 10 s! Having laid her egg, the cuckoo then abandons it, leaving all subsequent care to the hosts. Hosts sometimes reject the cuckoo egg, but often they accept it. The cuckoo chick usually hatches first, needing an unusually short period of incubation, whereupon, just a few hours old and still naked and blind, it balances the host eggs on its back, one by one, and ejects them from the nest (Fig. 4.11b). Any newly hatched host young suffer the same gruesome fate and so the cuckoo chick becomes the sole occupant of the nest and the hosts slave away, raising it as if it was their own (Plate 4.1 a−c, between pps 212−213).

To what extent has each party evolved in response to selective pressure from the other? This can be investigated experimentally, by testing host responses to variously coloured model eggs. Two

(a)   (b)

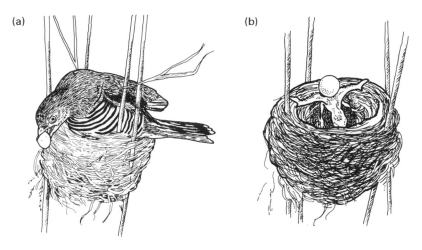

Fig. 4.11 (a) A female cuckoo, *Cuculus canorus*, laying in a reed warbler nest. She has removed one of the host eggs, which she is holding in her bill while she lays her own egg in its place. (b) The newly hatched young cuckoo ejects the host eggs from the nest.

sources of evidence show that cuckoos have responded to selection from hosts.

Cuckoos have adaptations to increase host acceptance of their eggs

1 The cuckoo's egg-laying tactics are specifically designed to circumvent host defences. Experimental 'parasitism' of reed warbler nests with model cuckoo eggs reveals that the hosts are more likely to reject the model egg if it is a poorer match of the hosts' own eggs, if it is laid too early (before the hosts themselves begin to lay), if it is laid at dawn (when the hosts themselves lay) or if the hosts are alerted by the sight of a stuffed cuckoo on their nest (Davies & Brooke 1988). Therefore the cuckoo's mimetic egg, timing of laying and unusually fast laying are all adapted to increase the success of parasitism.

2 The degree of host-egg mimicry exhibited by the various cuckoo gentes reflects the degree of discrimination shown by their respective hosts. For example, dunnock-cuckoos do not lay a mimetic egg, as expected from the fact that dunnocks, in contrast to most other favourite hosts, do not reject odd eggs from their nest.

Two sources of evidence show that hosts, in turn, have responded to selection from cuckoos.

Potential hosts that have never been exposed to cuckoos do not show egg rejection

1 Species which are unsuitable as hosts, either because they have a seed diet (young cuckoos need to be raised on invertebrates) or because they nest in holes (inaccessible to a female cuckoo), show very little, if any, rejection of eggs unlike their own. In contrast to favourite hosts, they also show little aggression to adult cuckoos near their nests (Davies & Brooke 1989a; Moksnes *et al.* 1991). Thus the rejection of odd eggs by favourite hosts and their intense mobbing of cuckoos have evolved specifically in response to cuckoo parasitism.

2 Populations of favourite hosts isolated from cuckoos also show much less rejection of non-mimetic eggs (Davies & Brooke 1989a; Soler & Møller 1990). For example, meadow pipits in Iceland, where there are no cuckoos, show much less egg discrimination than the parasitized pipit populations in Britain.

Given this evidence that each party has evolved adaptations in response to selection from the other, what is the expected outcome of the arms race?

EVOLUTIONARY EQUILIBRIUM OR CONTINUING ARMS RACE?

Figure 4.12 summarizes the results from experiments with model eggs on hosts of the cuckoo in Europe and those of the brown-headed cowbird, *Molothrus ater*, in North America. Unlike the cuckoo, this cowbird is a generalist parasite with females laying eggs in a huge variety of host nests; 216 host species have been recorded! Cowbird eggs do not show any close mimicry of their host eggs. Another important difference is that the cowbird young do not eject the host young, so the hosts may still gain some reproductive success from a parasitized nest though the cowbird chick often outcompetes the hosts' own young for food and so host success is usually severely reduced. Furthermore, cowbirds are much more abundant than cuckoos. Typically, less than 5 per cent of the host population suffers parasitism by cuckoos whereas in some cases over 50 per cent of the host nests are parasitized by cowbirds. In some cases this heavy parasitism is even driving hosts to extinction (May & Robinson 1985).

How can we explain the variation between hosts in rejection rate of eggs unlike their own (Fig. 4.12)? We discuss two hypotheses below, which raise some interesting questions about how 'well designed' we should expect animals to be (Davies & Brooke 1989b; Rothstein 1990).

Hosts of cowbirds and cuckoos show various degrees of egg rejection

Two hypotheses for the variation

(a) Continuing arms race

One possibility is that the variation reflects snapshots of different stages of a continuing arms race. Species which show little rejection may be recent hosts, which have not yet had time to evolve counter-adaptations to the parasite, while the more strongly rejecting species may be older hosts, at a later stage of the arms

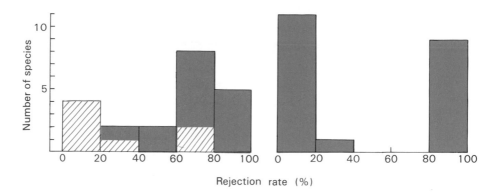

Fig. 4.12 Results of experiments with model eggs to test the rejection shown by various hosts to eggs unlike their own. (*Left*) Hosts (21 species) of the cuckoo, *Cuculus canorus*, in Europe. Hatched columns refer to current favourite hosts, black columns to suitable but rarely used hosts. From Davies & Brooke (1989a) and Moksnes *et al.* (1991). (*Right*) Hosts (21 species) of the brown-headed cowbird in North America. From Rothstein (1990).

Accepters may be more recent hosts ...

race. There is good evidence that brood parasites may change their use of hosts with time, so promoting arms races with new hosts. In Japan, the cuckoo has recently begun to parasitize a new host, the azure-winged magpie *Cyanopica cyana* (Yamagishi & Fujioka 1986; Nakamura 1990), while in North America the cowbird, originally confined to the short-grass plains in the central states, has undergone a remarkable increase in range and abundance within the last 200 years as forests have been cleared and tall-grass prairies ploughed for agriculture. This range expansion must have brought the cowbird into contact with many new hosts.

... while rejecters may have won the arms race

The continuing arms race hypothesis could explain two otherwise puzzling features of Fig. 4.12. First, current favourite hosts of the cuckoo, on average, show *less* rejection than other suitable but more rarely used hosts. Perhaps the latter were old favourites of the cuckoo which evolved strong rejection and forced the cuckoo to change to new hosts? Second, the cowbird hosts show a much stronger dichotomy into 'accepters' and 'rejecters'. This could reflect the much stronger selection exerted by cowbirds, with their extraordinarily high rates of parasitism. Once hosts begin to evolve rejection, it will sweep very fast through the population so at any one time few hosts will be in the intermediate stages. By contrast, rejection will take much longer to fix in cuckoo host populations due to the very low rates of parasitism, so more hosts will be at an intermediate stage.

(b) Evolutionary equilibrium

An alternative hypothesis is that the systems in Fig. 4.12 are at equilibrium with stabilizing selection for the degree of rejection now shown by each species. Although there is no evidence that more strongly rejecting hosts suffer higher rates of parasitism, it is possible that they face differing costs of rejection. First, some hosts may have problems in actually rejecting the parasite's egg. Rohwer and Spaw (1988) found that small-billed species were more likely to accept cowbird eggs and they suggested that this was because they were unable to puncture and eject the parasite egg, which has an unusually thick shell. Their options are then either to desert or accept. Acceptance may sometimes be better because the host often rears some of its own young along with the parasite and a repeat nesting may not be profitable if the season is short. This may explain some of the variation in Fig. 4.12(b) but the argument cannot apply to cuckoo hosts, which gain nothing from acceptance.

For some hosts acceptance may be better than rejection

The other problem hosts face is that of recognizing the parasite egg. Stephen Rothstein (1982) showed that cowbird hosts do not simply adopt the rule 'reject the egg type which is in the minority'. For two hosts, the American robin *Turdus migratorius* and the grey catbird *Dumetella carolinensis*, he replaced all but one of their eggs with model cowbird eggs, so that their own egg was now the minority type. The hosts rejected all the cowbird eggs and left their own intact. This implies that they know what their own eggs look like. There were also interesting differences between these two hosts in their tolerance of models which deviated from their own. Real cowbird eggs are white and spotted. The American robin's large pure blue eggs thus differ in three ways: size, colour and lack of spots. Experiments with a variety of models (Fig. 4.13) showed that the robins tolerated eggs which differed from their own in one way (e.g. same size and colour but with spots, or same colour and no spots but smaller) but they rejected models which differed in two ways. It may pay the robins to be tolerant of slightly deviant eggs to avoid the costs of mistakenly ejecting one of their own eggs, which sometimes get muddy in the nest lining and so become darker or spotty. But the robin should not be too tolerant or else it will end up accepting cowbird eggs. Thus there will be an optimal degree of discrimination to maximize the chance the parasite egg is detected without making a mistake. Catbirds also have blue eggs, but they are smaller, about the same size as a cowbird egg. Catbirds, therefore, cannot be as tolerant as the robin and Rothstein's experiments showed that

Variation in the degree of rejection may be related to mechanisms of egg recognition

Fig. 4.13 Top row, left-hand egg is that of the American robin. Bottom row, left-hand egg is that of the grey catbird. The other eggs are models used by Stephen Rothstein to test host discrimination. The right-hand model in the middle row is like a real cowbird egg. From Rothstein (1982).

they rejected eggs which differed from their own in one way only (e.g. a white egg the same size as their own).

These experiments suggest that different species may have differing degrees of rejection depending on how different their own eggs are from those of the parasite. This may explain some of the variation in Fig. 4.12, with different species stabilizing at different degrees of rejection (Lotem *et al.* 1992). Alternatively there may be a continuing arms race, with hosts evolving finer degrees of discrimination in response to the evolution of better mimicry by the parasite.

Finally, how do the hosts know what their own eggs look like? There is some evidence that hosts learn the characteristics of their own eggs the very first time they breed. Thus young birds are more likely to accept foreign eggs (either models or real parasite eggs), especially if they are introduced into their nests early on during laying, presumably because they learn these as part of their own set (Rothstein 1974; Lotem *et al.* 1992). In one experiment Rothstein gradually replaced an entire clutch of catbird eggs, as they were laid, with model cowbird eggs. He then gave the hosts one of their own eggs and they rejected it!

CONCLUSION

We have discussed this example of coevolution in detail because it provides a good example of how field experiments can reveal the nature of adaptations and how they change during evolution. Furthermore the studies show the importance of understanding mechanisms (how hosts recognize eggs) if we are to understand the evolution of behaviour. Some of the points we have raised echo the discussion at the end of Chapter 3. If animals seem to be behaving maladaptively (in this case accepting an egg unlike their own) is this because of evolutionary lag to a new selective pressure or have we merely misunderstood the animal's constraints (problems .of recognizing and rejecting parasite eggs)? There is no easy answer to this problem but an exciting prospect for future studies will be to compare the egg recognition systems of different species to test whether host learning mechanisms change in response to exploitation by brood parasites.

Summary

The complex adaptations of predators and counter-adaptations by their prey reflect the result of an evolutionary arms race. Experiments on predation by birds on cryptic prey show that:
1 crypsis does indeed reduce predation, and polymorphism in the prey population is effective in reducing the efficiency of search image formation by the predator (blue jays versus underwing moths);
2 even slight crypsis, enough to impose just a few seconds extra discrimination time, can bring an advantage to prey and thus serve as the starting point for an arms race (great tits and conveyor belt experiment).
   Noxious prey are often brightly coloured. Experiments show that predators learn to reject noxious prey quicker if they are conspicuous rather than cryptic (chicks versus coloured breadcrumbs). Field and laboratory experiments with guppies show that male coloration reflects a balance between the advantages of bright colours for mating and the advantages of dull colours for avoiding predation. Experiments with model eggs reveal the adaptations of cuckoos and cowbirds and counter-adaptations by their hosts. Acceptance of non-mimetic eggs by some hosts may reflect evolutionary lag or the costs of recognizing and rejecting parasite eggs.

Further reading

We have not attempted to be at all comprehensive in this chapter, rather to take a few well-defined problems and show how these can be analysed by field and laboratory experiments. For general discussions of predator—prey coevolution see Endler (1991).

Rothstein (1990) reviews brood parasitism. For theoretical accounts of coevolution and arms races, see Slatkin and Maynard Smith (1979) and Dawkins and Krebs (1979). Baker and Parker (1979) discuss bird coloration in relation to the unprofitable prey hypothesis. For another well-studied example of coevolution see the papers on snakes versus ground-squirrels (Owings & Coss 1977; Goldthwaite *et al.* 1990; Poran & Coss 1990).

Topics for discussion

1   Why are warning colours conspicuous?
2   Why do cuckoo hosts show such fine discrimination of cuckoo eggs yet blindly accept a cuckoo chick which is clearly nothing like their own young?
3   Design a field experiment to test whether prey choice is influenced by searching images.

Chapter 5. Competing for Resources

Our discussion in Chapter 3 of how individuals exploit resources such as food omitted a crucial factor: competition. When many individuals exploit the same limited resources, they are competitors, and the decisions made by one competitor may be influenced by what others are doing. This is true, for example, of the male dungflies introduced in Chapter 3. Female dungflies (the limited resource for which males compete) occur at higher densities on cowpats than in the surrounding grass, but none the less it pays some males to search in the grass where competition is less intense.

In this chapter we will start by discussing the simplest form of competition, *exploitation*, which simply means 'using up resources', and then go on to describe another form of competition, *resource defence*, in which animals keep others away from resources by fighting or by aggressive displays.

Competition by exploitation: the ideal free distribution

Let us start with a model of competition. To keep things simple, imagine there are two places ('habitats'), a rich one containing a lot of resources and a poor one containing few, and that each individual chooses to exploit the habitat in which it can achieve the higher pay-off, measured as rate of consumption of resource. With no competitors, an individual would simply go to the better of the two habitats and this is what we assume the first arrivals will do. But what about the later arrivals? There is no territoriality or fighting and so there is no limit to the number of individuals that can go there. However the more competitors that occupy the rich habitat the more resource will be depleted, and so the less profitable it will be for further newcomers. Eventually a point will be reached where the next arrivals will do better by occupying the poorer quality habitat where, although the resource is in shorter supply, there will be less competition (Fig. 5.1). Thereafter the two habitats will be filled so that the profitability for an individual is the same in each one.

In other words, competitors adjust their distribution in relation to habitat quality so that each individual enjoys the same rate of acquisition of resources. This theoretical pattern of distribution of competitors between resources has been termed the 'ideal free' distribution by Stephen Fretwell (1972) because it assumes that animals are free to go where they will do best (there is no

A simple model of how competitors should be distributed between habitats or patches ...

... the ideal free distribution

102

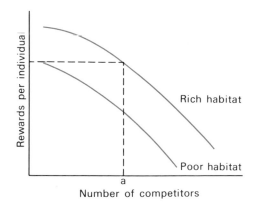

Fig. 5.1 The ideal free distribution. There is no limit to the number of competitors that can exploit the resource. Every individual is free to choose where to go. The first arrivals will go to the rich habitat. Because of resource depletion, the more competitors the lower the rewards per individual so at point *a* the poor habitat will be equally attractive. Thereafter the habitats should be filled so that the rewards per individual are the same in both. After Fretwell (1972).

exclusion of weaker competitors by stronger ones) and that the animals are ideal in having complete information about the availability of resources.

We can see an example of this in action when people stand in line at the counters of a supermarket. If all the serving clerks are equally efficient and all the customers are equal in the time they require for service, then the lengths of all the lines should end up being equal. If one line gets shorter, then customers would profit by joining it until its length becomes the same as the others. Because everyone is free to join whichever line they like, each person goes to the best place at the time, and the lines fill up in an ideal free way with the result that every customer should have the same waiting time for service.

Milinski's test of the ideal free model: sticklebacks and waterfleas

An equivalent in animals is Manfred Milinski's (1979) experiment with sticklebacks. Six fish were put in a tank, and prey (*Daphnia*) were dropped into the water from a pipette at either end. At one end prey were dropped into the tank at twice the rate of the other end. The best place for one fish to go depends on where all the others go. There was no resource defence and Milinski found that the fish distributed themselves in the ratio of the patch profitabilities, with four fish at the fast-rate end and two at the slow-rate end. When the feeding regimes were reversed, the fish quickly redistributed themselves so that four were again at the fast end (Fig. 5.2). This is the only stable distribution under ideal free conditions. With any other distribution it would pay an individual to move. For example, if there were three fish at each

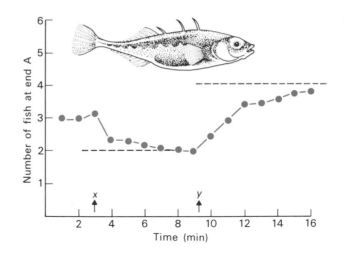

Fig. 5.2 Milinski's (1979) feeding experiment with six sticklebacks. At point *x*, end B of the tank had twice the amount of food as end A. At point *y* the profitabilities were reversed. The dashed lines indicate the number of fish predicted at end A according to ideal free theory, and the solid line is the observed numbers (mean of several experiments).

end then one fish would profit by moving from the slow to the fast-rate end. Once it had done so, it would not pay any of the other fish to move. In our supermarket analogy, this experiment is equivalent to what should happen if one clerk is twice as efficient at serving customers as another; the stable distribution would be for this line to be twice as long.

The stable distribution of searching individuals could be achieved in two ways. For example, if one habitat was twice as profitable as another then stability could come about by:

1 competitor numbers adjusting so that twice as many individuals go to the good habitat as to the poor one;

2 all individuals visiting both habitats, but each spending twice as much time in the good habitat as the poor one.

Three predictions of the ideal free model . . .

Milinski's experiment tested the ideal free distribution by examining the *numerical prediction* (number of predators at each end of the tank); as predicted the ratio of sticklebacks to food was equal. Two other predictions could also have been tested: the *equal intake prediction*, i.e. the intake rate of sticklebacks should be the same at both ends of the tank; and the *prey risk prediction*, i.e. the prey mortality should have been the same at both ends (Kacelnik *et al.* 1992).

. . . 'continuous input' and depletion

The example of Milinski's fish is what is usually called a 'continuous input' system, in which prey density does not change with time because prey arrive at a constant rate and are eaten as soon as they arrive. Whilst this may be a realistic representation of some natural foraging environments such as streams with insects drifting past waiting fish, more often prey are likely to be gradually depleted. The predictions of the ideal free model are more complicated in this situation (Kacelnik *et al.* 1992).

A field test of the numerical prediction for continuous input

Mary Power's (1984) study of armoured catfish (*Ancistrus spinosus*) in a stream in Panama shows that these alga-grazing fish are distributed among pools in an ideal free manner. The continuous growth of algae on the sides and bottom of the pools resembles the continuous input of food in Milinski's experiment. In sunny pools the algal film grew about six times as fast as in shady pools and the catfish were about six times as numerous in sunny pools, which would be expected from an ideal free distribution with continuous input.

Competition by resource defence: the despotic distribution

Consider the same situation as before: two habitats, one rich and one poor, but this time the first competitors to settle in the rich habitat defend resources by establishing territories (pieces of ground containing the resource) so that later arrivals are forced to occupy the poor habitat even though they do less well there than the individuals in the rich area. When the poor habitat fills up with territory-defending individuals the latest arrivals of all may end up being excluded from the resource altogether (Fig. 5.3). This kind of situation is very common in nature. In Wytham Woods, near Oxford, England, the best breeding habitat for great tits is in oak woodland. This is quickly occupied in the spring

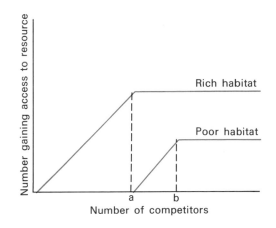

Fig. 5.3 Resource defence. Competitors occupy the rich habitat first of all. At point *a* this becomes full and newcomers are now forced to occupy the poor habitat. When this is also full (point *b*), further competitors are excluded from the resource altogether and become 'floaters'. After Brown (1969).

Removal experiments
show that territorial
behaviour may exclude
some competitors from
good habitats

and becomes completely filled with territories. Some individuals are excluded from the oak wood and have to occupy the hedgerows nearby where there is less food and consequently have lower breeding success. If great tits are removed from the best habitat then birds rapidly move in from the hedgerows to fill the vacancies (Krebs 1971). Similarly, in red grouse (*Lagopus lagopus scoticus*) territorial birds defend the richest areas of the heather moors as breeding and feeding territories. Excluded birds have to go about in flocks and exploit poor habitats where their chances of survival are low. Once again, if a territory owner is removed its place is quickly taken by a bird from the flock (Watson 1967).

In these examples the strongest individuals are despots, grabbing the best quality resources and forcing others into low quality areas or excluding them from the resource altogether.

Resource defence is one kind of competition that comes under the general heading of *interference*. Interference occurs whenever competitors interact with one another in such a way as to reduce their efficiency at searching for or exploiting resources. In addition to resource defence, interference might occur, for example, competitors bump into each other and waste time which would otherwise be spent searching, or as happens in some insect parasitoids, scent marks left by one competitor confuse another individual (Hassell 1971).

The ideal free distribution with unequal competitors

Most examples in nature will have features of both the simple models we have discussed above. Perhaps the commonest situation will be where the best place to search depends on where all the other competitors are, but within a habitat some individuals get more of the resource than others. In the fish experiment, for example, our population counts may show a stable, ideal free distribution of individuals but the chances are that some fish will be better competitors than others. At each end of the tank there could be one or two large fish grabbing most of the prey. The ideal free distribution could come about because of the way the subordinates distribute themselves in relation to the despots. In effect, the despots are part of the habitat to which the subordinates respond when deciding where to search (Milinski 1984a).

It is unlikely that there will be any populations where all individuals are of equal competitive ability. Even though male dungflies obey an ideal free distribution around cowpats, the larger males get more females than small males (Borgia 1979). In our supermarket there is probably no fighting between the customers but some will have more items of shopping than others and will require more time to be served. In a sense, they are greater competitors because standing in line behind them will impose greater waiting times.

Geoff Parker and Bill Sutherland (1986) pointed out that it might be difficult from the numerical prediction alone to distinguish between the simple ideal free distribution with equal competitors and one with unequal competitors, which they call the 'competitive unit' model, because it hypothesizes that the number of 'competitive units', rather than the number of individuals, is equalized across patches. If one individual can consume resources twice as rapidly as another it scores twice the number of competitive units. The difficulty in distinguishing the two versions of the ideal free distribution arises from the fact that, by chance alone, the competitive unit distribution will tend to look like the simple ideal free distribution (Fig. 5.4). This may be why many studies appear to support the numerical prediction of the ideal free distribution, even though competitors in these studies were unequal.

A good example which shows features of both the resource defence and ideal free models is the study by Thomas Whitham (1978, 1979, 1980) of habitat selection in the aphid (*Pemphigus betae*). In the spring, females known as 'stem mothers', settle on leaves of narrowleaf cottonwood (*Populus angustifolia*) to feed and they become entombed by expanding leaf tissue, so forming a gall. A stem mother reproduces parthenogenetically and the number of progeny she produces depends on the quantity and quality of the juices she can tap from the leaf. The largest leaves provide the richest supplies of vascular sap and result in the greatest reproductive success, with up to seven times the number of progeny that are produced by settling on a small leaf. As we would expect, all the large leaves are quickly occupied and so additional settlers have the problem of whether to settle on large leaves and share the resources or occupy smaller leaves alone.

Whitham made measurements of reproductive success which enabled him to plot a family of fitness curves for habitats of varying quality (leaves of different sizes) and with different densities of competitors (number of galls per leaf). Figure 5.5 shows the results, which enable us to draw three conclusions. First, for any competitor density, the average reproductive success increases with habitat quality. Second, within a habitat of a certain quality, success decreases as the number of competitors increases. This shows that stem mothers settling on the same leaf must compete with each other for resources. Third, if the *average* reproductive success is calculated for aphids who are alone on a leaf, those who share a leaf with one other and those who share with two

The 'competitive unit' model

Gall aphids: a test of the competitive unit model

Larger leaves correspond to better habitats

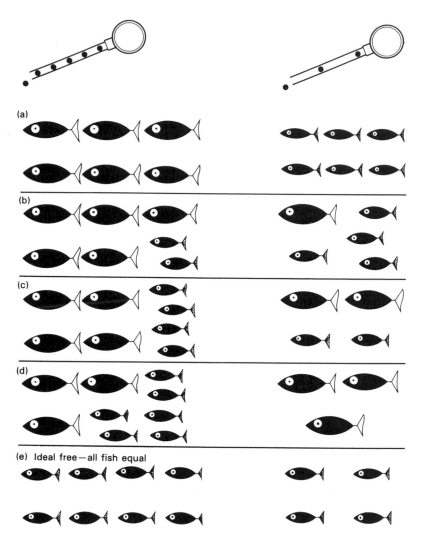

Fig. 5.4 An illustration of how in theory it is difficult to distinguish between a numerical distribution based on the simple ideal free distribution with equal competitors (e), and a distribution with unequal competitors (a–d). The left-hand patch has twice the input rate of that on the right, so the ideal free distribution (e) of 12 equal competitors is 8:4. If six of the fish (drawn as twice the size) are capable of eating twice as many prey per unit time than the other six there are four possible ways of distributing the 12 fish so that the average intake at the two ends is equal (a–d). However, the number of different ways in which each of these distributions can be achieved varies. Imagine each fish has a name. The 12 fish can be arranged in only one pattern to achieve distribution (a). For (b), (c) and (d), there are many ways of arranging the individual fish to achieve the distribution; the numbers of ways are 90, 225 and 20. In short, by chance alone, (c) is the most likely to be observed. Note that it has the same numerical pattern as (e). After Milinski and Parker (1991).

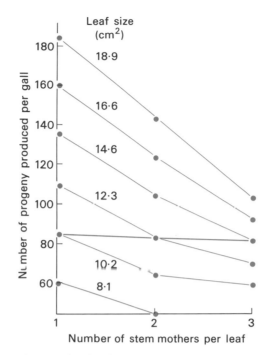

Fig. 5.5 The thin lines are a family of fitness curves for habitats of varying quality (leaf size) and competitor density (no. stem mothers per leaf) in the aphid *Pemphigus betae*. The solid horizontal line is the average success for one, two and three stem mothers per leaf. See text for explanation. From Whitham (1980).

Average reproductive success is equal on all leaves . . .

others, no significant differences are found. There was also no significant difference in average success on leaves with different numbers of competitors when other fitness measures were used such as body weight of the stem mother, abortion rate, development rate or predation. The results support the predictions of the ideal free model. The conclusion, therefore, is that the stem mothers settle on leaves of different sizes such that the average success in good habitats with a high density of competitors is the same as in poor habitats with fewer competitors.

. . . but individuals near the leaf base do better

However, although the results of average success on different sized leaves are in accord with ideal free predictions, within a habitat not all individuals get equal rewards. This is because a leaf is not a homogeneous habitat. The best place to be is on the mid-rib at the base of the leaf blade because everything translocated into and out of the leaf must flow past this point. Basal galls on a leaf give rise to more young than distal galls and the stem mothers spar with each other, like boxers in a ring, for occupancy of these prime positions (Fig. 5.6). As we would predict from the defence model, if a basal individual is removed her place is quickly occupied by another aphid from a distal site.

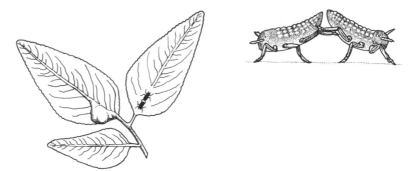

Fig. 5.6 Stem mother aphids *Pemphigus betae* fight for prime positions on a leaf by kicking and pushing. The winner will settle at the base of the mid-rib where food is richest. From Whitham (1979).

This example examines the distribution of unequal competitors within a species. Rosenzweig (1986) uses similar arguments to explain interspecific differences in habitat use.

The economics of resource defence

'Functions' of
territoriality

Some animals, as we have seen, compete for resources by exploitation, others by territoriality. Is it possible to predict when the latter form of competition should be adopted instead of the former? The first attempts to analyse this question did so by asking 'What is the function of territorial defence?' (Hinde 1956; Tinbergen 1957). This approach led to long lists of different functions for different species, emphasizing the diversity of resources defended. Hole nesting birds defend territories to secure nest sites, barnacles defend areas of rock (by chemical signals) to guarantee enough space in which to grow, bullfrogs defend bits of habitat which are attractive to females and so on. The 'functions' approach did not, however, answer the question of when it pays to compete by means of territorial defence. A more fruitful approach stems from the concept of economic defendability.

(a) Economic defendability

Jerram Brown (1964) first introduced the idea of economic defendability. He pointed out that defence of a resource has costs (energy expenditure, risk of injury and so on) as well as the benefits of priority of access to the resource. Territorial behaviour should be favoured by selection whenever the benefits are greater than the costs. This may seem a rather obvious conclusion and indeed as stated so far it is. However it led field workers to look in more detail at the time budgets of territorial animals, in particular the feeding territories of nectivorous birds. While Brown's idea is in principle applicable to any kind of territory, it has been most useful in looking at birds such as hummingbirds, sunbirds, and honeycreepers in which the costs and benefits can be measured in calories in the field. Frank Gill and Larry Wolf (1975), for example, were able to measure the nectar content of territories of the golden-winged sunbird (*Nectarinia reichenowi*) in East Africa, where it defends patches of *Leonotis* flowers outside the breeding season. They also calculated from time budget studies and laboratory measurements of the energetic costs of different activities such as flight, sitting and fighting, how much energy a sunbird expends in a day. When the daily costs were compared with the extra nectar gained by defending a territory and excluding competitors, it turned out that the territorial birds were making a slight net energetic profit. The resource was economically defendable (Box 5.1).

Time and energy budgets of sunbirds predict when resources should be defended

Box 5.1 *The economics of territory defence in the golden-winged sunbird (Gill & Wolf 1975).*

(a) The metabolic cost of various activities was measured in the lab:

Foraging for nectar	1000 cal/h
Sitting on a perch	400 cal/h
Territory defence	3000 cal/h

(b) Field studies showed that territorial birds need to spend less time per day collecting enough energy in the form of nectar to survive when the flowers contain more nectar:

Nectar per flower (μl)	Time to get energy (h)
1	8
2	4
3	2.7

(c) By defending a territory a bird excludes other nectar consumers and therefore increases the amount of nectar available in each flower. The bird therefore saves foraging

Continued on p. 112

Box 5.1 *Continued*

time because it can satisfy its energy demands more rapidly. It spends the spare time sitting on a perch, which uses less energy than foraging. For example, if defence results in an increase in the nectar level from 2 μl to 3 μl per flower, the bird saves 1.3 h per day foraging time (from (b)). It therefore saves:

$$(1000 \times 1.3) - (400 \times 1.3) = 780 \text{ cal}$$
$$\text{foraging} \qquad \text{resting}$$

(d) But this saving has to be weighed against the cost of defence. Measurements in the field show that the birds spend about 0.28 h per day on defence. This time could otherwise be spent sitting, so the cost of defence is:

$$(3000 \times 0.28) - (400 \times 0.28) = 728 \text{ cal}$$

In other words, the flowers are just economically defendable when the nectar levels are raised from 2 to 3 μl as a result of defence. Gill and Wolf found that most of their sunbirds were territorial when the flowers were economically defendable.

The idea of economic defendability has also been used to predict the levels of resource availability which could lead to territorial defence. If resources are very scarce, the gains from excluding others may not be sufficient to pay for the cost of territorial defence. Instead the animal might abandon its territory and move elsewhere. There may also be an upper threshold of resource availability beyond which defence is not economical. This upper boundary could arise for a number of reasons.

Resource defence in sunbirds does not pay when nectar is too scarce or too abundant

1 There may be so many intruders trying to invade rich areas that defence costs would be prohibitively high.

2 There may be no advantage of territoriality at high resource levels if the owner cannot make use of the additional resources made available by defence. In Gill and Wolf's sunbirds, one advantage of territorial defence was that it raised the amount of nectar per flower (by exclusion of nectar thieves) and hence saved foraging time (Box 5.1). But if nectar levels are already high, the extra increment resulting from territorial defence saves hardly any foraging time. This is because the bird's rate of food intake at high nectar levels is limited by the time it takes to probe its beak into the flower (the handling time). For example Gill and Wolf calculate that an increase from 4 μl to 6 μl per flower would save the birds less than 0.5 h of foraging time, while as shown in Box 5.1 an increase of from 1 μl to 2 μl saves 4 h. Thus when

nectar levels are high, territorial exclusion of nectar thieves does not pay for itself in savings of foraging time.

3 A third hypothesis which predicts an upper boundary was proposed by Carpenter and MacMillen (1976). They suggested that territory defence has associated risks such as increased conspicuousness to predators, so that whenever resource levels are high enough to allow an animal to satisfy its needs without excluding others, territories should be abandoned. As yet there is no critical evidence to test this idea.

(b) Optimal territory size

Although the use of Brown's concept to predict the range of resource levels over which animals should defend territories is a step forward from simply showing that territories are economically defendable, a more quantitative and powerful development is to predict the optimal amount of resource for an individual to defend.

Migratory hummingbirds adjust territory size to maximize weight gain

Rufous hummingbirds (*Selasphorus rufus*) studied by Lynn Carpenter (Carpenter *et al.* 1983) appear to defend territories of a size that maximizes energy gain (Fig. 5.7). These birds were studied on temporary territories around patches of Indian Paintbrush

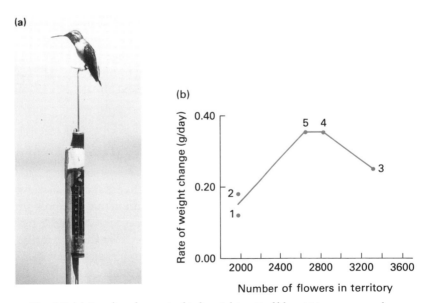

Fig. 5.7 (a) A rufous hummingbird weighing itself by sitting on a perch attached to a balance. Photo by Mark Hixon. (b) The daily weight gain of one territorial bird plotted against its territory size (measured as number of flowers defended) for five successive days. The bird started with a small territory on the first day, enlarged its territory on the third day and then gradually moved towards an intermediate size at which daily weight gain was maximal.

(*Castilleja linariaefolia*), defended during pauses on their southward migration through the Sierra Nevada mountains in California, where they put on weight in order to continue with their migration. By the ingenious method of putting out perches attached to a small balance, Carpenter *et al.* were able to measure the weight gain of birds defending territories of different sizes. At least some individuals appeared to adjust their territory size over a series of days towards the size that maximized rate of weight gain (Fig. 5.7).

One feature of hummingbird behaviour that is at first sight puzzling in view of their goal of maximizing weight gain, is the fact they spend about 75 per cent of their time perching and only 20 per cent of their time foraging. If maximizing energy gain is so important why do the birds not forage more often? A neat answer has been provided by Jared Diamond and colleagues (Diamond *et al.* 1986). They used isotope labelled glucose and polyethylene glycol to measure the time that it takes the hummingbird to absorb energy and the time required for a meal to pass through the digestive tract. They found that hummingbirds have an extremely rapid rate of active absorption of glucose through the intestinal wall, but that the limiting step in digestion is the time taken for food to pass from the crop, where it is stored when the bird first takes it in, to the stomach. It takes about 4 min for the crop to half empty after a 100-ml meal, which corresponds quite closely to the average interval between meals in the wild. In other words, while the bird is sitting it is in fact hard at work making room for the next meal! (Hixon 1982).

Why do hummingbirds rest? A digestive constraint

(c) Graphical models of optimal territory size: a caveat

Graphical models should be treated with caution

Brown's concept of economic defendability is often represented in the form of a graph of costs and benefits which can be used to predict the effects of changes in resource density and number of competitors on the amount of resource defended (Fig. 5.8). In a cautionary note, Tom Schoener (1983) has pointed out that the graphical model can lead to diametrically opposing predictions, depending upon the exact shape of the curves (Fig. 5.8). The message is that graphical models, in this book and elsewhere, should be handled with care!

SHARED RESOURCE DEFENCE

So far we have assumed that territories are defended by single individuals, but very often two or more competitors share the same territory. Typically, sharing individuals are a mated pair, as

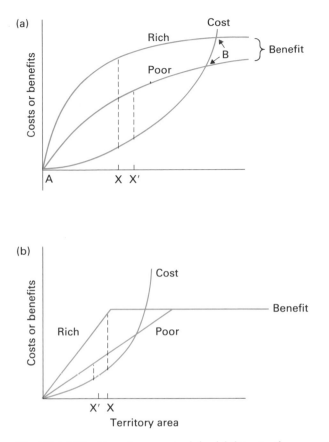

Fig. 5.8 (a) The idea of economic defendability. As the amount of resource defended (or territory size) increases so do the costs of defence. The benefits (e.g. amount of food available) are assumed to increase at first but level off as the resource becomes superabundant and the animal is limited in its capacity to process the resource. Two benefit curves are shown, one for a rich environment and one for a poor environment: the benefit curve rises more steeply in the former because the density of resources is higher. The resource is economically defendable between A and B. Within this range the optimal territory size depends on the currency: for maximizing net gain the optimal size is smaller for the rich environment (X) than for the poor one (X') (note that this is where the slopes of the cost and benefit curves are equal). (b) The same model, but with slightly different shaped curves. In this case the optimal territory size to maximize net gain is predicted to *increase* in a rich environment (X now greater than X'). After Schoener (1983).

in most territorial birds for example, and here the reasons for shared defence may be unrelated to economic considerations of costs and benefits of resource defence. The fact that European robins (*Erithacus rubecula*) share a territory in the breeding season but not at other times of year is more related to the necessity of the two birds to be together for mating and parental duties than because resource defence is more efficient when done by two birds. However, sometimes shared defence may occur because of economic considerations. An example which has been analysed in some detail is Nick Davies and Alasdair Houston's (1981) study of the pied wagtail (*Motacilla alba*).

In winter, these birds defend territories along stretches of the River Thames near Oxford. They feed on insects washed up by the river onto the bank. After a bird has foraged in a particular spot and depleted the insects, the numbers gradually build up as new insects are washed ashore. Eventually the rate of deposition and washing away is about equal and the insect abundance remains at a plateau. An efficient way to exploit this kind of renewing resource is to visit each place only after the food has had a chance to renew itself, and this is exactly what the wagtails do. The territory owner works systematically around its territory and revisits each spot on average about once every 40 min after the insect supply has been replenished.

It is not hard to see why territorial defence pays: without exclusive use of the river bank, a wagtail's strategy for harvesting the renewing food could easily collapse. It might turn up at a spot it has not visited for 40 min only to find that another bird has just been there, rather like the mushroom picker who shows up too late in the morning. One might expect, therefore, that wagtails would always defend exclusive territories, but this is not so. Sometimes the territory owner tolerates a second bird, a so-called 'satellite'. The two sharers move around the territory out of phase with one another, so that the average return time to a site is halved to 20 min, resulting in a lower feeding rate for the owner. To counterbalance this cost there is a benefit of sharing because the satellite chases away intruders. The owner therefore saves defence time leaving more time to feed. The net effect of these costs and benefits on maximizing feeding rate, the currency hypothesized by Davies and Houston, depends on the food supply. On days when the rate of renewal of food is high the cost of sharing is relatively low, and on days when intrusion rate is high, the benefit of sharing is relatively large. Intrusion rate is closely related to asymptotic food levels, so sharing will tend to pay on days with a high rate of food renewal and a high asymptotic level. By calculating the exact values of the costs and benefits in terms of feeding rate Davies and Houston were able to predict on which days it would pay a territory owner to share and on which days it would not. Their predictions were right for 34 out of 40 days (Fig. 5.9).

The wagtail study illustrates two general points. First it is an example of how apparently different kinds of costs and benefits (defence and feeding) can sometimes be reduced to a single

When should wagtails share a territory? When renewal of food is rapid and intrusion rate is high

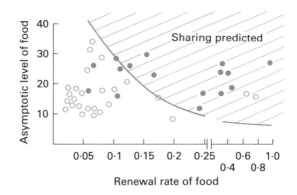

Fig. 5.9 Wagtails were predicted to share their territories when the rate of renewal of food and the asymptotic abundance of food were above the curve. These combinations represent instances where the costs of sharing are outweighed by the benefits. The observed outcomes are shown as dots, each one representing a single day; solid dots — shared territories, open dots — not shared.

currency — feeding rate in this case. Second, it shows that one advantage of group living is shared resource defence. The wagtail groups were never larger than two, but the same argument could be generalized to larger groups (Brown 1982). In Chapter 6 we will come back to the costs and benefits of group living.

INTERSPECIFIC TERRITORIALITY

Although competition for resources is usually more intense within than between species, individuals of different species with similar ecological requirements may compete and sometimes they do so by defending interspecific territories. In the past there has been some debate over whether interspecific territoriality arises because individuals of one species misidentify individuals of another species and treat them as competitors even though they are not, or whether it is a genuine case of economic resource defence. The instances that have been studied in detail favour the latter view. A nice example is Tim Reed's work (1981) on interspecific territoriality between chaffinches (*Fringilla coelebs*) and great tits (*Parus major*). The species have very different songs and are quite distinct in appearance. Over most of the British Isles they occur in the same habitats and do not interact at all, but on the small island of Eigg off the west coast of Scotland they defend interspecific territories. Reed showed this in two ways. First he carried out playback experiments on the island and on the nearby mainland of Arisaig; these showed that neither species responds at all to the other's song on the mainland, but that they respond

Interspecific competition may lead to interspecific territoriality

by aggressive approach and counter-singing on the island. Secondly, Reed did a removal experiment on the island: he captured chaffinches and removed them from their territories. Great tits soon moved in to occupy the empty spaces, indicating that before the removals they have been kept at bay by chaffinch territorial defence.

Why do the two species defend interspecific territories on the island but not on the mainland? The likeliest explanation is that on the island the wooded habitats which both species occupy are structurally much simpler than on the mainland. Because of the simplicity of the habitat the two species are forced to use the same resources (especially food) which they would not normally do.

Summary

Animals may compete for resources by pure exploitation or by resource defence, or by a mixture of both. A simple model of exploitation is the 'ideal free' distribution of competitors between resources. This pattern is seen in a number of examples. A useful concept in thinking of when it pays to compete by resource defence is that of economic defendability. This can be used in combination with studies of time budgets to predict conditions under which defence should occur and to formulate hypotheses about the optimal size of territories.

Further reading

Hixon *et al.* (1983) describe experiments on optimal territory size in hummingbirds. Verner (1977) develops the idea that animals might defend more resources than they actually need to maximize their own success, simply in order to reduce the success of others (superterritories). Appealing though it is, Verner's idea has been criticized on theoretical grounds by several authors (e.g. Rothstein 1979; Parker & Knowlton 1980). The essence of the criticism is that defending a superterritory might pay when the trait is rare, but it will not pay when most of the population adopts this strategy. Harper (1982) tests how ducks on a pond divide up between patches of different profitability by throwing pieces of bread at different rates. Inman (1990) tests whether starlings distribute themselves between patches according to the ideal free or the competitive unit model. Milinski and Parker (1991) and Kacelnik *et al.* (1992) review the ideal free distribution.

Topics for discussion

1 Do animals ever defend more resources than they need?

2 How can the effects of predation differences between patches be incorporated into the ideal free distribution? (see Abrahams & Dill 1989).

3 How would you apply the idea of economic defendability to resources other than food (e.g. nest sites, mates)?

Chapter 6. Living in Groups

Show anyone 10 000 flamingos nesting beak by jowl in a colony and the chances are that sooner or later they will ask 'Why on earth are they all nesting so close together?'. In this chapter we will look at why animals live in groups: why flamingos flock, horses herd and sardines shoal. Using the methods described in Chapter 2 (comparison between or within species and experimental studies of costs and benefits) we will show how ecological pressures might favour living in groups.

Food and predators influence the costs and benefits of group living

Comparisons between species suggest that the two main environmental influences on group size are food and predators (Chapter 2) and comparisons between populations within a species also emphasize their importance (Fig. 6.1). In many studies either costs or benefits related to feeding and predation have been measured and we will describe some of these in the first part of the chapter, before moving on to consider whether different kinds of cost and benefit can be combined to predict optimal group size. Animals which do not live in groups (and also some which do) often defend resources from which they exclude other members of the same species. Therefore the question 'Why live in a group?' is a natural complement to the question 'Why defend resources?' that we discussed in Chapter 5.

Living in groups and avoiding predation

The guppies in Fig. 6.1a live in groups when they are in streams where predators are common, which suggests that being in a group might help an individual to avoid becoming a meal. This could happen in several different ways.

INCREASED VIGILANCE

For many predators success depends on surprise: if the victim is alerted too soon during an attack, the predator's chance of success is low. This is true, for example, of goshawks hunting for pigeon flocks (Fig. 6.2). The hawks are less successful in attacks on large flocks of pigeons mainly because the birds in a large flock take to the air when the hawk is still some distance away. If each pigeon in the flock occasionally looks up to scan for a hawk, the bigger the flock the more likely it is that one bird will be alert when the hawk looms over the horizon. Once one pigeon takes off the others follow at once.

Many eyes are better than one

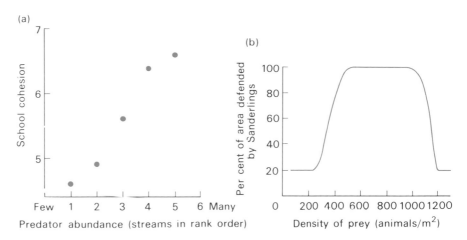

Fig. 6.1 Intraspecific variation in group size may be related to predators and food. (a) Guppies (*Poecilia reticulata*) from different streams in Trinidad: guppies from streams with many predators live in tighter schools than those from streams with few predators. Each dot is a different stream and 'cohesion' was measured by counting the number of fish in grid squares on the bottom of a tank. From Seghers (1974). (b) Sanderlings (*Calidris alba*) in Bodega Bay, California. The birds defend stretches of beach in some parts of the intertidal zone and feed in roving flocks on other parts of the beach. Whether or not the birds defend territories depends on the density of the major prey, an isopod called *Excirolana linguifrons*. Territories are mainly defended in areas of intermediate prey density. At very low densities there are not enough prey to make defence worthwhile and at very high densities there are so many sanderlings trying to feed that defence would not be feasible because of high intruder pressure. In the area where birds defend territories there is an inverse correlation between territory size and food density. From Myers *et al.* (1979).

The precise way in which vigilance changes with flock size depends on how individuals in the group spend their time. In ostrich flocks, for example, Brian Bertram (1980) found that each individual spends a smaller proportion of its time scanning than when alone but that the overall vigilance of the group (proportion **Ostriches scan at random** of time with at least one bird scanning) increases slightly with group size (Fig. 6.3). Therefore each bird in the flock has more time to feed and enjoys greater awareness of approaching lions (a potential predator of ostriches). The increase in vigilance with group size is as predicted if each bird raises its head independently of the others. The ostriches also raise their heads at random time intervals which makes it impossible for a stalking lion to predict how much time it has to creep forward undetected between look-ups by its victim. Any predictable pattern of looking could be exploited by the lion in its tactics of approach.

The problem of how individuals in a group scan is complicated **In scanning groups it may** by the fact that in a large group, where overall vigilance is at the **pay to cheat . . .** maximum value of 100 per cent, it would pay an individual to

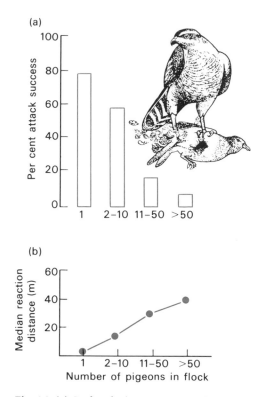

(a)

Per cent attack success

1   2–10 11–50 >50

(b)

Median reaction distance (m)

1   2–10 11–50 >50
Number of pigeons in flock

Fig. 6.2 (a) Goshawks (*Accipiter gentilis*) are less successful when they attack larger flocks of wood pigeons (*Columba palumbus*). (b) This is largely because bigger flocks take flight at greater distances from the hawk. The experiments involved releasing a trained hawk from a standard distance. From Kenward (1978).

'cheat' and spend all its time with its head down feeding. The cheater loses nothing in terms of vigilance because others are busy scanning and it gains extra time to feed. It is not known how this kind of cheating is prevented from evolving, but one suggestion is the following. Although the 'innocent' strategy of scanning regularly regardless of what others do is susceptible to cheating, a flock made up of more canny individuals, which do not scan unless they have seen their neighbours doing the same thing, might be resistant to cheaters (Pulliam *et al.* 1982). The general point is that even when there is an overall benefit of being in a group, each individual will be expected to try to get more benefit than the others. In groups of Thompson's gazelle, the individual that happens to be scanning when a predator approaches is more likely to escape (Fitzgibbon 1989). Here there is a direct benefit to the scanning individual, so no selection for cheating.

... unless the scanning individual gets a higher pay-off

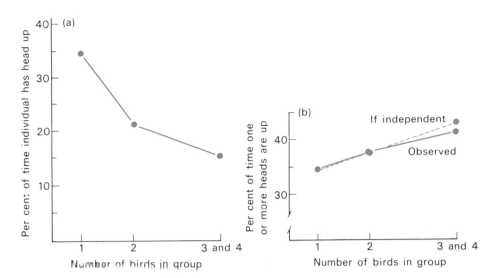

Fig. 6.3 Vigilance in groups. (a) An ostrich (*Struthio camelus*) spends a smaller proportion of its time scanning for predators when it is in a group. (b) The overall vigilance of the group increases slightly with group size (solid line), as predicted if each individual looks up independently of the others (broken line). From Bertram (1980).

DILUTION AND COVER

Although there is only a slight increase in vigilance with increasing group size in ostriches, the chances that any one individual will be eaten during an attack by lions decreases rapidly with group size, because the lions can kill only one ostrich per successful attack. By living in a group the ostrich dilutes the impact of a successful attack because there is a good chance that another bird will be the victim. To some extent this dilution effect may be offset by the increased number of attacks on larger and more conspicuous groups, but usually the net effect probably favours living in a group, as the following hypothetical example illustrates. An individual antelope in a herd of a hundred has (all things being equal) only a one in a hundred chance of being the victim in a single attack and the herd is not likely to attract more than a hundred times as many attacks as a solitary antelope (see also Fig. 6.4). In fact if the herd is more vigilant it may pay the predator to concentrate its attacks on small groups and solitary individuals.

The dilution effect may explain . . .

One study in which the survival rate of individuals in different sized groups was measured showed an overall benefit of group living from dilution. The monarch butterfly (*Danaus plexippus*) migrates from North America to spend the winter in warmer

. . . communal roosts of butterflies . . .

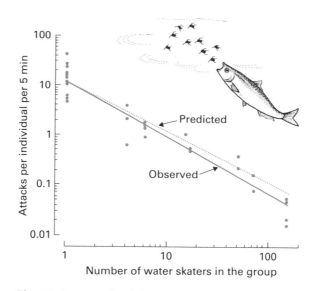

Fig. 6.4 An example of the dilution effect. The prey are insects called water skaters (*Halobates robustus*) that sit on the water surface; their predators are small fish (*Sardinops sagax*). The fish snap the insects from below, so there is little possibility that vigilance increases with group size. The attack rate by the fish was similar for groups of different sizes, so the attack rate per individual varies only because of dilution. The 'predicted' line is what would be expected if the decline in attack rate with group size is entirely caused by dilution; this line is very close to the observed. From Foster and Treherne (1981).

places such as Mexico. They assemble into enormous communal roosts in which the trees over an area of up to 3.0 ha may be clothed in resting butterflies. The monarch is not a very palatable butterfly, but some birds attack them in the winter roosts. Counts of the remains of predated butterflies showed that predation rate is inversely related to colony size, so the advantage of dilution seems to outweigh any disadvantage of greater conspicuousness in a large roost (Calvert *et al.* 1979).

. . . stealing young . . .   The dilution effect is probably a very widespread advantage of being in a group and it might explain the strange behaviour of birds such as ostriches and goosanders when they have young. When two females meet, each appears to try and steal the other's young and incorporate them into its own brood. Usually caring for someone else's young doesn't pay, but if predation pressure is severe it might, because of dilution. A more concrete example of the dilution effect comes from a study of semi-wild horses in the Carmargue, a marshy delta in the south of France. In the summer months the horses are plagued by biting tabanid flies and during this period they are more likely to cluster together in large groups.

. . . herding in horses . . .   Measurements of the number of flies per horse in large and small

groups showed that horses in a large group are less likely to be attacked. An experiment in which horses were transferred from large to small groups and vice versa confirmed that living in a group gives protection by the dilution effect (Duncan & Vigne 1979).

... and cicada cycles

In some animals dilution is achieved by synchrony in time as well as in space, and this might explain the remarkable 13- and 17-year life cycles of certain species of cicada. These insects live as nymphs underground and the adults emerge after 13 or 17 years depending on the species and location. In the 17-year cicada studied by Dybas and Lloyd (1974) millions of adults (of three species) emerge in synchrony over a wide area, effectively 'flooding the market' so that the chances of any one individual falling victim to a predator is reduced. Lloyd and Dybas (1966) and others have speculated on why the cycle should be 13 or 17 years long and not, for example, 15 or 18. The advantage of a very long dormant stage between emergence periods is that it forces specialist predators and parasites out of business. When there are no cicadas around for 13 or 17 years the predators have either to die or to switch to other prey or to become dormant themselves. The very long cycle could have evolved as a result of an 'evolutionary race' (Chapter 4) in which both cicadas and their predators gradually extended their life cycles until the cicadas eventually 'won'. The significance of the 13- and 17-year periods is that these are prime numbers which means that a predator could not regularly fall into synchrony with the cicadas if it had a short life cycle of which the cicada cycle is a multiple. If, for example, cicadas had a 15-year cycle, predators with 3- or 5-year life cycles would fall into step with their prey every fifth or third generation.

This idea remains an interesting speculation, but synchrony is certainly an advantage. Field evidence shows that cicadas emerging at the peak of the cycle have a lower chance of succumbing to predators than those emerging early or late (Simon 1979). Selection therefore acts to maintain synchrony once it is established.

Individuals in the middle of a group may be safer than those at the edge

Just as a cicada in the middle of the emergence period is safer than one at either end, individuals in the middle of a flock, school or herd may enjoy greater security than those at the edge. If the predators pick off victims from the edge, each member of the group should jockey for a central position and, in effect, seek cover behind the others (Hamilton 1971). This may explain why starling flocks, for example, bunch together in a tight group when a predator approaches. Why should predators attack the edge of the group? The old trick of throwing three tennis balls to a friend at the same time shows how difficult it is to track one of

Confusion effect

a number of rapidly moving objects in the visual field for long enough to catch one. There is some evidence that predators suffer from the same type of confusion when attacking a dense group of prey (Neill & Cullen 1974) and this may provide an explanation of why attacks should be directed at the edge of a group.

GROUP DEFENCE

Prey animals are often not just passive victims and by living in a group they may be able to defend themselves against the unwelcome attentions of a predator. In colonies of black-headed gulls nesting pairs will mob a crow when it flies near their nest and in the centre of a dense colony many gulls mob the crow at the same time because it is close to many nests. The effect of this is to reduce the success of the crows in hunting for gulls' eggs (Kruuk 1964) (see also Fig. 6.5a).

COSTS OF BEING IN A GROUP

Communal mobbing in fieldfares

As we mentioned earlier, one of the costs of group living might be increased conspicuousness. This cost was studied experimentally by Malte Andersson (Andersson & Wicklund 1978) using artificial nests of the fieldfare, a thrush-like bird which breeds colonially in Scandinavian boreal forests. The bulky nests are quite conspicuous and a colony of artificial nests attracted more predators than did solitary nests. However, fieldfares vigorously mob and defecate on crows and other predators, and Andersson and Wicklund found that artificial nests placed near a colony of fieldfares survived better than those placed near solitary fieldfare nests. They concluded therefore that the benefit of group mobbing by members of a colony more than offsets the disadvantage of being conspicuous. This is supported by Volker Haas's (1985) observation that nesting success is higher for colonial than for solitary fieldfares.

Living in groups and getting food

FINDING GOOD SITES

The comparative studies described in Chapter 2 revealed that species which feed on large ephemeral clumps of food such as seeds or fruits often live in groups. For these animals, the limiting stage in feeding is the problem of finding a good site: once the patch has been found there is usually plenty of food, at least for a

(a)

(b)

Fig. 6.5 (a) Group living and avoiding predators. In dense colonies of guillemots (*Uria aalge*) like this one, breeding success is higher than in sparse colonies because of more effective defence against nest predators such as gulls. From Birkhead (1977). Photo by T.R. Birkhead. (b) Group living and hunting for food. Hyenas (*Crocuta crocuta*) can successfully attack prey which are larger than themselves because they hunt in a group. From Kruuk (1972). Photo by Hans Kruuk.

short while. Peter Ward and Amotz Zahavi (1973) developed the idea that communal roosts and nesting colonies of birds may act as 'information centres' in which individuals find out about the location of good feeding sites by following others. The idea is that unsuccessful birds return to the colony or roost and wait for the chance to follow others who have had more success on their last feeding trip. Unsuccessful birds might recognize successful ones by, for example, the speed with which they fly out from the colony on their next trip.

Colonies and roosts may act as information centres

It is perhaps unfortunate that Ward and Zahavi used the phrase 'information centre' which carries with it a connotation of mutual co-operation in the transfer of information, as for example in a honeybee or ant colony. As we shall see in Chapter 13 there are special reasons to expect co-operation in social hymenopteran colonies, but these do not apply to bird roosts or nesting colonies. 'Mutual parasitism' might be a more appropriate label here, since the successful foragers are in effect parasitized by unsuccessful birds. Each individual is out to maximize its own success and not the success of the colony as a whole. In some species the 'informer' might be unable to avoid being followed because it is conspicuous when it leaves the nest, for example seabirds leaving a colony on a cliff. The 'informer' might, however, benefit from being followed. The benefit could be long term: on a later trip the leader becomes a follower, or short term: there may be an advantage to feeding in a group because, for example, of the reduced risk of predation. If these benefits do not outweigh the disadvantage of competition at the feeding grounds which arises from being followed, a successful bird should conceal as much as possible information about its success.

Information transfer does not necessarily imply co-operation

The most direct experimental test of Ward and Zahavi's idea is a study of communally roosting weaver birds (*Quelea quelea*) by Peter de Groot (1980). *Quelea* nest in a colony and roost in groups sometimes estimated to contain over a million birds. They are serious agricultural pests in parts of central Africa and can devastate a grain field in a few hours. De Groot's experiments were done on a somewhat more modest scale (Fig. 6.6). Two groups of birds roosted together in the large aviary labelled X and had access to foraging areas in the small compartments labelled 1–4. The birds could not see into the foraging compartments from the roosting area but had to pass through small entrance funnels to explore them for food or water. In one experiment one of the groups (A) was trained to find water in one of the four compartments and the other group (B) was trained separately to find food in another compartment. The two groups were allowed to roost together and were deprived of food or water. When they

Information transfer in *Quelea*

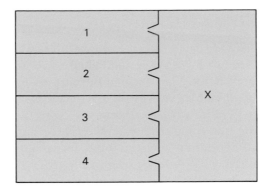

Fig. 6.6 An experiment to test the 'information centre' hypothesis with *Quelea*. The birds roost in the large area labelled X and feed in the smaller compartments labelled 1–4.

were thirsty, the birds of group B followed A to the drinking site, and when they were hungry A followed B to the feeding site. Somehow the 'naive' birds assessed that the other group was knowledgeable and followed it to the resource supply.

In a second experiment group A was trained to forage on a good food supply (pure seed) in one of the compartments while group B was separately trained to fly to another compartment for a poor food supply (seed in a bed of sand). When the two groups roosted together, the members of the second group followed the first group when they left the roost at dawn. It is not yet known how the birds recognize which individuals to follow.

In another study, however, the mode of information transfer has been identified. Geoff Galef and Stephen Wigmore (1983) trained rats (*Rattus norvegicus*) to search for food in a three-arm maze. Each arm had food with a different flavour, cocoa in one, cinnamon in another, and cheese in the third. In the first part of the experiment the rats learned that on any particular day only one of the three sites contained food, but the site was unpredictable. Then on the days of the actual experiment each of the seven test rats was allowed to sniff a 'demonstrator' rat in a neighbouring cage. The demonstrator had been allowed to feed on whatever randomly chosen food was available for that day, and four of the seven test rats, having sniffed the demonstrator, went to the correct site on their first choice of the day. 'Sniff' is the operative word, because other experiments showed that the cue the test rat picks up from the demonstrator is the smell of the food it has eaten, just as you can tell when your friend has had a garlic pizza.

Learning about potential food sources in a more direct way, by seeing others exploiting them, is important in both flocks of birds and shoals of fish (Krebs *et al.* 1972; Pitcher *et al.* 1982).

Information is transferred in rats by smell

CATCHING DIFFICULT PREY

Predators that chase
their prey

Individuals in groups may be able to capture prey which are
difficult for a single individual to overcome either because the
prey is too large for one predator to handle (e.g. lions hunting
adult buffalo) or because it is too elusive for one predator to catch
(e.g. killer whales hunting porpoise). When the prey themselves
are in a group the predators may, by hunting in a group, succeed
in separating a victim from its companions and subsequently
chasing it until they overtake it. This is how predatory fish such
as the jack (*Caranx ignobilis*) hunt for schooling prey. Individuals
in a group are more successful than single fish when hunting for
schools of Hawaiian anchovy (*Stolephorus purpureus*) (Fig. 6.7).
However the benefit is not shared equally between members of
the hunting group: fish at the front of the school when it is
chasing a prey get more than those at the back (Fig. 6.7). In fact
the fourth and fifth fish could do better by hunting alone, but it
may be that different individuals occupy the lead position during
different chases. This is a reminder of the general point that
benefits of being in a group may not be shared equally.

HARVESTING RENEWING FOOD

Goose flocks optimize
their return times

Suppose an animal eats food which renews itself continuously,
for example growing vegetation. The amount of food available in
a site increases with time since the last visit, so an individual
could get the maximum possible foraging returns by coming back
to the same site after the appropriate time interval. Returning
too soon means not finding enough food and returning too late
means missed opportunities to eat a plentiful food supply. As we
saw with the wagtails in Chapter 5, the problem with harvesting
a renewing food supply in this way is that it only works if there
is no interference by others with the renewal pattern. Individual
A's strategy of returning after 10 days would fail if B visited the
site after, for example, 9 or 8 days. One way to prevent interference
by others is to defend a territory (Charnov *et al.* 1976) and another
way is to visit sites in a group so that everyone returns at the
same time. Wintering flocks of Brent geese (*Branta bernicla*)
feeding on salt marshes in Holland seem to do the latter, where
territory defence would not be feasible because the marsh is
frequently inundated at high tide. Continuous observation of 40
1-ha plots from dawn to dusk for 24 days during the spring
showed that the flocks return to exactly the same site on the
marsh at regular 4-day intervals. Not only does this allow the sea
plantain (*Plantago maritima*) to recover between visits by the

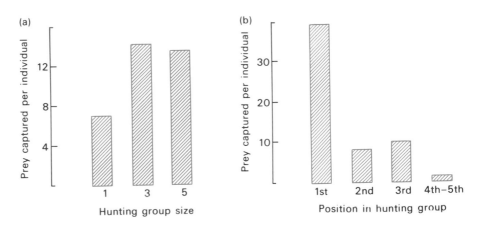

Fig. 6.7 The jack is a predatory fish which hunts in schools. (a) Each fish on average captures more prey per experiment when hunting in a group. (b) But fish at the front of the group benefit the most. From Major (1978).

geese, but also the regular cropping pattern actually stimulates the growth of young leaves which are rich in nitrogen. Experiments in which sea plantain was cut with scissors to simulate goose grazing at different time intervals suggested that given the average bite size, the geese may even return after a time interval which maximizes the growth of young shoots (Prins *et al.* 1980).

Geese at the front and back of flocks may obtain similar payoffs

Do individuals at the back of the group fare less well than those at the front, as in the hunting schools of fish described earlier? The answer is not yet known, but it is possible that the overall benefit in terms of food intake is similar for birds in different parts of the flock. The front birds eat most of the vegetation, but the youngest and most nutritious parts of the *Plantago* plants are the base of the leaves close to the ground. These parts are exposed only after the older, taller, parts of the leaves have been cropped off, so it is reasonable to hypothesize that the first birds eat larger mouthfuls while the later birds eat food of higher quality. The overall effect may be that all birds obtain the same quantity of nutrients.

COSTS ASSOCIATED WITH FEEDING

The goose study suggests an important potential cost of feeding in a group: competition for food. Competition may take the form of direct exploitation as in the jack where fish at the front catch the prey and deprive those at the back of the school, or it may arise as a result of interference in which the availability of food

Disturbance of prey is a
cost of group feeding

to a group member is reduced as a result of the behaviour of
nearby companions. This occurs in the redshank (*Tringa totanus*),
a shorebird studied by John Goss-Custard on the coastal mudflats
of Britain. Redshank feed in tight flocks at night and more loosely
scattered or solitarily during the day: this difference seems to be
related to interference. In the daytime, redshank feed by sight on
small shrimps (*Corophium*) which live with their tails just sticking
out of the mud surface, while at night, when visual search is
impossible, the birds turn to feeding by touch on snails (*Hydrobia*),
sweeping their long beaks through the mud. When feeding on
shrimps, the birds are likely to interfere with one another because
the shrimps retreat into the mud and become unavailable as soon
as they detect the heavy clump of redshank feet (this was shown
in experiments with captive birds). Feeding rate is therefore greater
with increasing neighbour distance and the birds tend to space
out. At night there is no feeding interference because the birds do
not depend on seeing the prey, and the snails in any case do not
react quickly to disturbance. Feeding rate in these conditions is
not related to flock density and the birds crowd into tight groups
(Goss-Custard 1976). One interpretation of these results is that
there is an advantage to tight clumping in redshank (perhaps as a
consequence of seeking cover from predators in the middle of the
flock), and that during the night there is little cost to feeding
close together. During the day, however, feeding close together
causes interference and the birds spread out. The nearest neighbour
distance under these conditions appears to reflect a balance between the costs and benefits of group living.

Sparrows recruit other
individuals to divisible
food sources

Another study which illustrates how competition for food
influences group size is Mark Elgar's (1986) work on house sparrow
(*Passer domesticus*) flocks. The first sparrow to arrive at a feeding
site on the ground gives a 'chirrup' call from a perch before going
down to feed. The chirrup draws in other sparrows and the group
goes to the food source together. It may seem maladaptive for a
sparrow, having found a good source of food, to attract other
competitors to it, but it is known that sparrows spend less time
scanning in larger groups (Elgar & Catterall 1981) so there is
some benefit to the first sparrow in generating a flock in which
to feed. However, and this is the most interesting point, the first
sparrow does not give a chirrup to recruit others if the food
source is indivisible. The cost of recruiting in terms of competition
for food now outweighs the benefits derived from reduced scanning
time. Elgar's experiment was simply to provide the same amount
of food — a piece of bread — either in a lump (indivisible) or in
crumbs (divisible). The first sparrow to arrive was more likely to
chirrup when the food was in crumbs.

Weighing up costs and benefits — optimal group size

The message of the chapter so far is that there are many different costs and benefits of living in a group, some or all of which might be relevant to a particular species. Pigeons, horses, ostriches and cicadas do not necessarily get together for the same reasons, but they could do so for any or all of the reasons discussed. Our list has been by no means comprehensive; we could have gone on to describe costs of group living such as transmission of diseases, cannibalism, and cuckoldry, or benefits such as protection from the elements, co-operative territorial defence, and increased efficiency of locomotion. (Some of these are summarized in Table 6.1.) But rather than try to extend the list of possible costs and benefits to its exhaustive and exhausting conclusion, we will return to the more interesting question of whether different kinds of cost and benefit can be combined to predict an optimal group size.

Other costs and benefits of group living

Table 6.1 Examples of studies in which possible costs and benefits of group living other than those mentioned in the text have been measured

Hypothesis	Test	Reference
1 Warm-blooded animals save energy because of thermal advantage of being close together	Pallid bats (*Antrozous pallidus*) roosting in groups use less energy than solitary roosters	Trune and Slobodchikoff (1976)
2 Inferior competitor can overcome competitive advantage of another species by group foraging	Blue tang surgeon fish (*Acanthurus caeruleus*) are excluded from algal mats by dusky damsel fish (*Stegastes dorsopunicans*) when solitary but not when in a group	Foster (1985)
3 Hydrodynamic advantage for fish swimming in a school. They save energy by positioning themselves to take advantage of vortices created by others in the group	Measurements of distances and angles between individuals show that they are *not* correctly positioned to benefit according to the predictions of the theory	Weihs (1973) Partridge and Pitcher (1979)

Continued on p. 134

Table 6.1 *Continued*

Hypothesis	Test	Reference
4 Increased incidence of disease as a result of close proximity of others	Measure number of ectoparasites in burrows of prairie dogs (*Cynomys* spp.). There are more parasites per burrow in larger colonies	Hoogland (1979b)
5 Risk of cuckoldry by neighbours	In colonial nesting red-winged blackbirds (*Agelaius phoeniceus*) the mates of vasectomized males laid fertile eggs. They must have been fertilized by males other than their mates	Bray *et al.* (1975)
6 Risk of predation on young by cannibalistic neighbours	In colonies of Belding's ground squirrels (*Spermophilus beldingi*) females with small territories are more likely to lose their young to cannibalistic neighbours than are females with large territories around their burrows	Sherman (1981a)

COMPARATIVE STUDIES

A qualitative picture of how costs and benefits interact can be drawn from comparisons between species. For example, some species of shorebirds such as the knot feed in dense flocks while others such as the ringed plover feed in loose flocks or as solitary birds (Fig. 6.8). It is known that living in a flock confers protection on shorebirds against attacks by birds of prey (Page & Whitacre 1975) so why do not all species feed in tight flocks? The species which feed in dense flocks hunt by touch and walk slowly, probing or swinging their beaks through the mud, while solitary and loose flock feeders hunt by sight and move rapidly, picking prey off the surface of the mud or water. Perhaps, as with the redshank, the costs of feeding interference are so big in the latter species that the net benefit for an individual is higher when alone, even though the risk of predation is greater.

Shorebird species: touch feeders live in flocks; visual hunters tend to be solitary

Fig. 6.8 Knot (*Calidris canutus*) (top) feed by touch and live in dense flocks while ringed plovers (*Charadrius hiaticula*) (bottom) feed by sight and live in loose flocks or as solitary individuals. From Goss-Custard (1970). The difference can be interpreted in terms of the costs and benefits of flocking.

COLONIALITY IN CLIFF SWALLOWS: A CASE STUDY

Few studies have attempted to assess both the costs and the benefits of group living at the same time in order to weigh up whether one particular size of group results in a higher fitness than others. One example is the work of Charles and Mary Brown on colonial nesting in cliff swallows *Hirundo pyrrhonota*. In Nebraska, where the Browns did their work, cliff swallows nest in colonies ranging in size from 1 to 3000 pairs. They build mud-cup nests on cliff overhangs, and under man-made structures such as bridges. Swallows are aerial insectivores, gathering in large aggregations to feeding on swarms of insects. It was found that, on average, the birds feed in a particular patch for about 25 min, before they move on to a new patch, presumably because the insect swarm disperses or is depleted.

Foragers hunting for ephemeral large patches of prey in this
way seem to be likely candidates to reap the benefits of a colony
as an information centre (see above) and the Browns' (1986a)
observations support this idea. They found that all swallows,
when collecting food for their young, often followed other individuals from the colony to feeding sites. What is more, an individual's tendency to follow other birds depended on its success
during the previous foraging trip. A successful bird came back to
its nest with a large ball (bolus) of tiny insects squashed together
in a lump bound with spittle, whereas an unsuccessful bird came
back with nothing or with a tiny ball of insects. If a bird had been
successful on its last trip, it followed another individual on the
next trip on only 17 per cent of occasions, whilst if it had been
unsuccessful, it followed 75 per cent of the time (most often
following a neighbour or its nest mate). Looking at the data the
other way round, to ask which birds were likely to be followed
by others, the Browns found that birds that had just been successful
were followed on 44 per cent of their trips whilst unsuccessful
birds departing on their next trip were followed on only 10 per
cent of occasions. In short, unsuccessful foragers follow successful
foragers to feeding sites. All individuals played the role of follower
about equally often, so all obtained about the same benefit from
learning by following others, as result of which the overall rate of
delivery of food to the nest was greater in larger colonies (Brown
1988).

Colonially nesting cliff swallows therefore appear to benefit
from learning about rich ephemeral food patches. But coloniality
also has a cost (Brown & Brown 1986), namely ectoparasitism.
Nestlings are often attacked by a blood-sucking hemipteran, the
swallow bug *Oeciacus vicarius* (Fig. 6.9a). These bugs spend most
of their lives in swallow nests apart from a brief dispersal phase
when they cling to the feathers of an adult bird. Larger colonies
of swallows have more bugs per nest (Fig. 6.9b) and the bugs have
a significant negative effect on nestling growth, as Fig. 6.9(c)
illustrates. In order to test whether or not the blood-sucking
ectoparasites actually cause the decrease in nestling growth, Brown
fumigated some nests with an insecticide: he found that nestlings
in fumigated nests grew much more rapidly than those in control

Fig. 6.9 (*Opposite*) (a) Two 10-day-old cliff swallow chicks. The left-hand
chick was from a control nest, the right-hand one from a fumigated nest.
Fumigation with insecticide killed the ectoparasitic swallow bugs (inset)
which cause reduced nestling growth. From Brown (1986b). (b) Number of
swallow bugs per nest increases with increasing colony size. (c) The body
mass of 10-day-old chicks decreases with increasing ectoparasite load. From
Brown and Brown (1986).

(a)

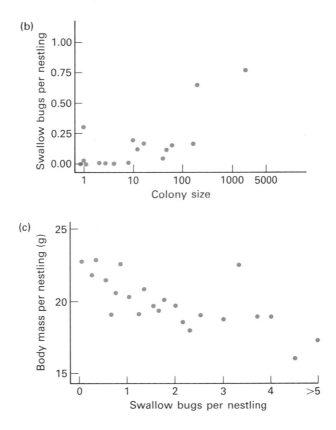

Experimental tests of
costs and benefits

Costs and benefits
may balance out in
cliff swallows

An optimality approach
to group size

nests (Fig. 6.9a) so that by 10 days of age, fumigated chicks were 3.4 g heavier than the 22 g non-fumigated controls.

How do these benefits (enhanced foraging) and costs (ecto-parasitism) balance out in different sized colonies? Brown (1988) found that if he fumigated all the nests in a colony to eliminate the effects of ectoparasitism, chicks in larger colonies grew faster (Fig. 6.10a): at 10 days of age, nestlings in colonies of 400 weighed about 1.5 g more than chicks in colonies of less than 10 birds. In order to eliminate the possibility that this difference was simply because larger colonies occurred near richer feeding sites, Brown experimentally reduced the size of some colonies and showed that the chicks in these colonies grew at a rate characteristic of their reduced size rather than their original size. Thus the ben-eficial effects of being in a larger colony on growth rate of the chicks was probably as a result of enhanced foraging success through copying other individuals, as explained above. Brown (1988) also compared nestling growth in colonies of different sizes without fumigation: this comparison includes both the ef-fects of enhanced foraging efficiency and the negative effect of ectoparasitism. As Fig. 6.10(b) shows, there is now no correlation between colony size and nestling growth. Apparently the costs and benefits cancel each other out. Does this mean that there is no single optimal colony size in cliff swallows? Of course nestling weight at 10 days is only one estimate of reproductive success, and perhaps more accurate estimates such as number of surviving young (Chapter 1) might tell a different story. In addition, there are other costs and benefits associated with coloniality in this species, including intraspecific nest parasitism (Brown & Brown 1988) which may also affect the picture. Therefore it may be still premature to decide whether or not there is an optimal colony size.

TIME BUDGETS

In order to predict more precisely how different costs and benefits might combine to determine group size we will return to the approach of Chapter 3. Ultimately the costs and benefits influence survival and reproduction, but as with the optimality models described in Chapters 2 and 3 it is more useful to think in terms of a proximate currency that ultimately relates to fitness.

Ron Pulliam (1976) and Tom Caraco (1979a) have used time as a currency and developed a model of optimal group size based on time budgets. The model is meant to illustrate the factors influencing winter flocks of small birds. The survival of birds in a flock is considered to be dependent on two main risks, starvation

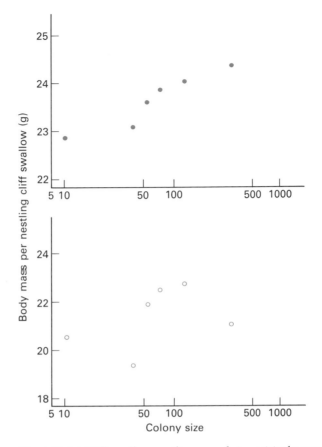

Fig. 6.10 (a) Cliff swallow nestling growth is positively correlated with colony size when ectoparasites are killed by fumigation. (b) However, without fumigation there is no correlation. From Brown (1988).

Junco flocks: scanning, fighting and feeding

and predation, and the birds' time budget is divided into three types of behaviour associated with these risks: scanning (for predators), feeding, and fighting (for food). Based on their observations of yellow-eyed junco flocks, Pulliam and Caraco divide fighting into two categories: short-term squabbles over access to pieces of food and attacks in which dominant birds attempt to evict subordinates from good feeding sites in order to ensure a supply of food for the rest of the winter. The three activities in the time budget are assumed to be mutually exclusive, that is a bird cannot, for example, scan and feed at the same time. In order to scan it has to point its head upwards, while pecking involves facing towards the ground. Finally they assume that scanning for predators takes precedence over feeding, since failing to see an approaching predator is more dangerous than failing to eat a seed.

Dominant birds are assumed to give higher priority to satisfying their daily energy requirements than to long-term eviction of subordinates, while for a subordinate bird, aggression must take priority over feeding since a bird cannot feed while it is being attacked. Figure 6.11(a) shows a simplified version of Pulliam and Caraco's model. The main features are as follows.

The model assumes that scanning decreases and fighting increases with flock size

1 The proportion of time spent scanning by an individual is assumed to decrease with increasing group size. The basis for this assumption is that a given level of vigilance can be maintained with less scanning per individual as group size increases (p. 121).
2 As group size increases and encounters between birds become more frequent, the proportion of time spent in aggression increases.
3 The time spent feeding therefore is at a maximum in flocks of intermediate size.

Can this model of time budgets be used to predict the optimal group size? If the only benefit of flock feeding is to increase the time available for feeding while maintaining a certain level of vigilance, the optimal flock size is the one indicated in Fig. 6.11a. If there are other benefits of flocking such as dilution and increased vigilance (see pp. 120–26), the optimal flock size may be larger than the one shown in Fig. 6.11a. Therefore the model can be used to test whether or not maximizing food intake is the only benefit of flocking. However the picture is probably more complicated than Fig. 6.11a suggests, because the optimal flock size may be different for dominant and subordinate birds. Dominant birds obtain a long-term benefit from evicting subordinates, so they should prefer to be in smaller groups.

Factors ignored in the model

Caraco (1979b) and Caraco *et al.* (1980) have tested some of the assumptions of their model by recording the time budgets of yellow-eyed juncos in winter flocks in Arizona. They found that the proportion of time spent by individuals in scanning and fighting change with flock size in the directions assumed by the model. However the decrease in scanning time was much greater than the increase in fighting time over the range of flock sizes they studied, so that feeding time increased with flock size, as it does to the left of the peak in Fig. 6.11a.

In order to test whether time budgets influence flock size in the way suggested by the model, Caraco *et al.* predicted the effect of various environmental changes on flock size. The predictions were as follows.

Group size decreased with increasing temperature

1 As average daily temperature increases, dominant birds should have more time to evict subordinates because they can satisfy their energy requirements more rapidly. Flock size should therefore decrease (Fig. 6.11b). This prediction was supported by observations: at 2°C flocks contained an average of seven birds, at

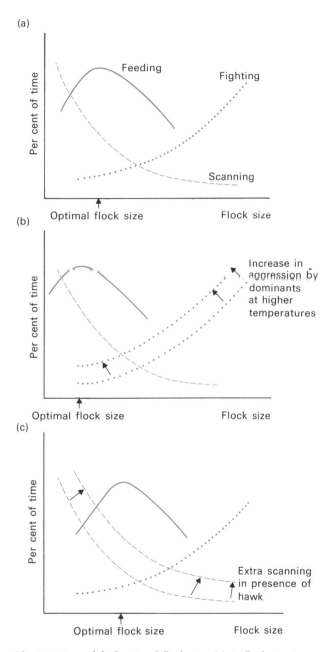

Fig. 6.11 A model of optimal flock size. (a) As flock size increases birds spend more time fighting and less time scanning. An intermediate flock size gives the maximum proportion of time feeding. (b) At higher temperatures (or when food is more plentiful) dominant birds can afford to spend more time attacking subordinates. The optimal flock size for the average bird therefore decreases. (c) When predation risk is increased by flying a hawk over the flock, the scanning level should go up and the optimal flock size is increased. Based on Pulliam (1976) and Caraco *et al.* (1980).

10°C they contained two birds. This decrease coincided with an increase in time spent fighting by dominant birds.

2 By a similar argument, an increase in food supply should produce a decrease in flock size and an increase in the proportion of time spent fighting by dominant birds. Again the results of field observations supported the prediction. When food was scattered in the canyon, the birds fed in smaller flocks.

3 An increase in the risk of predation should have exactly the opposite effect of the previous two changes. This is because a high risk of predator attack should cause the birds to spend more time scanning; they therefore have to feed in larger flocks to maintain a given rate of food intake (Fig. 6.11c). Caraco *et al.* (1980) allowed a tame hawk to fly over the canyon, and as predicted the birds spent more time scanning and the mean flock size increased. It was 3.9 birds without the hawk and 7.3 with the hawk.

4 Finally, Caraco predicted that by adding more cover to the canyon in the form of a bush, the effective risk of predator attack would be reduced because the juncos would have easier access to a safe hiding spot. The birds should therefore spend less time scanning, which allows more time for feeding and fighting. This is exactly what happened when an experimental bush was placed near one of the favoured feeding sites, and as expected the flock size decreased.

What can we conclude from these results? First they show that flock size is influenced by time budgets as hypothesized in the model shown in Fig. 6.11a. Flocking allows more time for feeding because less time is spent scanning, and the maximum flock size depends on the time available for dominant birds to evict subordinates. Second, the results allow us to reject the simplest hypothesis about optimal flock size. The birds did not feed in flocks of the size that would have maximized feeding time: under normal conditions the average flock in the canyon contained 3.9 birds, but measurements showed that the time available for feeding would have been higher in a flock of 6 or 7. As we have mentioned already, the optimal flock size for dominant and subordinate birds probably differs since dominant birds benefit by evicting subordinates. The observed flocks may have been a compromise between the optimum for dominant and subordinate birds. A further complication is that birds in larger flocks benefit by the dilution effect and increased vigilance, as described earlier for ostriches (Caraco *et al.* 1980).

The model in Fig. 6.11 is clearly too simple, but the study shows that time budgets can be used to analyse the effects of different costs and benefits on flock size. It also reminds us of the

Extra food caused a decrease in flock size

Extra predation hazard caused an increase . . .

. . . and extra cover a decrease

Pay-offs differ for dominant and subordinate birds

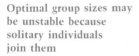

Flocking and territoriality as two sides of the same coin

idea that flocking and resource defence may be two ends of a continuum. The model could be viewed as one which predicts the conditions under which it pays dominant birds to exclude subordinates and defend a territory. When food is plentiful or predation risk is low, dominants can afford the time to maintain a defended area, or in other words the territory is economically defendable.

ARE GROUPS OF OPTIMAL SIZE STABLE?

Richard Sibly (1983) has pointed out that the groups of optimal size may rarely be found in the wild because if there was a group of this size, it would pay any solitary individuals around to join the group and therefore push the group above the optimal size. The idea is illustrated in Fig. 6.12. The curve shows a possible relationship between individual fitness (perhaps measured as feeding rate or probability of escaping predation) as a function of group size. The optimal group size, in which average fitness is maximal, is seven individuals. Imagine that individuals are free to join the group or forage alone as they arrive in the area. Clearly, if each individual chooses the option that maximizes its fitness, newcomers will join the group up to the point where group size is 14, twice the optimum! At this point the pay-off for foraging alone is the same as the pay-off for joining the group (note that this model is an example of the 'ideal free' distribution discussed in Chapter 5). The same principle would apply if one

Optimal group sizes may be unstable because solitary individuals join them

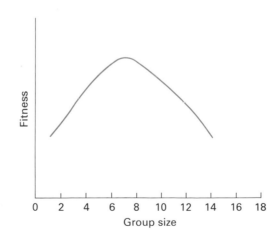

Fig. 6.12 Sibly's model of optimal and stable group size. Each individual joins the group that maximizes its fitness, so that the optimal size of seven is not necessarily stable — it will be joined by solitary individuals for example. After Sibly (1983).

imagined groups splitting and reforming as smaller units, but the argument is more complicated (Kramer 1985). A group of 12, for example would split into two units of 6, but then one individual would migrate to form a 7 and a 5-membered group, the seven would be joined by another because 8 is better than 5, and so on. The exact outcome depends on the shape of the curve in Fig. 6.12, and it is possible to draw fitness curves which will result in a stable optimal group size (Giraldeau & Gillis 1985). The general point, however, is that one should expect to find stable groups rather than optimal groups in nature and very often, observed groups will be larger than the optimum (Pulliam & Caraco 1984).

INDIVIDUAL DIFFERENCES IN A GROUP

We have already emphasized the point that individuals in a group may benefit from group living to different extents. In foraging groups of jack, individuals at the front do better than those at the back. Similarly in starling flocks, birds at the edge spend more time scanning than those in the middle (Jennings & Evans 1980). How are such differences between individuals maintained? In many cases they are simply a reflection of dominance relationships. Within the group, older, more experienced, or bigger individuals are able to commandeer the best positions and force others to take what is left. Subordinate individuals put up with less pay-off as long as they could not do better by moving elsewhere (Vehrencamp 1983).

Producers and scroungers

An alternative view of individual differences is that different individuals achieve similar pay-offs but in different ways (Chapter 10). In sparrow flocks, for example, some birds are good at locating new sources of food while others are good at stealing food once it is found. These two strategies, producer and scrounger, may coexist in the flock with equal pay-offs for the two (Barnard & Sibly 1981) (Fig. 6.13).

Evolution of group living: schooling in guppies

The studies described so far in this chapter are based on measurement of short-term costs and benefits of group living such as food intake, attack rate by predators and risk of disease. In one study, it has been possible to show how these costs and benefits translate into evolutionary change. In 1957, the American ichthyologist C.P. Haskins moved 200 guppies (*Poecilia reticulata*) from a predator-rich river system (the Caroni system) to the almost predator-free headwaters of another river system, the Oropuche (similar transfer experiments, carried out by John Endler, were

Rapid evolution of schooling

Fig. 6.13 Some mixed species flocks involve complex relationships. Barnard and Thompson (1985) studied golden plovers, *Pluvialis apricaria* (two in the foreground), lapwings, *Vanellus vanellus*, and black-headed gulls, *Larus ridibundus*, which often associate in winter feeding flocks in pastures in lowland Britain. Lapwings favour pastures where earthworm density is high and golden plovers select fields on the basis of the presence of lapwings, using them, in effect, as food finders. Black-headed gulls are kleptoparasitic and steal worms from the lapwings and golden plovers, but also bring benefits because they give early warning of the approach of predators. After a painting by Donald Watson.

described in Chapter 4). The transferred fish subsequently colonized the downstream parts of their new river system and again encountered predators. In 1989–91, more than 30 years after the original transfer, Anne Magurran (Magurran *et al.* 1992) collected guppies from several streams including the original predator-rich source from which Haskins took the guppies, the predator-free site where they were first introduced, and the predator-rich streams they later colonized from the introduction site. In guppies, schooling is an antipredator adaptation, giving the various forms of benefit discussed earlier in this chapter (Magurran 1990; see Fig. 6.1). Schooling also has a cost: Magurran and Seghers (1991) show that guppies with a high propensity to school are less good competitors over food. Apparently, selection for schooling carries with it a decrease in competitiveness. By placing single guppies in an aquarium with the choice of an empty beaker at one end and a 'school' in a beaker at the other, Magurran *et al.* showed that the fish transferred by Haskins first reduced their tendency to school in the predator-free streams but secondarily increased it

Laboratory test of the change in schooling behaviour

when they re-invaded areas with predators. These differences were shown by individuals bred and raised in standard conditions in the laboratory. Thus in guppies, the short-term costs and benefits of schooling translated into an evolutionary response to decrease and later an increase in schooling behaviour. These changes took place within 100 generations (guppies probably have about three generations per year).

Summary

Two of the most important selective advantages of living in a group are increased protection from predators and increased likelihood of finding or capturing food. To be weighed against these benefits are costs of group living such as increased competition for food and increased conspicuousness to predators. The size of group in which animals live may reflect a compromise resulting from these various costs and benefits. One way to analyse the trade-off between them is to use the common currency of time to predict optimal group sizes.

Two limitations of the notion of optimal group size are (a) that individuals in a group may attain different pay-offs and may have different optima. If observed groups are a compromise between the optima for different individuals, they may not be optimal for any one individual. (b) Optimal-sized groups may be unstable because they tend to be joined by individuals from smaller groups.

Further reading

Pulliam and Caraco (1984) is a good review of why animals live in groups. Hoogland (1979a,b) is a study of prairie dogs in which both the costs and the benefits of group living are investigated. The book edited by Barnard (1984) discusses producers and scroungers in groups. Ian Patterson's (1965) paper is a classic study of the benefits of synchronous and colonial breeding.

Topics for discussion

1 How can different kinds of costs and benefits of group living (e.g. feeding, predation, disease) be combined to predict group size?

2 Why do subordinates remain in groups?

3 How would you test the hypothesis that groups are stable rather than the optimal size?

Chapter 7. Fighting and Assessment

In the last two chapters we saw that individuals often have to compete with others for scarce resources such as food, territories or mates. We now consider the evolution of contest behaviour. What factors determine which will be the most successful fighting strategies? It might at first be supposed that we could use simple optimality models to analyse this problem, for example assessing the costs and benefits and working out the strategy which results in greatest net benefit. We cannot do this, however, because the optimal way for one individual to behave turns out to depend on what other competitors in the population are doing. The appropriate technique to analyse situations like this is game theory, which was originally developed for the study of economics but which has become a very useful theoretical tool for the analysis of animal behaviour, particularly due to the work of Maynard Smith (1982). It is easiest to describe this method if we begin with an example.

Game theory models for analysing contest behaviour

The war of attrition, or 'the waiting game'

(a) The problem

Female dungflies, *Scatophaga stercoraria*, come to fresh cowpats in order to lay their eggs. Swarms of males are waiting for them on and around the cowpats (Fig. 7.1) and whenever a female arrives, the first male to encounter her copulates with her and then guards her while she lays (p. 55). Females prefer to lay in fresh dung and as the pat gets older, and a crust forms over it (thus making it less suitable for egg laying), fewer females arrive. The male's problem is, what is the optimum time to spend waiting for females at each cowpat? The answer is that the best waiting time for one male depends on what all the other males are doing. For example, if most males remain for short times then a male who stayed a little longer would have high mating success because he could claim all the late arriving females. If, on the other hand, most males were staying a long time then it would pay our male to move quickly to a new pat to claim the early arriving females there. We have to analyse, therefore, a competition among the males where the pay-offs for different waiting times are frequency dependent.

Male dungflies compete for females

Fig. 7.1 A male dungfly guards the female while she lays her eggs in a fresh cowpat. Photo by G.A. Parker.

(b) The theory

This kind of game, where winners are decided simply by a contest involving waiting, or displaying, for different lengths of time is known as the war of attrition and the solution is as follows (Maynard Smith 1974). Imagine two individuals contesting for a resource of value V. The longer an individual waits, the greater the cost, m. Let animal A select a time x_A and animal B select a time x_B. The costs associated with these times are m_A and m_B. If $x_A > x_B$ then A wins the resource and so we have

> Pay-off to A = $V - m_B$
> Pay-off to B = $\quad - m_B$

Note that A only pays the cost of waiting for time x_B because B gave up first. We suppose now that waiting times are heritable and that individuals have offspring with identical waiting times to themselves. We suppose too that individuals reproduce in proportion to their pay-offs in the game.

How would evolution proceed in this game? Obviously B would have done better to wait for longer than x_A so it could have won. For example, a strategy 'wait for 1 min' would be beaten by a strategy 'wait for 1.1 min'. This might suggest that

If contests are settled by persistence, there is no single optimum

longer and longer waiting times would evolve (1.2 beats 1.1, 1.3 beats 1.2, and so on) but eventually these waiting times would become so long that the costs would exceed the value of the resource. It is clear that no 'pure' strategy can be an evolutionarily stable strategy, or ESS. (*An ESS is a strategy which when adopted by most members of the population cannot be beaten by any other strategy in the game.*) If an individual always played the same fixed waiting time, then other competitors could take advantage of this and play a time which would always beat it. In fact, it can be shown that the ESS for this game is to play a randomly chosen waiting time, in other words be unpredictable.

We have to seek an evolutionarily stable strategy (ESS)

This stable solution to the waiting game could come about in two ways.

1 Every individual in the population could play the same variable strategy, or mixed ESS. Every individual would sometimes wait for a long time and sometimes for a short time, and the times would be selected from the appropriate random distribution.

2 There could be a polymorphism in the population with every individual playing a fixed waiting time, or pure strategy, but with the frequencies of the strategies following the random distribution. There would then be a mixture of pure strategies, with most individuals playing short waiting times and fewer playing longer times.

The stable solution is for waiting times to follow a random distribution

The first prediction from this model is therefore that *waiting times will be variable*, either within or between individuals. The second prediction is that *at the ESS, the pay-offs for the different waiting times will be equal*. Equality of success will come about through frequency-dependent selection. Imagine, for example, that individuals with short waiting times are enjoying greatest reproductive success. Their progeny will inherit this successful strategy and the proportion of individuals adopting short stay times will increase in the population. As it does, however, the success of short stay times will decrease because of increased competition. As more and more male dungflies, for example, play short stay times, there will be more and more competition among these males for the early arriving females. The increase in proportion of the short stay times in the population will come to a stop when this strategy's reproductive success equals that of the other strategies. On the other hand, if a certain waiting time experienced lower reproductive success the frequency of this strategy would decrease in the population. As it did, its reproductive success would increase, because of reduced competition, and frequency-dependent selection would again stabilize the proportion at that which gave equal reproductive success to the other waiting times. The solution to the game described above is *stable* in the sense that once the population adopts this strategy then no individual who played a different strategy could gain higher reproductive success.

The model's assumption of asexual reproduction ('like gives rise to like') is not too restrictive. Provided the ESS phenotype can be produced by a genetic homozygote it will be uninvadeable (stable) in a sexually reproducing species with diploid inheritance just as in an asexual one (Maynard Smith 1982).

(c) Testing the theory

What do dungflies do? Parker (1970) measured the lengths of time that males spent waiting for females on cowpats and the results showed that, as predicted by the war of attrition model, the 'stay times' fitted a negative exponential (random) distribution (Fig. 7.2a). It is not yet known whether different individuals play different stay times or whether individuals play a mixed strategy. Parker also measured the reproductive success of males and showed that the distribution of male stay times was such that different waiting times resulted in similar mating success (Fig. 7.2b), again as predicted by the model.

As predicted, dungflies persist for random lengths of time

The evolution of conventional fighting

The war of attrition model is likely to be most applicable to situations, like the dungflies, where individuals are engaged in exploitation competition (p. 102) or 'playing the field' (Maynard Smith 1982). Many animal conflicts, however, are examples of contest competition where two individuals meet and compete for a resource. Here we would not expect the contest to be solved simply by a waiting game! Indeed, in cases like this there tends to be direct confrontation in the form of fights. For example, after a male dungfly has paired with a female other males may approach and attempt to take her over. There is often a struggle before one eventually wins and the other gives way (Fig. 7.3).

Why are contests settled by displays and not fights?

Although fights like these can sometimes be vicious in animals (see below), in general disputes are often settled before there is serious injury. Why do animals often settle contests by display rather than by all-out fighting? The long-accepted answer to this question was that escalated contests would result in many animals getting seriously injured and this would militate against the survival of the species (e.g. Lorenz 1966; Huxley 1966). This is a group selection argument and does not explain how natural selection acting on individuals can give rise to the evolution of conventional fighting. Since the mid 1960s many people realized

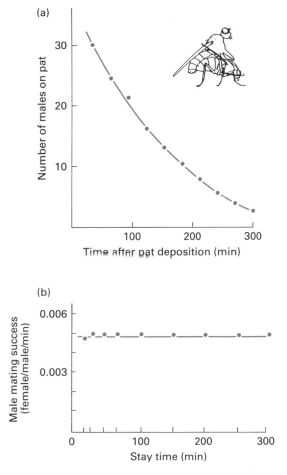

Fig. 7.2 (a) The number of male dungflies on a cowpat declines exponentially with time. This could come about through the population having a mixture of males with different fixed stay times or through all males being alike with flexible stay times; for example each may have the same constant probability of leaving per unit time. (b) Given this distribution of stay times, the result is that the mating success of males adopting different stay times is about equal, as predicted by an ESS model (see text). From Parker (1970).

that the answer must be framed in terms of costs and benefits of fights to individuals. Once again game theory models developed by Maynard Smith and others are useful for exploring the factors influencing the evolution of pair-wise contests.

HAWKS AND DOVES

Consider a game where there are just two strategies. 'Hawks' always fight to injure and kill their opponents, though in the process they will risk injury themselves. 'Doves' simply display

Fig. 7.3 A struggle between two male dungflies for the possession of a female. The attacking male (left) is attempting to push the paired male (right) off the female (hidden from view, underneath). Photo by G.A. Parker.

and never engage in serious fights. Although simple, these two strategies are chosen to represent the two possible extremes we may see in nature.

In this evolutionary game, let the winner of a contest score +50, and the loser 0. The cost of a serious injury is −100 and the cost of wasting time in a display is −10. These pay-offs are some measure of fitness and we will assume, for simplicity, that Hawk and Dove reproduce their own kind faithfully in proportion to their pay-offs. (The exact values do not matter and are chosen simply because this game is easier to explain with numbers rather than algebra.) The next step is to draw up a two by two matrix with the average pay-offs for the four possible types of encounter. These calculations are explained in Table 7.1.

How would evolution proceed in this particular game? Consider what would happen if all individuals in the population are Doves. Every contest is between a Dove and another Dove and the pay-off is on average +15. In this population, any mutant Hawk would do very well and the Hawk strategy would soon spread because when a Hawk meets a Dove it gets +50. It is clear that Dove is not an ESS.

However, Hawk would not spread to take over the entire population. In a population of all Hawks the average pay-off is

The Hawk−Dove game helps us to think about the evolutionary stability of displays

Table 7.1 The game between Hawk and Dove. After Maynard Smith (1976b)
(a) Pay-offs: Winner   +50   Injury   −100
               Loser   0   Display   −10
(b) Pay-off Matrix: average pay-offs in a fight to the attacker

	Opponent	
Attacker	Hawk	Dove
Hawk	(a) $\frac{1}{2}(50) + \frac{1}{2}(-100)$ $= -25$	(b) +50
Dove	(c) 0	(d) $\frac{1}{2}(50 - 10) + \frac{1}{2}(-10)$ $= +15$

Notes:
1 When a Hawk meets a Hawk we assume that on half of the occasions it wins and on half the occasions it suffers injury.
2 Hawks always beat Doves.
3 Doves always immediately retreat against Hawks.
4 When a Dove meets a Dove we assume that there is always a display and it wins on half of the occasions.

−25 and any mutant Dove would do better because when a Dove meets a Hawk it gets 0 (which is not very good, but still better than −25!). The Dove strategy would spread if the population consisted mainly of Hawks. Therefore Hawk is not an ESS either.

Nevertheless, a mixture of Hawks and Doves could be stable. The stable equilibrium will be when the average pay-offs for a Hawk are equal to the average pay-offs for a Dove. If the population moved away from this equilibrium then either Dove or Hawk would be doing better and so the population would not be stable. Each strategy does best when it is relatively rare and the tendency in this evolutionary game will be for frequency-dependent selection to drive the frequencies of Hawk and Dove in the population so that they each enjoy the same success. For the values in Table 7.1, the stable mixture can be calculated as follows.

Let h be the proportion of Hawks in the population. Therefore the proportion of Doves must be $(1 - h)$. The average pay-off for a Hawk is the pay-off for each type of fight multiplied by the probability of meeting each type of contestant. Therefore,

$$\bar{H} = -25h + 50(1 - h).$$

Similarly, for Dove the average pay-off will be

$$\bar{D} = 0h + 15(1 - h).$$

At the stable equilibrium (the ESS), $\bar{H}$ is equal to $\bar{D}$. Solving the two equations above by setting $\overline{H} = \overline{D}$ gives $h = 7/12$, and therefore, by subtraction, the proportion of Doves $(1 - h)$ must be $5/12$.

Just as in the war of attrition, the ESS could be achieved in two distinct ways.

1 The population could consist of individuals who played pure strategies. Each individual would either be Hawk or Dove, and the ESS would come about with 7/12 of the population being Hawks and 5/12 Doves.

2 The population could consist of individuals who all adopted a mixed strategy, playing Hawk with probability 7/12 and Dove with probability 5/12, choosing at random which strategy to play in each contest.

A mixture of fighting and display is evolutionarily stable

Both these would produce stability in the game. If the population consisted of a different mixture of individuals from that in (1) or of individuals playing their strategies with different probabilities from those in (2), then there would be no equilibrium; either Hawk or Dove would enjoy a temporary increase in success until the population evolved to the ESS once more.

It is instructive to note that at the ESS, the average pay-off is 6.25 per contest. This is less than the average pay-off which individuals would enjoy if they all agreed to fight as Doves, namely 15! This makes clear the point we made in the opening paragraph to this chapter. The optimal strategy to maximize everyone's fitness would be an agreement among all to play Dove. However, this would not be stable because in a population of all Doves, Hawk mutants would invade. We expect evolution to lead to stable strategies because, in the words of Richard Dawkins, 'They are immune to treachery from within'.

The ESS solution does not maximise every individual's fitness

HAWK, DOVE AND BOURGEOIS

Now imagine another strategy in this game, 'Bourgeois'. With this strategy the individual plays 'Hawk, if owner' and 'Dove, if intruder'. In other words, it fights hard if it's the owner but always retreats if it's the intruder. Let us keep the same pay-offs as before and, for simplicity, imagine that a Bourgeois individual finds itself owner half the time and intruder half the time. With three strategies in the game, the pay-offs are indicated in Table 7.2.

In this game, Bourgeois is an ESS. If all the population are playing this strategy, no one ever engages in escalated fights because when two individuals contest for a resource, one is owner and the other is intruder; the result is that the intruder always gives way. With everyone playing the Bourgeois strategy the average pay-off for a contest is +25. This is stable against invasion by Hawks, who would only get +12.5, and also stable against

Table 7.2 The Hawk–Dove–Bourgeois game. After Maynard Smith (1976b)
(a) Pay-offs (as in Table 7.1)

Winner	+50	Injury	−100
Loser	0	Display	−10

(b) Pay-off Matrix: average pay-offs in a fight to the attacker

	Opponent		
Attacker	Hawk	Dove	Bourgeois
Hawk	−25	+50	+12.5
Dove	0	+15	+7.5
Bourgeois	−12.5	+32.5	+25

Notes:
1 The top left four cells are exactly the same as in Table 7.1.
2 When Bourgeois meets either Hawk or Dove we assume it is owner half the time and therefore plays Hawk, and intruder half the time and therefore plays Dove. Its pay-offs are therefore the average of the two cells above it in the matrix.
3 When Bourgeois meets Bourgeois on half the occasions it is owner and wins while on half the occasions it is intruder and retreats. There is never any cost of display or injury.

'Bourgeois' can be an ESS

Doves, who would only get +7.5. In fact Bourgeois is the only ESS in the game in Table 7.2. If all the population played Hawk, both Dove and Bourgeois could invade and do better. If all the population played Dove, then both Hawk and Bourgeois could invade.

SIMPLE MODELS AND REALITY

These models are so simple in comparison with what actually happens in nature that we may well ask what use they can possibly be in helping us to understand real animal contests. There are three main conclusions.
1 The most important point is that the best fighting strategy for any one individual must depend on what other competitors are doing, because the pay-offs for employing a strategy will be frequency dependent. Is Hawk a good strategy? The answer is yes if the population consists mainly of Doves and no if the population consists mainly of Hawks. Instead of asking whether a strategy is a *good* strategy, we should really instead be asking is it a *stable* strategy, or ESS?
2 The ESS will depend on the strategies in the game. In the first example, where there were just two strategies, the ESS was a mixture of Hawk and Dove. However, when we introduced another strategy, Bourgeois, into the game the ESS solution changed; Bourgeois turned out to be a pure ESS.

These three strategies are undoubtedly too simple to represent in detail the strategies that animals adopt in the wild. Nevertheless they are plausible alternatives which we could regard as simplified versions of strategies we might see in nature. It is interesting to find that neither Hawk nor Dove is an ESS in our simple games, but some mixture of Hawk and Dove behaviour can be stable. This is exactly what we see in real animals, a mixture of display and fighting. Because of its simplicity, we don't expect to actually see Hawk and Dove strategies in nature or to test the model in Table 7.1. The model should be used simply for gaining some insights into how contest behaviour might evolve.

In nature, contests are more complex than Hawk−Dove

More detailed models will have to consider a greater variety of more complicated strategies and we will have to rely on biological intuition to help us define the range of possibilities. ESS models cannot tell us what strategies will evolve, only what will be stable given a defined set of alternatives. There is no doubt that an animal with a machine gun would invade the Hawk−Dove game and soon spread, but until we see real animals behaving like this there is little point in putting the strategy into our models.

3   The ESS will also depend on the values of the pay-offs in the game. If we changed the pay-offs in Table 7.1 then the ESS mixture of Hawks and Doves would change. In fact, in this game it can be shown that as long as the cost of injury exceeds the value of winning then the ESS is a mixed one. The relative amount of Hawk-like behaviour we would expect will depend on the costs and benefits in the contest. If the value of the resource was greater than the cost of injury then pure Hawk is an ESS.

Therefore if game theory models are going to make precise predictions about contest behaviour, we will not only need to know the range of possible strategies but also the costs and benefits for the pay-off matrix. In the real world, ecological circumstances such as resource abundance and competitor density will determine the pay-offs in the evolutionary game. In practice it is not going to be easy for the field worker to go out and measure on a common scale (effect on fitness) the cost of displaying, the cost of serious injury and the value of winning. However we can see in an intuitive way that fighting strategies in nature vary depending on the value of the resource.

Examples of animal contests

(a) Serious fights

Conventional display does not always occur; sometimes fights are fierce and there is injury or death. Some weapons, such as horns and antlers, have evolved in relation to their efficiency in attack and defence (Geist 1966). In the musk ox, from 5 to 10 per cent of the adult bulls may die each year from fights over females (Wilkinson & Shank 1977) and figures for mule deer indicate that up to 10 per cent of males more than 1.5 years of age show signs of injury each year (Geist 1974). Narwhals (*Monodon monoceros*) use their tusks for fighting and in one study over 60 per cent of the adult males had broken tusks; some had tusk tips embedded in their jaws and most adult males were covered in head scars (Silverman & Dunbar 1980). Smaller animals may also fight viciously. Some male fig-wasps have large mandibles which are

Severe fights occur when the pay-off for winning is high

able to chop another male in half. When several males occur inside the same fig fruit they may engage in lethal combat for the opportunity to mate with females in the fig (Hamilton 1979). One such fruit was found to contain 15 females, 12 uninjured males and 42 damaged males who were dead or dying from fighting injuries. Damage included legs, antennae and heads completely bitten off, holes in the thorax, and eviscerated abdomens.

In all these examples the value of the resource probably exceeds the cost of injury. We would therefore expect Hawk-like strategies because failure in a contest could mean failure to pass on genes to future generations; individuals are really fighting for genetic life or death. Where, on the other hand, the resource is less valuable, or where the costs of fights are very high, then would we expect Bourgeois (owner wins, intruder retreats) to evolve as a strategy for settling contests?

(b) Respect for ownership

One possible example comes from a study of lions by Packer and Pusey (1982). A coalition of several males (up to 7) defends the pride (Fig. 7.4a). Sometimes the males are close relatives, often brothers, but 42 per cent of the coalitions contain unrelated males. Competition among males within the pride for access to oestrus females consists of competition for temporary ownership of a female. When a female comes into oestrus, one male forms a consortship with her, guarding her from the approaches of other males and copulating with her frequently for a period of about 4 days (see Fig. 7.4b and Chapter 1).

What determines which males get consortship with a particular female? Contests between unrelated males were no more intense than between related males. Two important factors in settling contests were age and size; small males and very old or very

(a)

(b)

Fig. 7.4 (a) A group of male lions with a pride of females. Photo by Jonathan Scott, Planet Earth Pictures. (b) Lions mating. Photo by © Craig Packer.

young (less vigorous) males were less likely to gain access to an oestrus female. Among prime males, however, the main factor influencing who won a contest was simply respect for ownership. Usually the first male to encounter an oestrus female guarded her and other males then gave way. Most serious fights occurred when ownership was undecided, for example when two males simultaneously came into the vicinity of an unconsorted and potentially oestrus female, or when ownership was unclear, for example when the consorting male moved further from the female than a rival male. There was then usually a race between the males to arrive first at a female. On arrival the loser would defer to the winner, even though in subsequent consortships the roles might be reversed with the previous consorting male now a rival to the new consort. Lions may use this Bourgeois-like strategy to settle contests because the costs of serious fights are so high and the benefits relatively low (p 7)

A study of the damselfly, *Calopteryx maculata*, provides an especially neat example of how escalated contests occur when ownership of a territory is unclear. Males defend small clumps of vegetation in streams where females come to lay eggs. Males usually expel intruders and neighbours with brief chases lasting less than 15 s but longer contests involving back and forth chases and spiralling flights, which could go on for an hour or more, occur when ownership is confused, for example when an interloper occupies a territory while an owner is temporarily absent. Ownership confusion was induced experimentally by attaching floating pieces of the defended vegetation to fishing line and then gradually moving two males' territories together until they fused. Escalated contests always ensued. In this case the asymmetry of owner and intruder was absent and both contestants behaved like Hawks. What determines who wins these prolonged encounters? Pairs of males which had engaged in long contests were caught to test whether victory was correlated with any physical characteristics. Size itself was not a predictor of which male won but winners had greater energy reserves in the form of fat. This suggests that escalated contests are energetic wars of attrition in which males with the lowest energy reserves are forced to give up because they cannot afford to continue the contest (Waage 1988; Marden & Waage 1990).

In both lions and damselflies, males sometimes respect ownership

When ownership is confused, escalated contests occur

(c) The influence of resource value

The observation that owners tend to win contests versus intruders does not, of course, necessarily mean that the Bourgeois strategy is being adopted. Owners may win for at least three reasons.

Three hypotheses for why owners win

Hypothesis 1. They are better fighters. This may be the reason why they got ownership in the first place!

Hypothesis 2. They have more to gain from a fight and so are prepared to fight harder. If you were attacked in the street you would probably fight harder if you had a lot of money in your pocket than if you had very little. Similarly, owners may value the resource more than intruders. For example, in disputes over a territory the owner will be familiar with the territory boundaries and will know the location of the good feeding and nesting sites.

Hypothesis 3. Arbitrary asymmetry of ownership settles the contest. Ownership is simply a conventional settlement, as in the Hawk−Dove−Bourgeois game, similar to the conventions we use in deciding positions in a queue at a supermarket or occupancy of a seat on a bus.

Krebs (1982) tested these three hypotheses in a study of territorial defence by male great tits *Parus major*. He removed owners and kept them in a cage. Newcomers took over the vacancy and then, after various time intervals, the original owner was re-released onto the territory. The prediction from hypothesis 1 is that the original owner will win back his territory, through his superior fighting ability. Hypothesis 2 predicts a gradual reversal of dominance in favour of the newcomer with increased time on the territory, which allows the newcomer to learn the territory characteristics and so gain greater value from winning a contest.

In great tits owners fight harder because they have more to gain

Hypothesis 3 predicts that the newcomer will win simply because it is now in the role of owner. The data supported the second hypothesis, with newcomers escalating and being more likely to win after they had had time to learn the territory characteristics and settle boundaries with the neighbours.

(d) Contests of strength

The Hawk−Dove game showed how ritualized displays, rather than dangerous fighting, could evolve as a means of settling contests. As we pointed out earlier, the game is best viewed as a useful illustration of the logic of evolutionary game theory rather than having direct relevance to real contests in nature. Real contests tend to involve sequences of various displays in which individuals appear to assess each other's fighting ability (Parker 1974), something which was not incorporated in the Hawk−Dove model.

Assessment is a key feature of many contests

As an example, consider how red deer stags, *Cervus elaphus*, compete for females in the autumn rut. A male's reproductive success depends on fighting ability; the strongest stags are able to command the largest harems and enjoy the most copulations. Although fighting brings great potential benefits, it also entails serious costs. Almost all males suffer some slight injuries, and between 20 and 30 per cent of stags will become permanently injured sometime during their lives, through broken legs or being blinded by an antler point, for example. Competing stags minimize fighting costs by assessment of each other's fighting potential and so avoid contests with individuals they are unlikely to beat (Clutton–Brock & Albon 1979).

In the first stage of the display, the harem holder and challenger roar at each other (Fig. 7.5a). They start slowly at first and then escalate in rate. If the defender can roar at a faster rate then the intruder usually retreats. Roaring is a good signal of fighting ability because to roar well a stag has to be in good physical condition. At the peak of the rut, harem holders may be roaring at intruders throughout the day and night and, because they do not feed much during this time, they show a steady decline in body weight. Some stags literally become 'rutted out' and are unable to roar strongly. Their females may then be taken over by other stags. In the second stage of a contest if a challenger outroars the defender, or matches him, then he approaches and both stags engage in a parallel walk (Fig. 7.5b). This presumably enables them to assess each other more closely. Many fights end at this stage, but if the contestants are still equally matched then a serious fight ensues where they interlock antlers and push against each other. Body weight and skillful footwork are important determinants of victory, but there is a chance that even the

Assessment may involve trials of strength or reliable signals of fighting ability

winner may get injured. The important point is that these escalated fights are rare, and most contests are settled at an earlier stage by displays.

Many animal contests proceed like this and are direct or indirect trials of strength. Buffalos charge at each other and assess their fighting potential in head-on clashes (Sinclair 1977). Beetles engage in pushing contests in which the larger one emerges as victor (Eberhard 1979). Male frogs and toads have wrestling matches where the large ones win the best territories or the most females (Chapter 8).

In many species of frogs and toads the pitch of a male's croak is closely related to his body size; the larger the male, the larger the vocal cords and so the deeper the croak. Common toads (*Bufo bufo*) assess the body size, and hence fighting potential, of their rivals by the pitch of their croaks. An attacker is much less likely to attempt a take-over of a female when deep croaks are broadcast from a loudspeaker next to the pair than when high-pitched croaks are played (Fig. 7.6).

Fig. 7.5 Stages of a fight between two red deer stags. The harem holder roars at the challenger (a). Then the pair engage in a parallel walk (b). Finally they interlock antlers and push against each other (c). Photos by Tim Clutton-Brock.

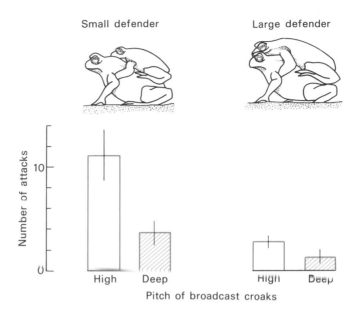

Fig. 7.6 An experiment on fighting assessment in toads, *Bufo bufo*. Medium-sized males attack either small or large paired males, which are silenced by means of a rubber band passing through their mouths. During an attack, tape recorded croaks are broadcast from a loudspeaker next to the pair. For both sizes of defender there are fewer attacks when the deep croaks of a large male are played than when the high-pitched croaks of a small male are broadcast. Therefore croak pitch is used to assess a rival's body size, which is a good predictor of fighting ability because large males are more difficult to displace. Croaks cannot be the only assessment cue, however, because for either croak pitch there are fewer attacks at large defenders. The strength of a defender's kick may also be important. From Davies & Halliday (1978).

Honest assessment

Three conclusions emerge from these observations of how individuals assess each other. First, the displays involved in assessment appear to be reliable signals of size and strength, such as visual or vocal displays and pushing contests which are correlated closely with fighting ability. This honest signalling makes good sense, otherwise weak individuals would be able to mimic the signal and so gain an advantage by bluffing. Second, the displays often involve a degree of co-operation between the contestants; rivals run parallel to each other or engage head on. Opponents will have a common interest in obtaining information about each other and co-operation is likely to be stable when it is better to accept an offer of safe assessment rather than to start a dangerous fight. Third, the contests often proceed through a set pattern, with a sequence of repeated displays divided into phases which increase in intensity (e.g. roar, parallel walk, pushing). Why should this be so?

SEQUENTIAL ASSESSMENT

Magnus Enquist and Olof Leimar (1983, 1987, 1990) suggest that information accumulates during a contest in a way similar to statistical sampling. The outcome of a single bout contains a random error and to get a more accurate estimate contestants must 'increase the sample size' by repeating the behaviour. On this view we would expect the least costly displays to be used first of all, followed by the more costly but more accurate means of assessment. An individual will give up when it assesses its fighting ability to be less than that of its opponent. With more closely matched opponents, fights should last longer and escalate further simply because it will take longer to assess which is the strongest, just as a statistician needs to sample more to detect a small difference than a large one.

Assessment is partly information gathering

This idea makes good sense of contests involving sequences of displays, such as those of the South American cichlid fish *Nannacara anomala*. In nature, males compete for females who are about to spawn and contests can be readily staged in laboratory tanks. The order of displays in a contest is very consistent, with constant rates of behaviour within each phase (Fig. 7.7). (a) Initially the males orient laterally with erect fins. (b) Then they engage in tail-beating, pushing a stream of water towards the flank of an opponent. This involves alternating roles in which each fish takes turns between performing a tail beat and manoeuvring into a lateral position to receive the stream from its opponent's beat. (c) Then biting increases in frequency and the contestants begin to orient frontally. (d) This is followed by mouth-wrestling, where the males get a firm grip of each other's jaws and have a pushing and pulling contest allowing more accurate assessment of relative strength. (e) Finally, circling occurs where both fish swim fast in a tight circle and try to bite the back of their opponent. At the end of the contest, losing fish signal their defeat by folding their fins, changing colour and retreating. Larger males were more likely to win and with very large differences in size the smaller fish gave up more or less immediately at stage (a). As the difference in size decreased, however, fights became longer and escalated through the sequence with increasing risks of injury at successive stages (Fig. 7.8).

The more closely matched the contestants, the longer the assessment

Contests with differences in both resource value and fighting ability

So far we have identified both resource value and fighting ability as important factors influencing the outcome of contests. In many

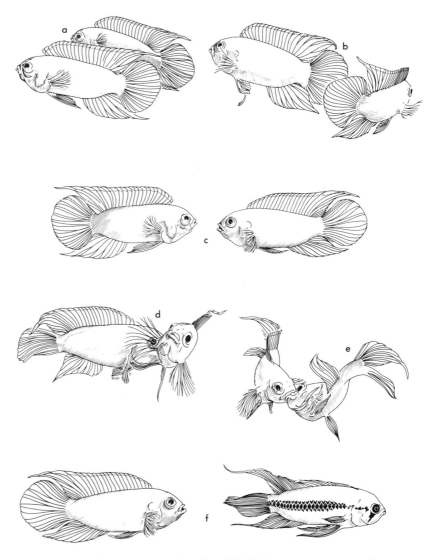

Fig. 7.7 Fighting sequence of male cichlid fish, *Nannacara anomala*.
(a) Lateral orientation. (b) Tail-beating. (c) Frontal orientation. (d) Biting.
(e) Mouth wrestling. (f) The loser (right) gives up. Drawing by Bibbi
Mayrhofer. From Jakobsson *et al.* (1979).

contests the opponents will vary both in fighting ability and how
much they value the resource. How will fighting strategies evolve
in these cases?

THEORY: THE ASYMMETRIC WAR OF ATTRITION

The question of whether we expect to see contests in nature
settled simply by the arbitrary convention of ownership, as opposed

(a)

(b)

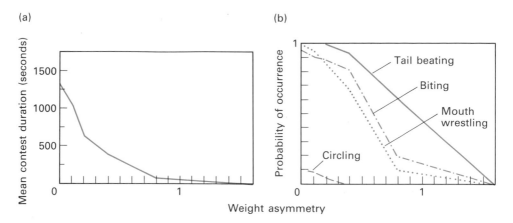

Fig. 7.8 Fighting in the cichlid fish, *Nannacara anomala*, is more prolonged (a) and escalates to more dangerous stages (b) the more closely the contestants are matched in size. Weight asymmetry is measured as the logarithm of the weight of the heavier fish divided by the weight of the lighter fish (0 = weight ratio of 1, i.e. equal weights). From Enquist *et al.* (1990).

to real differences in fighting ability, depends on how individuals can vary their degree of risks of injury in a fight. In the Hawk–Dove–Bourgeois game, contestants can engage in conflicts at only two levels of risk, namely display (low risk) and escalated fighting (high risk). This discontinuity is crucial for the generation of the Bourgeois strategy as an 'uncorrelated' asymmetry to settle contests. (Uncorrelated means ownership need have no correlation with fighting ability or resource value, for example; it is simply an arbitrary convention to settle disputes without escalation.) If the contestants can control their degree of risk along a continuum, however, and therefore adjust their escalation in relation to the value of the resource, then the uncorrelated Bourgeois rule turns out to be no longer evolutionarily stable.

A contest in which there are differences in fighting ability, and in which individuals can vary their cost of escalation along a continuum is referred to as an 'asymmetric war of attrition'. The theory behind this game is complex and brave readers are referred to Parker & Rubenstein 1981 and Hammerstein & Parker 1982. The ESS for this kind of contest is for individual A to give up and retreat when

$$\frac{V_A}{K_A} < \frac{V_B}{K_B}$$

where V is the value of the resource to contestants A and B, and K is the rate at which the two individuals accrue costs during a contest. K will be related to fighting ability; good fighters will accrue costs at a slower rate than poor fighters.

Therefore if ownership is used to settle contests of the war of attrition kind, it is because there are asymmetries in V and/or K which favour residents, and residents are willing to escalate further. A verbal summary of this rule for settling contests is 'Withdraw if you will run out of fitness budget before your opponent'. This is the common-sense solution to the game; the winner is always the one with the greatest benefit : cost ratio.

DATA: CONTESTS BETWEEN MALE SPIDERS

The outcome of contests
depends on resource
value and fighting ability

Steven Austad (1982, 1983) studied contests for females between male bowl and doily spiders, *Frontinella pyramitela*. Once mature, the males feed very little and spend most of their time going around female webs in search of mating opportunities. They only live for about 3 days in the final stage and so the way they compete for females is vital for their reproductive success. Laboratory experiments showed that there was first-male priority in sperm competition. These experiments involved allowing two males to mate with a female, one normal male and one irradiated male. The eggs fertilized by the normal male developed normally while the eggs fertilized by the irradiated male failed to develop because of genetic abnormalities induced by the irradiation. If the sequence of matings was first the normal, then the irradiated male, 95 per cent of the eggs developed (therefore 95 per cent were fertilized by the first male). If the sequence was first the irradiated and then the normal male, only 5 per cent of the eggs developed (therefore 95 per cent were again fertilized by the first male). Note that this is the reverse of what happens in dungflies, where the *second* male has the advantage (Chapter 3). The second finding was that there was no sperm displacement as a result of a take-over by another male; if the first male had transferred enough sperm to fertilize 30 per cent of the female's eggs he would fertilize these even if another male then took over and copulated. This means that if a male lost a fight half-way through the mating, he would lose only future possible gains; the take-over would not affect his past success.

(a) Measuring resource value, V

Copulation proceeds as follows. First there is a pre-insemination phase during which the male assesses the female, for example to discover whether she is sexually mature. Then copulation itself occurs, the male passing sperm from his pedipalps to the female. The proportion of eggs fertilized increases at a rapid rate early on in copulation and then at a slower rate as time goes on (Fig. 7.9).

Resource value: the expected number of eggs fertilized

The result is that the female is of different value to the male at different stages of the mating sequence. The average number of eggs fertilized, for all females encountered by a male, is 10 eggs, so when a male first meets a female her expected value is 10. Imagine this female is in fact a virgin, which is the most profitable female that a male encounters, being worth on average 40 eggs. During the pre-insemination phase the value increases to 40 as the male discovers that the female is a virgin. Copulation then begins and the value of the female declines as her eggs are fertilized. After 7 min of copulation, 90 per cent of the eggs have been fertilized so the female is now worth only four eggs. After 21 min of copulation, the value is less than one egg because 99 per cent of the eggs have now been fertilized (Fig. 7.9).

Imagine now that an intruding male comes along at different stages in this mating sequence by the resident male. Intruders are unaware of the state of the female unless they themselves are able to assess her. Therefore the intruder will always assume that the female is of average value, 10 (Fig. 7.9). This meant that Austad could vary V in a fight simply by introducing intruders at different stages. At the end of the pre-insemination phase, for example, V is higher for the resident, who has the knowledge that the female is a virgin and particularly valuable. After 7 min of copulation, V is higher for the intruder, because only the resident has the knowledge that most of the eggs are already fertilized. We may expect residents to fight harder for the female at the end of the pre-insemination phase than after 7 min of copulation.

(b) Measuring the costs of a fight, K

This is the expected reduction in lifetime reproductive success per minute of fighting. Austad found that costs, measured in terms of the probability of serious injury (which is a measure of loss of future reproductive success), were linearly related to the duration of the contest. In fights, the males grab each other and grapple with their jaws and legs. Long fights led to body wounds, loss of legs and eventually certain death. A male's fighting ability depended on his body size. Therefore it was possible to vary K by staging encounters between males of different size.

Fighting ability depends on size

(c) Results of staged encounters

Many contests were arranged, varying V and K as described above. The results were as follows.

1 When V was the same for both males (both introduced simul-

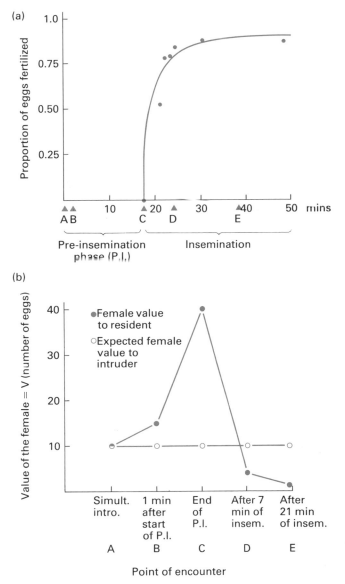

Fig. 7.9 (a) The proportion of eggs fertilized by a male bowl and doily spider during the process of copulation. The stages A to E were used for the introduction of intruders in staged fights for the female. (b) The value of the female to the owner and intruder varies depending on the stage at which the intruder is introduced. From Austad (1983).

taneously at the start of the pre-insemination phase) fights were settled by differences in K. Larger males won 82 per cent of the fights and grapple duration increased with closer matching of body size between the contestants, presumably because it then took longer to assess differences in fighting ability.

2 When K was the same (same size opponents), fights were settled by differences in V. Thus residents fought for longer and were more likely to win at the end of the pre-insemination phase, whereas they fought for less time and thus intruders were likely to win at the stage of 7 min of copulation (Fig. 7.10).

3 When residents were smaller than intruders, they persisted more in fights when V was greater. For example, at the end of the pre-insemination phase, when V was at its highest for the resident, small residents would persist in fights for so long that 90 per cent of the grapples led to them sustaining serious injury. After 7 min of copulation, however, they gave up much sooner and only 30 per cent of the fights led to injury.

4 The most serious escalations occurred where V/K was identical for both contestants. In these cases neither male was prepared to give way and most of the fights ended in serious injury and death for one of them. This occurred, for example, in fights between small residents versus large intruders at the end of the pre-insemination phase. Here V and K were both large for the resident, and were both small for the intruder.

Austad's experiments show clearly that both V and K influence duration of contests and who wins. Similar results have been

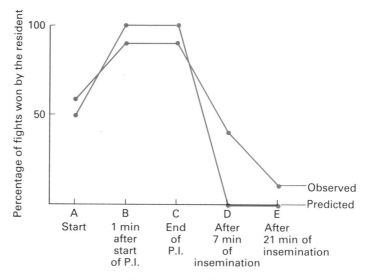

Fig. 7.10 Contests between two male bowl and doily spiders who have the same fighting ability (are the same size). Who wins tends to be determined by differences in resource value, V. For predicted curve, see values of V for residents and intruders in Fig. 7.9b. From Austad (1983).

Fig. 7.11 Plumage variability in the Harris sparrow. The darker males are dominant in the flocks and win most of the fights.

obtained by Sigurjonsdottir and Parker (1981) who studied struggles between male dungflies for the possession of females on cowpats.

Badges of status

Badges of status are a puzzle: why not cheat?

Some animals live in groups and individual differences in fighting ability determine who will have priority of access to food or mates. Many displays within groups probably involve reliable assessment of strength; however the exact link between the display and fighting ability is not always obvious.

In the Harris sparrow (*Zonotrichia querula*) there is enormous plumage variability in winter when the birds go around in flocks searching for food (Fig. 7.11). Individuals with the blackest plumage are the dominants and they always displace pale birds from food supplies (Rohwer & Rohwer 1978). If blackness signals dominance then why don't the subordinates grow blacker feathers and enjoy a rise in status? In the spring all the birds come into breeding plumage and even the pale males develop black coloration, so the dark feathering is not an obviously uncheatable signal of strength like the roar of a rutting stag or the deep croak of a large male toad.

The Rohwers attempted to create 'cheats' in the flock by experimental treatment of subordinate birds (Table 7.3). In the

Table 7.3 Summary of the experiments on signals of dominance in flocks of the Harris sparrow. After Rohwer and Rohwer (1978)

Experimental treatment of subordinates	Look dominant	Behave as dominant	Rise in status?
1 Paint black	Yes	No	No
2 Inject testosterone	No	Yes	No
3 Paint black and inject with testosterone	Yes	Yes	Yes

Cheating is prevented
because the badge is
checked by escalated
fights

first experiment they simply painted subordinates black. These treated birds were attacked by others and failed to rise in status. Secondly, some subordinates were injected with testosterone but their pale plumage was left unaltered. These birds behaved more aggressively and attempted to assert their dominance but they did not rise in status because their opponents did not retreat during disputes. Finally, subordinates were painted and injected so that they both looked and behaved like dominants. This time the cheating worked: the birds won more fights and were respected by others in the flock.

Therefore the earlier attempt to create a cheat by painting alone did not fail simply because the other birds didn't like the paint! It failed because although the darkened birds looked dominant, they did not behave so. The conclusion is that plumage alone is not a passport to high status in the flock; the signal must be backed up by dominant behaviour as well. Nevertheless the dark plumage still acts as a badge to reduce the amount of fighting in the flock. In another experiment dominant birds were bleached so that they signalled low rank. They were attacked a lot by others who tried to displace them. They didn't yield however and eventually exerted their superiority but only after a lot of squabbling.

It is clear from the experiments above that cheating is prevented because a subordinate cannot become dominant through darker plumage alone. But why can't a low ranking bird increase its testosterone level as well? This same question arises from work on red grouse (Watson 1970) where males were injected with testosterone and immediately doubled their territory size and increased the number of females they attracted. Presumably the answer to this problem is that an increase in androgen levels would push the behaviour of a subordinate bird beyond its real capabilities. Although it may gain a short-term increase in success, perhaps the long-term effect would be that all its strength is sapped and it suffers a decrease in fitness (Silverin 1980; Røskaft et al. 1986).

Several other studies have shown experimentally that plumage variability signals status. For example, male scarlet-tufted malachite sunbirds, *Nectarinia johnstoni*, on Mount Kenya compete to defend territories containing *Lobelia* flowers, on whose nectar they feed. The males display their scarlet pectoral tufts in territorial disputes and males with larger tufts defend more inflorescences. When tuft size was manipulated experimentally, by cutting and gluing feathers, birds with enlarged tufts gained control of more inflorescences while those with reduced tufts spent increased time in aggressive interactions and lost a number of

Experiments on pectoral
tuft size in sunbirds

inflorescences to neighbours (Evans & Hatchwell 1992). As with the Harris sparrows, it appears that cheating is controlled because the badges have to be backed up occasionally by real signs of strength such as chasing.

Not all cases of plumage variability are concerned with status signalling. For example, a wading bird, the turnstone (*Arenaria interpres*) has very variable plumage. Whitfield (1986) removed male territory owners and replaced them with models painted to either resemble the removed owner or to be lighter or darker. Neighbours were much less likely to attack the models if they looked like the removed owner than if they looked different. However, the attack intensity towards models that looked different was not related to whether the models were light or dark. In this species, plumage variability may simply facilitate individual recognition.

Plumage variability may facilitate individual recognition

Summary

Game theory is a useful way to analyse animal contests because the best fighting strategy for one animal to adopt depends on what others in the population are doing. Evolution is expected to lead to an evolutionarily stable strategy, or ESS. The war of attrition model applies to examples of scramble competition (e.g. among male dungflies waiting for females on a cowpat) where the pay-offs for different waiting times are frequency dependent. In pair-wise contests, however, there is often direct confrontation between opponents. The Hawk–Dove–Bourgeois model shows how conventional fighting may be an ESS. Most real contests involve assessment and are settled by differences in fighting ability (red deer), resource value (great tits) or both (spiders). Assessment often involves sequences of displays which escalate in intensity when opponents are more closely matched (cichlid fish). Plumage variability in birds may be concerned with status signalling or individual recognition.

Further reading

Parker (1974) discusses assessment strategies, and Caryl (1980) and Maynard Smith (1982) review the application of game theory to contests. Caldwell (1985) shows that mantis shrimps can distinguish individuals on the basis of odour and are more likely to avoid odours of opponents that had defeated them previously than odours of opponents they had beaten. Rubenstein and Hack (1992) show by experiment that horses assess opponents using both sound and scent.

Topics for discussion

1 How has game theory helped us to understand animal contests?
2 Why do territorial residents win?
3 Is assessment always based on reliable signals?

Chapter 8. Sexual Conflict and Sexual Selection

Ethologists used to view courtship rituals and mating as harmonious ventures in which male and female co-operated to propagate their respective genes. Admittedly some animals were obviously not co-operative, for example the praying mantis in which the female eats the male during mating, but on the whole courtship was seen as serving functions of common interest to male and female: to 'synchronize the sexual arousal of the sexes', to 'establish the pair bond' to 'allow species identification', and so on. However this view is no longer so widely held and more emphasis is placed on the idea that there are conflicts of interest between male and female in courtship and mating. The sexes are seen as forming an uneasy alliance in which each attempts to maximize its own success at propagating genes. They co-operate because both pass on their genes via the same progeny and therefore each has a 50 per cent stake in the survival of the offspring. But choice of mating partner, provisioning of the zygote with food, and caring for the eggs and young are all issues over which the sexes may disagree. The outcome of this sexual conflict is often more akin to exploitation by one sex of the other than to mutual co-operation.

Reproduction includes conflicts of interest between the sexes

In order to understand why sexual reproduction should be viewed in this way, we have to go right back to the beginning, to the fundamental difference between male and female.

Males and females

Sexual reproduction entails gamete formation by meiosis and the fusion of genetic material from two individuals. It almost always, but not invariably, involves two sexes called male and female. In higher animals the sexes are often most readily distinguished by external features such as genitalia, plumage, size, or colour, but these are not fundamental differences. In all plants and animals the basic difference between the sexes is the size of their gametes: females produce large, immobile, food-rich gametes called eggs, while male gametes or sperm are tiny, mobile, and consist of little more than a piece of self-propelled DNA. Sexual reproduction without males and females occurs in many protists such as *Paramecium* where the 'gametes' which fuse during sex are of the same size. This is referred to as *isogamous* sexual reproduction. The fusion of two gametes of unequal size, one large and one small is, however, much commoner and occurs in virtually all

Females: the sex that produces large gametes

175

sexually reproducing multicellular plants and animals. It is called *anisogamous* sex.

It is thought that anisogamy evolved from isogamy by an evolutionary process in which smaller than average gametes successfully parasitized those that were larger than average. This eventually led to the two distinct specializations we see today: small active sperm and large passive eggs (Parker *et al.* 1972)..

As we shall show in the rest of this chapter, the fundamental asymmetry in gamete size and associated investment in offspring has far-reaching consequences for sexual behaviour. Because females put more resources than do males in each offspring, male courtship and mating behaviour is to a large extent directed towards competing for and exploiting female investment. Where the usual difference between the sexes in parental investment is reversed, males providing more care than females, so the roles in sexual competition are reversed and females compete for access to males.

FEMALES AS A SCARCE RESOURCE

Anisogamous sexual reproduction, then, involves parasitism of a large egg by a small sperm. Females produce relatively few large gametes and males produce many small ones. In addition, females often invest more than males in other forms of care. Because of this males can potentially fertilize eggs at a faster rate than they are produced (illustrated by the fact 5 ml of human semen contains enough sperm to fertilize in theory eggs amounting to twice the population of the USA), and females are therefore a *scarce resource* for which males compete. Even in species where males temporarily deplete their sperm supply when offered a surfeit of females, their potential for producing offspring is greater than that of females (Nakatsuru & Kramer 1982).

Male reproductive success is often limited by access to females ...

A male can increase its reproductive success by finding and fertilizing many different females, but a female can only increase her success by turning food into eggs or offspring at a faster rate (Fig. 8.1). This point is graphically demonstrated by mammals such as man, in which a female spends many months producing a single child, during which time a male could potentially fertilize hundreds of other mates. Only by speeding up her production of

... whilst females are limited by resources

young can the female have more children in a lifetime. The same argument holds whenever females invest more than males in each offspring, whether the investment is in the form of food in the egg or care of the eggs and young later on.

This point was neatly summarized by Robert Trivers (1972), who was the first person to emphasize the relationship between

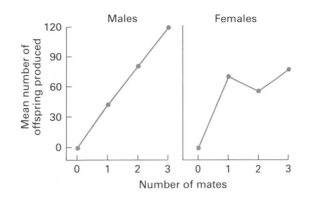

Fig. 8.1 A.J. Bateman (1948) put equal numbers of male and female fruit flies (*Drosophila melanogaster*) in bottles and scored the number of matings and offspring produced by each individual, using genetic markers to assign parentage. For males reproductive success goes up with number of matings, for females it does not, beyond the first mating. After Trivers (1985).

Parental investment, potential rate of reproduction and sexual competition

the investment of resources in gametes and other forms of care, and sexual competition. He wrote 'Where one sex invests considerably more than the other, members of the latter will compete among themselves to mate with members of the former'. The term 'investment' was used by Trivers to refer to the effort put into rearing an individual offspring from the parent's limited pool of resources. The sum of parental investment in all offspring during a parent's lifetime is referred to as 'parental effort'. Females generally put most of their reproductive effort into 'parental effort' while males put most of theirs into 'mating effort' (Fig. 8.2).

The consequence is that males usually have a much greater potential rate of reproduction than do females (Table 8.1) (Clutton-Brock & Vincent 1991) and are therefore under strong selection to be good at seeking out and competing for females: the pay-off for a successful male in terms of offspring fathered is enormous. Much of male reproductive behaviour can be understood with these ideas in mind.

THE SEX RATIO

If one male can fertilize the eggs of dozens of females why not produce a sex ratio of, say, one male for every 20 females? With this ratio the reproductive success of the population would be higher than with a 1:1 ratio since there would be more eggs around to fertilize. Yet in nature the ratio is usually very close to 1:1 even when males do nothing but fertilize the female. As we saw in Chapter 1, the adaptive value of traits should not be viewed as being 'for the good of the population', but 'for the good

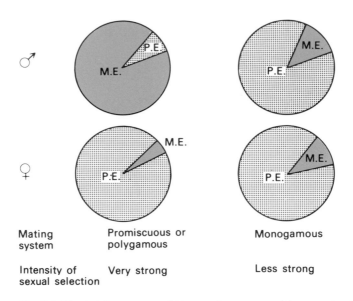

Fig. 8.2 The total resources of time and energy used by an animal in reproduction is referred to as reproductive effort. This is represented by a circle. Reproductive effort can be partitioned into parental effort (provisioning and rearing offspring) and mating effort (acquiring mates). These are represented by the stippled and coloured areas of the circles respectively. In general, males put relatively more into mating effort than do females, but this varies between species. The intensity of sexual selection (see p. 183 for a definition) therefore also varies. The differences in relative parental effort of the sexes is often related to the mating system. In monogamous species male and female effort is more similar than in polygamous and promiscuous species (see Chapter 9). After Alexander and Borgia (1979).

of the individual', or more precisely 'for the good of the gene'. As R.A. Fisher (1930) first realized the 1:1 sex ratio can readily be explained in terms of selection acting on the individual; his argument is simple but subtle.

The sex ratio is usually 50:50

Suppose a population contained 20 females for every male. Every male has 20 times the expected reproductive success of a female (because there are on average 20 mates per male) and therefore a parent whose children are exclusively sons can expect to have almost 20 times the number of grandchildren produced by a parent with mainly female offspring. A female-biased sex ratio is therefore not evolutionarily stable (p. 149) because a gene which causes parents to bias the sex ratio of their offspring towards males would rapidly spread, and the sex ratio will gradually shift towards a greater proportion of males than the initial 1 in 20. But now imagine the converse. If males are 20 times as common as females a parent producing only daughters will be at an advantage. Since one sperm fertilizes each egg, only one in every 20 males can contribute genes to any individual offspring

Table 8.1 In polygamous or promiscuous species males have a much higher potential reproductive rate than females. The data for man came from the Guinness Book of Records: the male was Moulay Ismail the Bloodthirsty, Emperor of Morocco, the woman had her children in 27 pregnancies. The data for elephant seals are from Le Boeuf and Reiter (1988), for red deer from Clutton-Brock *et al.* (1982). In the monogamous kittiwake, where male and female invest similarly in each offspring, the difference in maximum reproductive output is negligible. From Clutton-Brock (1983)

Species	Maximum number of offspring produced during lifetime	
	Male	Female
Elephant seal	100	8
Red deer	24	14
Man	888	69
Kittiwake gull	26	28

and females therefore have 20 times the average reproductive success of a male. So a male-biased sex ratio is not stable either. The conclusion is that the rarer sex always has an advantage, and parents which concentrate on producing offspring of the rare sex will therefore be favoured by selection. Only when the sex ratio is exactly 1:1 will the expected success of a male and a female be equal and the population stable. Even a tiny bias favours the rarer sex: in a population of 51 females and 49 males where each female has one child, an average male has 51/49 children. This *average* value is the same whether one male does most of the fathering or whether fatherhood is spread equally among the males.

The argument that the sex ratio should be 1:1 can be refined by re-phrasing it in terms of investment. Suppose sons are twice as costly as daughters to produce because, for example, they are twice as big and need twice as much food during development. When the sex ratio is 1:1 a son has the same average number of children as a daughter. But since sons are twice as costly to make they are a bad investment for a parent: each of its grandchildren produced by a son is twice as costly as one produced by a daughter. It would therefore pay the parents to concentrate on making daughters. As the sex ratio swings towards a female bias, the expected reproductive success of a son goes up until at a ratio of two females to every male an average son produces twice the number of children produced by an average daughter. At this point sons and daughters give exactly the same return per unit investment; a son costs twice as much to make but yields twice the return. This means that when sons and daughters cost different amounts to make, the stable strategy in evolution is for the

More precisely, the ratio of investment in males and females is equalized

parent to *invest* equally in the two sexes and not to produce equal numbers. An example to illustrate this point is Bob Metcalf's (1980) study of the sex ratio of two species of wasp: *Polistes metricus* and *P. variatus*. In the former females are smaller than males, while in the latter they are similar in size. As predicted, the population sex ratio is biased in *P. metricus* and not in *P. variatus*. In both species the investment ratio is 1 : 1.

The prediction that parents should invest equally in sons and daughters does not always hold, and demonstrations of these deviations from 1 : 1 investment are among the most convincing pieces of evidence that the sex ratio has evolved in the way suggested by Fisher. We will pick out some examples in the following paragraphs.

(a) Local mate competition

If brothers compete for mates, the sex ratio should be female-biased

Fisher's theory predicts a different outcome when brothers compete with each other for mates (so-called 'local mate competition'). Suppose, for example, that two sons have only one chance to mate and that they compete for the same female. Only one of them can be successful in mating, so from their mother's point of view one of them is 'wasted'. This is an extreme example, but it illustrates the general point that when sons compete for mates their value to their mother is reduced. The mother should therefore bias her ratio of investment towards daughters. The exact degree of bias predicted by Fisher's theory depends on the degree of local mate competition. Extreme competition is to be expected in species with limited powers of dispersal (because brothers will stay together in the same place) and therefore it tends to be associated with inbreeding. In the extreme case of inbreeding, a mother 'knows' that all her daughters will be fertilized by her sons. The best sex ratio in this instance is to produce just enough sons to fertilize the daughters, since any other males will be wasted. The crucial difference between this and the earlier argument for a 1 : 1 sex ratio is that here the ratio of males to females in the rest of the population does not matter. A female-biased ratio within a brood will not give other parents a chance to benefit by concentrating on sons. An example which supports this prediction is the viviparous mite, *Acarophenox*, which has a brood of one son and up to 20 daughters. The male mates with his sisters inside the mother and dies before he is born (Hamilton 1967).

Jack Werren (1980) has tested the prediction that the degree of bias depends on the extent of local mate competition. He studied the parasitoid wasp, *Nasonia vitripennis*, which lays its eggs inside the pupae of flies such as *Sarcophaga bullata*. If one female parasitizes a pupa, her daughters are all fertilized by her sons and as predicted the sex ratio of her clutch of eggs is biased towards females. Only 8.7 per cent of the brood is male. If a second female lays her eggs in the same pupa, what should her sex ratio be? If she lays few eggs she should produce mainly sons, since the first female has laid predominantly female eggs. But as the proportion of the total number of eggs in the pupa that come from the second female increases, the chance that sons of the second female will compete for mates also increases. Therefore her brood should have a female-biased sex ratio. Werren found exactly this pattern: when the second female's clutch was 1/10 the size of the first female's it contained only males, but when it was twice as large as the first female's it contained only 10 per cent males, and the quantitative details of the change in sex ratio with relative brood size were much as predicted

(b) Local resource competition or enhancement

The sex that does not compete with parents may be more valuable

Anne Clark (1978) found that the South African prosimian, *Galago crassicaudatus*, has a male-biased investment ratio among its offspring. She pointed out that this could be explained by the species' life history. As with most mammals, female *Galago* disperse less far than males, and often end up competing both with their mother and with each other for rich sources of food such as gum and fruit trees in the mother's home range. This local resource competition among females reduces their value as offspring: in the extreme case only one daughter might be able to survive on the food available near home, and so investment in other daughters would be wasted.

Exactly the opposite effect could arise if the sex that stays at home, rather than hindering one another or their parents, actually helps. As we shall see in Chapter 12, in some bird species, males but not females stay at home and help. The consequence of this is to make males slightly more valuable than females as an investment (since they help the parent in its future reproduction) and hence a male-biased investment ratio might be expected (Emlen *et al.* 1986).

If one sex helps its parents, the sex ratio may be biased in favour of that sex

(c) Maternal condition

We saw in Chapter 7 that male red deer compete for females by prolonged roaring and antler wrestling contests. In these contests it is an advantage for a male to be big, and size depends among other things on how well fed the male was as a youngster, which in turn depends on his mother's ability to compete for good sources of food and hence produce a plentiful supply of milk. In other words there is a direct link between a mother's competitive ability while lactating and her son's expected reproductive success. Now if a mother 'knew' that her sons would be highly successful harem holders it would pay her to invest heavily in sons rather than daughters: the pay-off in terms of grandchildren would be much greater. Similarly a mother who 'knew' that her sons would not grow up to be big and strong would do better to have daughters, since a daughter's future reproductive success does not depend so much on her mother's milk. Exactly this pattern has been found in red deer: dominant females who are able to gain access to good feeding sites while lactating and hence produce strapping sons, tend to have sons, while subordinate females have daughters (Clutton-Brock *et al.* 1984). It is not known how the sex ratio is adjusted by the mother in the red deer or the galagos studied by Clark, but the fact that they are adjusted is striking because agricultural geneticists have failed to select for adjustment of the sex ratio or to separate male and female sperm of domestic mammals (imagine the value of a female-biased sex ratio to the milk farmer!), and it has often been concluded that mammalian sex ratios are very inflexible (Williams 1979; Maynard Smith 1980). Adjustment of the sex ratio in Hymenoptera such as the wasps studied by Werren is not at all a problem because the mother can determine whether an egg becomes male or female by whether or not she fertilizes it (Chapter 13).

Female red deer in good condition tend to produce more male offspring

(d) Population sex ratio

When the population ratio of investment deviates from $1:1$, a compensatory bias in favour of the rarer sex should occur. In Metcalf's study of *P. metricus* he found that some nests produced only male offspring. As explained in Chapter 13 these offspring are the product of unfertilized eggs and are produced by workers in nests where the queen has died. In the remainder of the nests in the population Metcalf found a female-biased sex ratio, so that the ratio of investment in the population as a whole is $1:1$.

Biased sex ratios in paper wasp nests may be influenced by the population sex ratio

Finally it is worth pointing out that the theory of sex ratios discussed here is one example of a more general theory of sex *allocation* (Charnov 1982). Other examples of the problem of allocation of resources to male and female reproduction include the division of resources into eggs and sperms by simultaneous hermaphrodites and the timing of sex change in sequential hermaphrodites (see Chapter 10).

Sexual selection

The combination of females investing more than males and $1:1$ population sex ratio means that males usually compete for females. The potential pay-off for male success is high, so selection for male ability to acquire matings is very strong. Selection for traits which are solely concerned with increasing mating success is usually referred to as *sexual selection*. It can work in two ways: by favouring the ability of one sex (usually males) to compete directly with one another for fertilizations, for example by fighting (*intra-sexual selection*), or by favouring traits in one sex which attract the other (*inter-sexual selection*). Often the two kinds of selection act at the same time.

Sexual selection: selection for traits that increase mating success

The intensity of sexual selection depends on the degree of competition for mates. This in turn depends on two factors: the difference in parental effort between the sexes (Fig. 8.2) and the ratio of males to females available for mating at any one time (referred to as the *operational sex ratio*). When parental effort is more or less equal, as for example in monogamous birds where both male and female feed the young, sexual selection is less intense than in species with very different levels of parental effort. This follows from the point made earlier that the sex making little investment has a higher potential rate of reproduction (p. 177). If equal numbers of the two sexes come into breeding condition at the same time, the degree of sexual selection is reduced because there is less chance for a few males to control access to very large numbers of females. In contrast, when females come into breeding condition asynchronously there is a chance for a small number of males to control many females one after the other. With such high potential pay-offs, sexual competition is very intense (see Chapter 9). The relationship between parental investment, operational sex ratio and sexual selection is discussed in more detail by Clutton-Brock & Vincent (1991).

Sex differences in parental investment and the operational sex ratio influence the strength of sexual selection

ARDENT MALES

The most dramatic and obvious way in which males compete for mates is by fighting and ritualized contests, and often males have evolved weapons for fighting. Males may dispute over direct access to females or over places where females are likely to go, as for example when male damselflies defend clumps of vegetation (Chapter 7). Fighting is often a risky business, as illustrated by the injuries sustained by red deer stags referred to in Chapter 7. The most intense fights in many species occur when females are

Males fight for access to females . . .

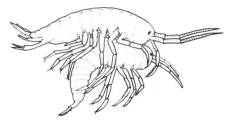

(a)

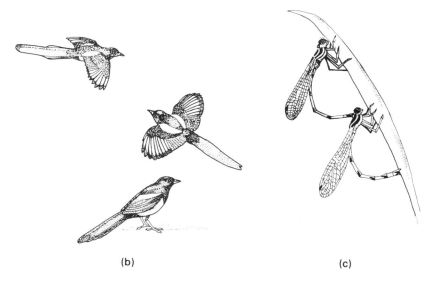

(b)   (c)

Fig. 8.3 Mate guarding as a form of sexual competition. (a) Precopulatory
mate guarding in the freshwater amphipod *Gammarus*. The mature female in
this species is ready to be fertilized immediately after she moults. Males
guard females in the few days preceding this moult. From Birkhead and
Clarkson (1980). (b) Male European magpies (*Pica pica*) assiduously guard
their mates against intruding males just before and during the period of egg
laying. From Birkhead (1979). (c) After copulation the male damselfly guards
the female while she lays her eggs, by clasping her thorax with the tip of his
abdomen in the 'tandem' position. From Corbet (1962).

ready to be fertilized and once a male finds a female he often
guards her (Fig. 8.3).

  Males often compete in ways which are less conspicuous than
fights, but are no less effective and often more bizarre. The
invertebrates are a particularly rich seam of examples. Female
dragonflies, as with many other insects, mate with a number of
males and store the sperm in a special sac (the spermatheca) in
the body for use at a later date. The males compete for fertilizations
by trying to ensure that previous sperm is not used by the female.

... or compete by sperm
competition

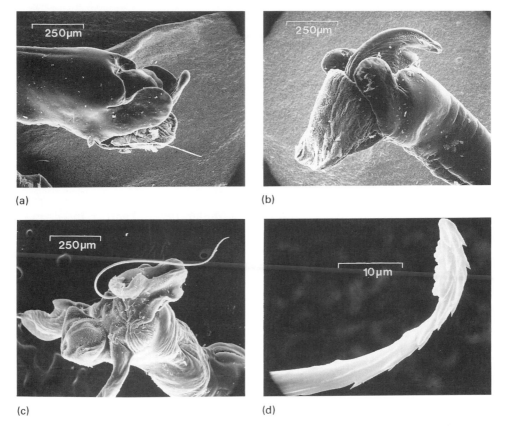

(a)

(b)

(c)

(d)

Fig. 8.4 Two sperm displacement mechanisms in Odonata. Photos by M. Siva-Jothy. *Crocethemis erythraea*: (a) Uninflated penis. (b) Inflated penis. The horn-like structure repositions sperm of previous males in the spermatheca. *Orthetrum cancellatum*: (c) The whip-like flagellum is everted during copula. (d) It carries barbs which remove sperm from the narrow ducts of the spermatheca (Siva-Jothy pers. comm.).

The penis of a male *Orthetrum cancellatum* is equipped with a barbed whip at the end which is used to scrape out of the female any sperm left by previous males before he injects new sperm into the sperm sac. *Crocethemis erythraea*, another dragonfly, uses an inflatable penis with a horn-like appendage to pack the sperm of previous males into corners of the spermetheca (Fig. 8.4; see also Waage 1979).

In some invertebrates (especially insects) the male cements up the female's genital opening after copulation to prevent other males from fertilizing her. The males of *Moniliformes dubius*, a parasitic acanthocephalan worm in the intestine of rats, produces a chastity belt of this kind but in addition to sealing up the female after copulation, the male sometimes 'copulates' with rival males and applies cement to their genital region to prevent

Copulatory plugs

Homosexual mating

them from mating again (Abele & Gilchrist 1977). No less remarkable are the habits of the hemipteran insect *Xylocoris maculipennis*. In normal copulation of the species the male simply pierces the body wall of the female and injects sperm, which then swim around inside the female until they encounter and fertilize her eggs. As with the acanthocephalan worms, males sometimes engage in homosexual 'copulation'. A male *Xylocoris* may inject his sperm into a rival male. The sperm then swim inside the body to the victim's testes, where they wait to be passed on to a female next time the victim mates (Carayon 1974).

Competition between males to prevent each other's sperm from fertilizing eggs is sometimes referred to as 'sperm competition' (Birkhead & Møller 1992). Another insect example was described in Chapter 3: the sperm of a second male displaces that of the first male to mate with a female dungfly. Sperm competition also occurs in vertebrates. For example, during courtship male salamanders and newts deposit little sperm-capped rods of jelly (spermatophores) on the bottom of the pond and then try to manoeuvre the female onto the spermatophore to achieve fertilization. In the salamander *Ambystoma maculatum*, males compete by depositing their spermatophores on top of those of other males. The top spermatophore is the one that fertilizes the female's eggs (Arnold 1976).

Anti-aphrodisiac scents

A fourth example of the arcane methods of male–male competition found among invertebrates is the use of anti-aphrodisiac smells. Larry Gilbert (1976) noticed that female *Heliconius erato* butterflies always smell peculiar after they have mated. He was able to show experimentally that the scent does not come from the female herself, but is deposited by the male at the end of mating. Gilbert also found that the scent discourages other males from mating with the female, perhaps because it resembles a scent used by males to repel one another in other contexts.

RELUCTANT FEMALES

Since females in the great majority of species are the chief providers of resources for the zygote, they might be expected to choose their mates carefully in order to get something in return. To put it another way, each egg represents a relatively large proportion of a female's lifetime production of gametes when compared with a sperm, so the female has more to lose if something goes wrong. Mating with the wrong species could cost a female frog her whole year's supply of eggs, but would cost the male very little apart from lost time — he could still go on to mate successfully with a member of the correct species the next day. Not surprisingly therefore, females are on the whole choosier than males during courtship. Choosiness extends not only to discriminating between species, but also to discriminating between males within a species. Females often select males on the basis of material resources they can offer and perhaps sometimes to obtain genetic benefits for their offspring.

(a) Non-genetic benefits: good resources and parental ability

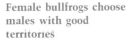

In many animal species males defend breeding territories containing resources which play a crucial role in the survival of a female's eggs or young (see also Chapter 9). For example male North American bullfrogs (*Rana catesbeiana*) defend territories in ponds and small lakes where females come to lay their eggs (Fig. 8.5). Some territories are much better for survival of eggs than others and these are the ones which females prefer. One factor which has an important influence on survival of eggs is predation by leeches (*Macrobdella decora*). Two environmental features of a territory influence leech predation: if the water is warm the eggs develop faster and are therefore exposed to predation for fewer days, and if the vegetation in the water is not too dense the eggs can form into a ball which the leeches find hard to attack. In territories with a dense mat of vegetation the eggs lie in a thin film on top of the plants and are more easily attacked. The bullfrogs also show that female choice and male–male

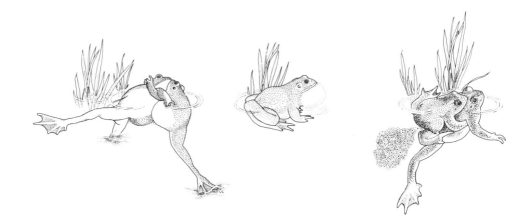

Fig. 8.5 Sexual selection in male bullfrogs. Males compete by wrestling and calling (left and middle) for good territories, in which the females prefer to lay their eggs (right). The good territories have high survival of eggs because they are warm and because the vegetation is not too dense. From Howard (1978a,b).

competition may go hand in hand. The preferred territories are hotly contested by males and the largest, strongest frogs end up in the best sites.

Food is a resource which often limits a female's capacity to make eggs and during courtship females may choose whether or not to mate with a male on the basis of his ability to provide food. In some birds and insects for example, males may provide food for the female during courtship ('courtship feeding') which makes a significant contribution to her eggs. Female hanging flies (*Hylobittacus apicalis*) will mate with a male only if he provides a large insect for her to eat during copulation. The larger the insect, the longer the male is allowed to copulate and the more eggs he fertilizes (Fig. 8.6). The female gains from a large insect by having more food to put into her eggs. Gifts provided by insects during courtship may help to protect, rather than nourish the eggs. In the moth *Utethesia ornatrix*, the male transfers protective alkaloids to the female during mating. Further, the same alkaloids are used by the male as a pheromonal attractant. The female is able to assess the quantity of poison she will receive by the concentration in the pheromone (Dussourd *et al.* 1991). In birds, the male usually helps to feed the young and courtship feeding may play the additional role of indicating to the female how good the male is at bringing food for the young.

In the common tern (*Sterna fuscata*) there is a correlation between the ability of the male to bring food during courtship feeding and his ability to feed the chicks later in the season. Pairs often break up during the courtship feeding period and it is possible that females are assessing their mates and rejecting poor quality partners (Nisbet 1977).

(b) Genetic benefits

If some males have 'better' genes than others, could a female improve the success of her progeny by choosing males with good genes? Good genes are ones which increase the ability of her offspring to survive, compete and reproduce. One of the few studies attempting to test this experimentally by Linda Partridge (1980). She took groups of female fruit flies (*Drosophila*) and either allowed them to mate freely with a population of males or forced each female to mate with a randomly chosen partner.

The offspring of the 'choice' and 'no choice' females were then tested for their competitive ability by rearing the larvae in bottles with a fixed number of standard competitors (these were distinguishable by a genetic marker). Partridge found that the offspring of the 'choice' group did slightly but consistently

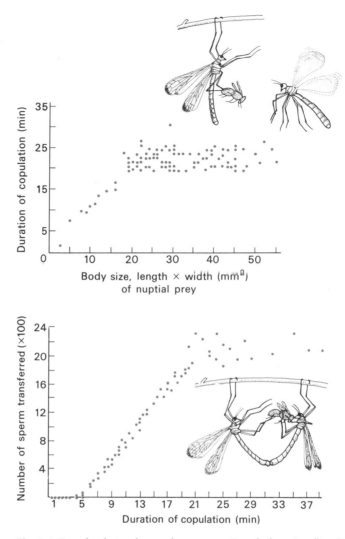

Fig. 8.6 Female choice for good resources. Female hanging flies (*Hylobittacus apicalis*) mate with males for longer if the male brings a larger prey item to eat during copulation. The male benefits from long copulation because he fertilizes more eggs. From Thornhill (1976).

better than those of 'no choice' females in the larval competition experiments. This experiment suggests that females are able to increase the survival of their offspring by choosing good genes in their mates, but it must be borne in mind that the results could also be in part explained by intra-sexual competition: in the 'choice' experiment, the males that mated may have been superior competitors against other males.

ELABORATE ORNAMENTS: FISHER'S HYPOTHESIS AND THE
HANDICAP HYPOTHESIS

The theory of sexual selection is most famous as an attempt to
explain the evolution of excessively elaborate adornments and
displays of male peacocks (Petrie *et al.* 1991), pheasants, birds of
paradise and so on (Plate 8.1, between pp. 212–213). Some
elaborate displays may have evolved for use in contests
between males, but some have almost certainly evolved as
a result of selection by females for genetic benefits. There are
two competing hypotheses to explain how selection for genetic
benefits might produce elaborate traits: *Fisher's hypothesis*
(sometimes called the 'runaway process' because it postulates
runaway positive feedback between female preference and male
displays), and the *handicap hypothesis*. In the following section
we will first describe two examples of studies in which females
have been shown to prefer elaborate male traits, then consider
how the two hypotheses might explain these results. Finally, we
consider whether there is any evidence for one or both hypotheses.

*Two hypotheses for
elaborate adornments*

*(a) Examples of female preference for elaborate
male displays*

There are many studies which have looked for correlations
between male mating success or female preference, and male
sexual displays. However studies based purely on observation or
correlations do not allow us to demonstrate a cause–effect
relationship. One of the classic experimental studies is that of
Malte Andersson (1982), who showed that females of the long-
tailed widow bird (*Euplectes progne*) in Kenya prefer males with
long tails. This highly polygynous species is an ideal candidate
for sexual selection; the male is a sparrow-sized bird with a tail
up to 50 cm long. The female's tail is about 7 cm long, presumably
close to the optimum for flight purposes. Andersson studied 36
males which he divided into four groups. In one group he docked
the tails to about 14 cm, while in another group he attached the
severed bits of group I tails with Superglue. This increased the
tail length of group II males by an average of 25 cm. The remaining
two groups were controls: one lot were left untouched and the
others had their tails cut and glued without altering the length.
By counting the number of nests in each territory, Andersson
showed that before his experimental manipulations there was no
difference in mating success of the different groups, while afterwards the long-tailed males did significantly better than the
controls or the shorter tailed birds (Fig. 8.7).

*Females prefer elaborate
traits*

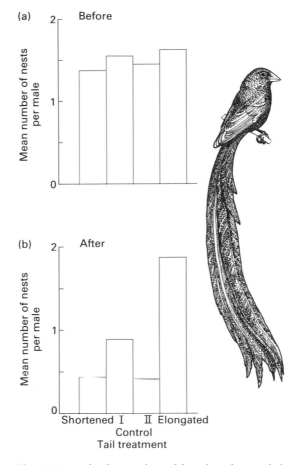

(a)

Before

Mean number of nests per male

(b)

After

Mean number of nests per male

Shortened I   II Elongated
Control
Tail treatment

Fig. 8.7 Sexual selection for tail length in long-tailed widow birds. The top line shows that there was no difference between the four groups before the tails were altered. The bottom line shows that after the tails were cut and lengthened the mating success went down and up respectively. The two kinds of control birds were (I) unmanipulated, and (II) cut and glued back without altering length. Mating success is measured as the number of active nests in each male's territory. From Andersson (1982).

Another nice experimental study of a sexually selected elaborate display is that of Clive Catchpole (1980; Catchpole *et al.* 1984) on the song of the European sedge warbler. The song consists of a long stream of almost endlessly varying trills, whistles and buzzes and is sung by the male after arriving back on the breeding territory from the winter quarters: as soon as a male pairs, it stops singing. Catchpole's measurements showed that the males with the most elaborate songs are the first to acquire mates (Fig. 8.9). Further, when female warblers were brought into the laboratory and treated with oestradiol to make them sexually

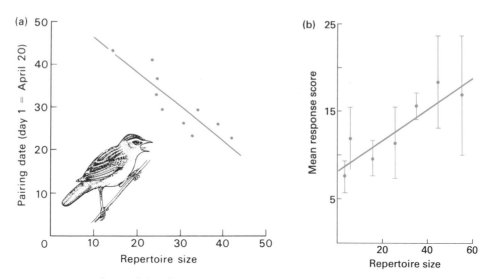

Fig. 8.8 (a) Male sedge warblers with the largest song repertoires are the first to acquire females in the spring. The size of song repertoire is estimated from sample tape recordings of each male. The results were collected in such a way as to control for the possibilities that older males, or males in better territories, both mate first and have larger repertoires. From Catchpole (1980). (b) The mean ± s.e. response score of five females to repertoires of different sizes. The response score measures sexual behaviour. From Catchpole *et al.* (1984).

active, they were more responsive to large than to small repertoires. In contrast to the long tail of the widow bird it is not obvious what might be the counter-selection limiting the elaboration of song; one possibility is that elaborate songs are more readily detected by predators, as has been found in the leopard frog (Ryan *et al.* 1982).

(b) Fisher's hypothesis

Selection for attractiveness alone

R.A. Fisher (1930) was the first to clearly formulate the idea that elaborate male displays may be sexually selected simply because it makes males attractive to females. This may sound circular, and indeed it is, but that is the elegance of Fisher's argument. At the beginning, he supposed, females preferred a particular male trait (let us take long tails as an example) because it indicated something about male quality. Perhaps males with longer tails were better at flying and therefore collecting food or avoiding predators. An alternative starting point is to suppose that larger tails were simply easier to detect (Arak 1983) or that females had a pre-existing sensory bias to respond to certain stimuli (Ryan *et al.* 1990; see Chapter 14). If there is some genetic basis for differences between males in tail length the advantage will be passed on to the female's sons. At the same time, a gene which causes females to prefer longer than average tails will also be favoured since these females will have sons better able to fly or more readily detected by potential mates. Now once the female preference for longer tails starts to spread, longer tailed males will gain a double advantage: they will be better at flying and be more likely to get a mate. The female similarly gets a double advantage from choosing: she will have sons that are both good fliers and attractive to females. As the positive feedback between female preference and longer tails develops, gradually the benefit of attractive sons will become the more important reason for female choice, and the favoured trait might eventually decrease the survival ability of males. When the decrease in survival counter-balances sexual attractiveness, selection for increasing tail length will grind to a halt. Box 8.1 describes some aspects of Fisher's hypothesis in more detail.

Box 8.1 *Sexual selection for nose length: the importance of genetic covariance for Fisher's hypothesis. After Lande (1981).*

1 Imagine that at the start there was a range of nose lengths and of female preferences in the population. Females with a preference for slightly longer than average noses would be mated to males with longer noses and vice versa. The crucial fact to note is that offspring of these matings would have *both* the nose and preference genes: either genes for long nose plus long preference or short nose and short preference. The preference is expressed only in females and the nose in males, but everyone carries both kinds of gene. In short, there will arise an association or *covariance* between nose and preference genes. You could look at a female's preference and predict what kind of nose genes she carries to give to her sons (Fig. a).

2 How will evolution proceed, given this covariance? If equal numbers of females have preference above and below the mean nose length (x), there will be no change. But if by chance there was a slight predominance of females on one side of the mean (it could be long or short but let us take long), then positive feedback will start. This is shown by the arrows on Fig. a. Females select for long noses (long-nosed males have a higher chance of mating) and thereby, *because of the covariance*, select for long preference. This

Continued on p. 194

Box 8.1 *Continued*

(a)

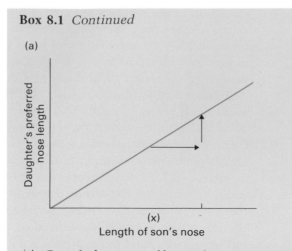

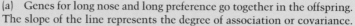

(a) Genes for long nose and long preference go together in the offspring. The slope of the line represents the degree of association or covariance.

in turn produces a further push to long noses and hence an increase in preference.

3 The final outcome of sexual selection in quantitative models of this hypothesis depends on the exact assumptions made in the model, for example whether or not there is a cost of female choice (Pomiankowski *et al.* 1991). However, the important general point is that covariance between the male trait and female preference underlies Fisher's hypothesis.

(c) The handicap hypothesis

Amotz Zahavi (1975, 1977) suggested an alternative view of elaborate male sexual displays. He pointed out that the peacock's long tail is a handicap in day-to-day survival, a view which few would dispute. He then went on to suggest that females prefer long tails (or other equivalent traits) precisely *because* they are handicaps and therefore act as a reliable signal of a male's genetic quality. The tail demonstrates a male's ability to survive in spite of the handicap, which means that he must be extra good in other respects. If any of this ability is heritable, then the tendency to be 'good' at surviving will be passed on to offspring. Thus females select for good genes by selecting to mate only with males whose displays honestly indicate their genetic quality. Note that in this hypothesis the 'good genes' are genes for the utilitarian aspects of survival and reproduction, rather than genes purely for attracting females, as assumed in Fisher's hypothesis. When it was first

The handicap hypothesis

published, Zahavi's idea was not accepted, but subsequent theoretical papers (Kodric Brown & Brown 1984; Nur & Hasson 1984; Grafen 1990a,b) have led to the view that the handicap hypothesis is a plausible explanation for the evolution of elaborate sexual displays, and perhaps of animal signals in general (Zahavi 1987; see Chapter 14). The most important feature of theoretical models of the handicap principle that 'work' (i.e. show that females could benefit from choosing males because of their handicaps) is that males only express the handicap, in other words develop the full sexual display, when they are in good condition. This gets around the difficulty some critics saw in Zahavi's original idea, that males were forced to carry the handicap whether or not they could afford it, because it was viewed as a fixed trait. There are different variants of the flexible handicap idea (some authors refer to 'revealing handicaps' that reveal a male's current vigour, others to 'condition dependent handicaps' expressed in proportion to the male's condition), but the essential feature of all these models is that the degree of expression of the male sexual display tells the female about his genetic quality.

Selection for male quality may work if handicaps are flexible traits

The best studied version of the handicap hypothesis is the one first proposed by Bill Hamilton and Marlene Zuk (1982), namely that sexual displays are reliable indicators of genetic resistance to disease. If males are able to show the full development of their secondary sexual characteristics only when they are free of diseases, females, by choosing for elaborate displays, might in effect be acting as diagnostic veterinarians. This particular version of the handicap hypothesis has two features which make it especially attractive. First of all, disease is a very widespread agent of selection, so the idea could have general application as a reason for selecting handicapped males. Second, diseases have the property of continually evolving new varieties, hence continually presenting new selective hurdles. The importance of this can be illustrated with a hypothetical example.

Disease resistance and the handicap hypothesis

Suppose a farmer wants to select for larger body size in a population of pigs. He takes the heaviest males and females to start the next brood and repeats this procedure for several generations. What will happen? Assuming that there is some genetic variance for body weight, selection will at first be fairly effective, but soon the stock will become less variable with respect to genes for body weight, because only a few genotypes (the heaviest) have been allowed to breed. When the genetic variance is 'used up' selection will cease to be effective in changing body size. In the same way, females cannot improve the genetic quality of their offspring indefinitely by choosing males for good genes. Further, if choosing has a cost, for example time spent searching,

Exhausting genetic variation

females should stop choosing when genetic benefits no longer accrue. The fact that diseases continually evolve new forms might get around this problem: selection for disease resistance never continues for long enough in one direction for all the genetic variation to be used up.

(d) Evidence for the Fisher and handicap hypotheses

Both the Fisher and the handicap hypotheses are attempts to explain why females should prefer elaborate or extreme male displays. Therefore experiments and correlations such as those on long-tailed widow birds and sedge warblers described above, whilst consistent with both hypotheses, do not discriminate between them. In order to demonstrate that a trait had evolved by Fisher's process, it would be necessary to show that there is genetic variation for both female preference and the male trait, and that the genes tend to covary (Box 8.1). Because Fisher's hypothesis assumes that the only benefit of the selected trait is increased mating success, it would also be necessary to show that expression of the male trait did not correlate with any inherited 'utilitarian' aspect of fitness such as disease resistance or ability to gather scarce resources, as proposed by the handicap hypothesis. Two ways to examine this prediction would be (a) to see whether or not, within a population of males, more extreme expression of the trait was correlated with viability, and (b) to examine the offspring of males with extreme traits. According to Fisher's hypothesis they should have no enhanced viability, only enhanced male mating success. The difficulty with both of these tests is that in order to support Fisher's hypothesis one would need to see a negative result. Negative results could arise for many reasons, including not having a large enough sample or not measuring the appropriate variables.

Among the few studies to demonstrate a genetic correlation between male display and female preference is the work on guppies, *Poecilia reticulata*, by Anne Houde (Houde 1988; Houde & Endler 1990). Guppies occur in many different stream systems in Trinidad, and males from different populations differ greatly in the extent to which they develop bright orange and blue spots, which are a stimulus for females during courtship. These differences between streams are correlated with the presence or absence of predators, including other species of fish and prawns. In streams without predators males have large spots, in predator-rich streams the spots are smaller (Chapter 4). Females from streams with large-spotted males have stronger preferences for males with large orange spots (Houde 1988) than do females from streams with

Choice for elaborate traits does not discriminate between the hypotheses

Covariance of male display and female preference

Size of male orange spots
and female preference in
guppies

Disease resistance and
displays

The two hypotheses are
not incompatible

small-spotted males. Furthermore, the differences between populations in both male sexual colour pattern and female preference are genetic: they persist in the laboratory for many generations when offspring are reared in standard conditions. Thus, in guppies, there is a genetic correlation between male display and female preference, which is essential for the Fisher process to operate. The fact that the differences persist under standard laboratory conditions suggests that the expression of the spot size does not depend on, for example, ability to gather food or on disease resistance, but as already pointed out, it would be difficult to prove that the trait is totally unrelated to viability. Therefore the guppy results, whilst consistent with Fisher's hypothesis, cannot rule out the handicap hypothesis. In contrast to the genetic differences between streams in spot size, the brightness of the orange pigment in the spots may be influenced by the environment (Kodric Brown 1989; Houde & Torio 1992).

Tests of the handicap hypothesis have mainly focused on the version proposed by Hamilton and Zuk, that male displays reveal their degree of resistance to parasites (Milinski & Bakker 1990; Clayton 1991). However, most of these studies are incomplete tests: they do not test the preference for the trait, the correlation between expression of the trait and disease resistance, and the heritability of resistance. The most detailed study to date to meet all these criteria is that of Møller on swallows (Box 8.2).

To summarize, there is now convincing evidence that females in a variety of species prefer extreme male displays. In some species, such as the guppy, we know that variation in the trait and the preference is heritable. In others, such as the swallow, we know that variation in the trait is also correlated with another aspect of fitness, as in the handicap hypothesis. We do not yet, however, have sufficient evidence to say which of the two hypotheses discussed in this section is more generally applicable. In fact, they are not necessarily incompatible with one another (Iwasa et al. 1991). Given a genetic correlation between preference and the trait, Fisher's process has the potential to operate even if the trait is also a handicap.

MALE INVESTMENT

We have so far assumed that females are investors (i.e. have low potential reproductive rate) and males are competitors. While this picture describes most animal species, there are exceptions. In many birds, some amphibians, and arthropods, both male and female invest about equally in the eggs or young by feeding, guarding or brooding.

Box 8.2 *A case study: the tail of the swallow.*

Testing the Hamilton–Zuk hypothesis is by no means straightforward. For example, it is not sufficient to show merely that females prefer males with lower parasite burdens. They may do this not because they are shopping for good genes for their offspring but simply because they want to avoid infection during the act of mating, or because they want a partner able to provide efficient parental care (heavily parasitized males may be debilitated). The four key assumptions that need testing are: (a) parasites reduce host fitness; (b) parasite resistance is genetic; (c) parasite resistance is signalled by the elaboration of sexual ornaments; (d) females prefer males with the most elaborate signals. All four factors have been demonstrated in a detailed field study of the barn swallow, *Hirundo rustica*, in Denmark by Anders Pape Møller (1988, 1989, 1990).

The barn swallow is a monogamous insectivorous bird which feeds on the wing and often nests in colonies in farm buildings. There is little difference between the sexes, except that males have more elongated outer tail feathers which they display, either in the air or while perched, in an attempt to attract a female (Fig. a). Although sexual selection is

(a) Male barn swallow, in flight, showing elongated outer tail feathers. Courtesy FLPA. Photo by Hugh Clark.

expected to be less intense in monogamous species than in polygynous species like the widow birds (see above), there will still be competition between males for mates because there is an advantage to pair up early. This leads to greater reproductive success not only because there is often more food available early in the season but also because early breeding increases the chance of raising several broods in the year. Males compete, therefore, to pair up as early as possible.

First of all, Møller showed that females preferred males with longer tails. Males with experimentally elongated tails paired up more quickly and were also preferred by females seeking extra-pair matings (Fig. b). As a result of pairing early, males with elongated tails were more likely to have two broods in the season and so enjoyed greater reproductive success. Why then do males not grow such extra-long tails naturally? The answer is that there is a cost. Males with experimentally elongated tails were handicapped in their foraging; they caught smaller, less profitable, prey and grew poorer quality feathers and shorter tails at the next moult, probably as a result of food deficiency. As a result, they were slower to attract a mate the following year and suffered reduced reproductive success.

(b)

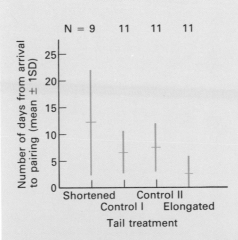

(b) Male barn swallows with experimentally elongated tails pair up sooner than controls (I, feathers cut and re-glued; II, unmanipulated), and males with shortened tails pair up last of all. From Møller (1988).

Continued on p. 200

Box 8.2 *Continued*

Why do females prefer males with longer tails? Could the ornament signal a male's genetic quality in terms of his ability to resist parasites? The most obvious parasite on swallows is a blood-sucking mite, *Ornithonyssus bursa*, which infects both adults and nestlings. The life cycle of the mite, from egg to adult, lasts just 5−7 days so one reproductive cycle of the swallow provides time for 8−10 generations of mites. This means that numbers of mites in a nest can build up rapidly and the maximum recorded was 14 000! Møller showed that nestlings reared in nests with lots of mites were lighter and smaller and suffered increased mortality. Experiments in which mites were either added or removed (by spraying with pyrethrin solution) confirmed that mites were the cause of the reduced growth. The precise cause of the harm is not yet known; birds may have suffered simply from loss of blood or the mites could have been vectors of blood parasites such as haematozoa or viruses.

There was large variation in the population in the degree of parasite infection. To test whether parasite resistance was heritable, Møller exchanged half the nestlings between pairs of nests soon after hatching. He found that a nestling's parasite burden was correlated with that of its parents, even when the nestling was reared in another nest (Fig. c), but not with that of its foster parents. Thus the genetic origin

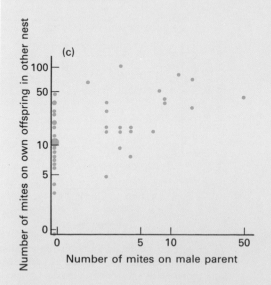

(c) Cross-fostering experiments showed that offspring parasite loads were correlated with those of the male parent.

of an offspring was a predictor of its parasite levels, not the site of rearing. This suggests that parasite resistance is partly genetic.

And now finally to the link with the swallow's tail. In the cross-fostering experiments Møller found that parents with longer tail ornaments had offspring with smaller mite loads, even when their offspring were raised in another nest. This relationship was strongest for male parent tail length and offspring mite loads (Fig. d). This suggests that the length of a male's tail signals his degree of parasite resistance. In conclusion, the female's preference for the male ornament makes good sense under the Hamilton–Zuk hypothesis that they are choosing males able to pass on 'good genes' to the offspring.

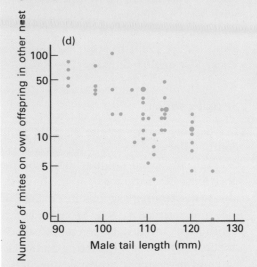

(d) Males with longer tails fathered offspring with lower parasite loads. From Møller (1990).

Sometimes the usual sex roles are completely reversed so that males do the investing and females the competing (Chapter 9). The ideas about sexual conflict and sexual selection can still be applied in modified form to species with equal or primarily male investment. When both sexes care equally for the offspring, for example, courtship may involve assessment and choice by males as well as by females. Males of species with internal fertilization can never be absolutely sure that they have fathered the children of their mate, and one role of courtship may be as an insurance against cuckoldry. A prediction of this idea is that courtship

Courtship as an insurance against cuckoldry

allows males to assess whether or not females have previously mated with others. This was tested by Erickson and Zenone (1976). They found that male barbary doves (*Streptopelia risoria*) attack a female instead of courting her if she performs the 'bow posture' (an advanced stage of courtship) too quickly. Since the females which responded in this way had been pre-treated by allowing them to court with another male, the reaction of the test males in rejecting eager females is adaptive if courtship plays a role in assessing certainty of paternity, before investing in offspring. It would not have been predicted by the older view that male courtship serves to sexually arouse the female!

When males invest, they are choosy

In species with high male investment, females tend to be the competitive sex and males may be choosy. In moorhens (*Gallinula chloropus*) males do almost three-quarters of the incubation and females play an active role in competing for the chance to mate with good incubators. These ideal husbands are small and fat: well equipped to survive on their reserves during long incubation stints (Petrie 1983). In other species, investing males may actually reject low quality females (Fig. 8.9).

Sexual conflict

Let us now return to the starting point of this chapter, sexual conflict. Recall the view of the origin of anisogamy as the primeval example of sexual conflict. The conflict was one about mating decisions. Macrogametes might have done better had they been able to discriminate against microgametes, but in the evolutionary race microgametes won. Similar, but more directly observable, conflicts of interest between the sexes are still apparent today, not only with respect to mating decisions but also in the contexts of parental investment, multiple matings and infanticide.

(a) Mating decisions

As we have emphasized earlier in the chapter, females have more to lose and therefore tend to be choosier than males. Thus for a given encounter it will often be the case that males are favoured if they do mate and females if they do not (Parker 1979). An extreme manifestation of this conflict is enforced copulation as exemplified by scorpionflies (*Panorpa* spp.). Male scorpionflies usually acquire a mate by presenting her with a nuptial gift in the form of a special salivary secretion or a dead insect (this is very similar to the *Hylobittacus* described earlier). The female feeds on the gift during copulation and turns the food into eggs. However, sometimes a male enforces copulation: he grasps her

Enforced copulation in scorpionflies

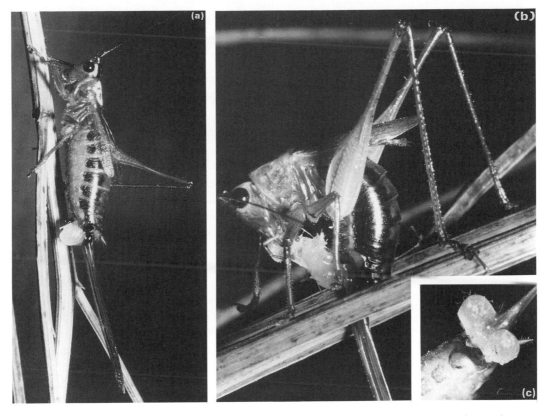

Fig. 8.9 A female katydid (bush cricket, *Conocephalus nigropleurum*). In some species of crickets the male produces a huge, protein-rich spermatophore (it may weigh up to 27 per cent of the male's body weight) which is eaten by the female and used to make eggs. With this large investment the male can afford to be choosy. Females prefer large males because they make large spermatophores while males reject small females because they are capable of laying fewer eggs than are large females. (a) Mated female with spermatophore. (b) Female eating spermatophore. (c) Close up of spermatophore. From Gwynne (1982).

with a special abdominal organ (the notal organ) without offering a gift (Thornhill 1980). Enforced copulation appears to be a case of sexual conflict. The female loses because she obtains no food for her eggs and has to search for food herself, while the male benefits because he avoids the risky business of finding a nuptial gift. Scorpion flies feed on insects in spiders' webs and quite often get caught up in the web themselves, so foraging is certainly risky (65 per cent of adults die this way). Why do not all males enforce copulations? The exact balance of costs and benefits is not known, but it appears that it results in a very low success rate in fertilizing females, so perhaps males adopt this strategy only when they cannot find prey or make enough saliva to attract a female.

(b) Parental investment

This is a topic to which we will return in the next chapter. Here it is sufficient to note that in species with investment beyond the gamete stage, each sex might be expected to exploit the other by reducing its own share in the investment. The outcome of this sexual conflict may depend on practical considerations such as which sex is the first to be in a position to desert the other. When fertilization is internal, for example, a male has the possibility of deserting the female immediately after fertilization and leaving her with the eggs or young to care for.

(c) Infanticide

Infanticide may benefit males and not females

As we saw in Chapter 1, male lions may slaughter the cubs in a pride shortly after they take over as group leaders. This behaviour (which is also seen in some primates) probably increases male reproductive success, as explained in Chapter 1, and clearly decreases female success. This seems to be a case of sexual conflict in which the males have won, but it is perhaps surprising that females have not evolved counter-adaptations. They could, for example, eat their own young once they have been killed in order to recoup as much as possible of their losses (Chapter 12).

(d) Multiple matings

As Bateman's experiments with *Drosophila* showed (p. 177) females may often gain little by mating with more than one male (but see p. 229). However, because of sperm competition males may gain by mating with already fertilized females. Multiple matings are likely to be costly to the female at the same time as being advantageous to the male. This is dramatically illustrated by the dungflies described in Chapter 7. When two males struggle for possession of a female, the female is sometimes drowned in cowdung by the fighting males on top of her!

Conflicts of interest between the sexes will lead to an evolutionary race of the sort envisaged by Parker *et al.* for sperms and eggs. There is no simple answer to the question 'Which sex is more likely to win the chase?' As we discussed earlier, factors such as the strength of selection and the amount of genetic variation will determine how fast the two sexes can evolve adaptation and counter-adaptation, but it is not possible to make any more specific statements about the outcome of sexual conflict races.

The significance of courtship

As we have mentioned earlier in the chapter, some aspects of courtship behaviour can be interpreted in terms of sexual conflict and sexual selection. However this is not true of all courtship signals: many are designed for species identification, and here the interests of the two sexes are similar because both benefit by mating with a member of the same species. Some of the clearest examples of this role of courtship come from studies of frog calls.

Reproductive isolation

When several species of frogs live in the same pond, each has a characteristic and distinct mating call given by the male, and females are attracted only to calls of their own species. In some frogs (e.g. the cricket frog *Acris crepitans*) it has been shown that the female's selectivity of response results from the fact that the auditory system is tuned to the particular frequencies in the male call (Capranica *et al.* 1973).

Courtship displays may also play a role in competition between males within a species for mating opportunities. Often the same displays simultaneously serve to repel other males and attract females. An example for which this has been demonstrated experimentally is the mating call of the Pacific tree frog (*Hyla regilla*) (Whitney & Krebs 1975a,b). Males are repelled and females attracted by loudspeakers broadcasting the mating call and females select out of a group of loudspeakers the one which calls for the longest bouts. Females may choose between displays purely on the basis of sexual attractiveness, as explained by Fisher's theory of sexual selection, but there is also the possibility that differences in courtship between males may indicate habitat quality, for example males with territories containing a lot of food might be able to afford to spend more time displaying.

Intra-sexual competition

Assessment

A third role of courtship to which we have referred is assessment. In a species with male parental care females may assess the ability of the male to look after young and males may assess whether a female has previously been fertilized. Early work by ethologists on birds and fish showed that at the beginning of courtship males are often aggressive and females are coy or reluctant. Courtship was seen, therefore, as serving to synchronize sexual arousal of the partners. A possible explanation of *why* it should be necessary to overcome aggression and reluctance is that the early phases involve assessment by both partners before investing in offspring.

Throughout this chapter we have emphasized the role of

females as investors in the zygote and offspring, but we have also mentioned that sometimes males invest as much as or more than females. Why should this happen in some species but not in others? In order to answer this question we will turn in the next chapter to the influence of ecological pressures.

Summary

Conflict lies at the heart of sexual reproduction. The fundamental difference between male and female is the size of gametes. Males produce tiny gametes and can be viewed as successful parasites of large female gametes. Because sperm is cheap, males can increase their reproductive success by mating with many females. Females can only increase their success by making eggs or young at a faster rate. Females are a scarce resource for which males compete and much of male courtship can be understood in terms of competition for matings. Females may be reluctant to mate unless they can choose partners with resources or genetic benefits. Sometimes the general rule of high female investment is reversed and males are the main investors: here females are competitive and males are choosy.

The two major hypotheses for how genetic benefits may be gained by sexual selection are (a) Fisher's hypothesis that the benefits are purely aesthetic (genetically attractive sons) and (b) the handicap hypothesis that the benefits are to do with general viability, for example disease resistance.

Further reading

Malte Andersson's (1993) book is the definitive account of sexual selection. Clayton (1991) reviews the evidence for Hamilton and Zuk's hypothesis and Gibson and Höglund (1992) discuss the idea that females in leks (Chapter 9) may copy each other's choice of mate. Charnov's (1982) book is the definitive work on sex allocation, including sex ratio theory.

The papers by Burley et al. (1982) and Burley (1986a,b) present a remarkable set of data on mate choice in a monogamous species of bird, the zebra finch. Burley discovered that the coloured leg rings placed on the birds for individual identification in her captive colony influenced mating success. Males with red rings and females with black ones had a higher reproductive success than those with some other colours, green and blue being especially unattractive rings on males and females respectively. It appears that individuals with attractive rings get partners who are willing to do more parental work in return for mating with a desirable member of the opposite sex.

A study of a species with male investment is that of Smith (1979) on a giant water bug, in which the male carries the eggs on his back. To ensure paternity of the eggs he carries, the male repeatedly inseminates the female during egg laying, the record being 100 copulations in 36 hours.

McKinney *et al.* (1983) give an excellent review of forced copulation in waterfowl.

Topics for discussion

1 How might sexual selection have given rise to (a) antlers and (b) the peacock's tail?

2 What hypotheses might account for bright coloration in birds (see also Baker & Parker 1979, Lyon & Montgomery 1985 and references therein).

3 Why do males usually compete for females rather than vice versa? (Include in your discussion the concepts of parental investment, operational sex ratio and potential rate of reproduction.)

4 Is it possible to discriminate with empirical data between Fisher's hypothesis and the handicap hypothesis?

Chapter 9. Parental Care and Mating Systems

In the last chapter we explored the fundamental differences between males and females. A male has the potential to father offspring at a faster rate than a female can produce them. Therefore females are a limiting resource for male reproductive success and males are expected to compete to maximize their number of matings. On the other hand, because females put a lot of investment into the zygote we expect them to resist male ardour and to be selective in their choice of mate.

However, care for the zygote does not end with gamete investment. In many animals there is some form of parental care, for example guarding the eggs or feeding the young. Parental care may be done by both parents (e.g. starling, *Sturnus vulgaris*), by the female alone (e.g. red deer) or by the male alone (e.g. sea horse, *Hippocampus*, where the male carries the fertilized eggs around in his brood pouch).

Variation in mating systems

Often, differences between species in parental care are associated with differences in mating system. Mating systems can be categorized under the following broad headings.

1 *Monogamy.* A male and a female form a pair bond, either short or long term (part of or a whole breeding season or even a lifetime). Often both parents care for the eggs and young.

2 *Polygyny.* A male mates with several females, while females each mate with only one male. A male may associate with several females at once (simultaneous polygyny) or in succession (successive polygyny). With polygyny it is usually the female that provides the parental care.

3 *Polyandry.* This is exactly the reverse of polygyny. A female either associates with several males at once (simultaneous polyandry) or in succession (successive polyandry). In this case, it is often the male who does most parental care.

4 *Promiscuity.* Both male and female mate several times with different individuals, so there is a mixture of polygyny and polyandry. Either sex may care for the eggs or young. *Polygamy* is often used as a general term for when an individual of either sex has more than one mate.

These divisions are not hard and fast but they will help as a general guideline.

The ideal way for a male to maximize his reproductive success may be to go round copulating with lots of females, each of which stays at home to care for his offspring. For a female, the ideal may be to mate and then leave the male to care for the offspring while she forms reserves for more eggs. In practice, as we shall now see, there are two factors that will influence how the sexes resolve this conflict. First, different groups of animals have different physiological and life history constraints which may predispose one sex to do more parental care than the other. Second, ecological factors will influence the costs and benefits associated with parental care and mating behaviour.

Proximate constraints on parental care

These can be illustrated by contrasting three classes of vertebrates, fish, birds and mammals (Table 9.1). Are there any basic differences in the physiology and life histories of these three groups which could account for the differences in parental care and mating system? While considering this question we must bear in mind that both male and female will be selected to maximize their reproductive success and may behave at the expense of the other sex (Trivers 1972).

BIRDS

Birds often have biparental care

As we saw in the first chapter, reproductive success in birds can be limited by the rate at which food is delivered to the nest. At least in those species where parental feeding of the young is important, we could imagine that two parents will be able to feed twice as many young as a single parent. Therefore both male and female increase their reproductive success by staying together. If either sex deserted, they would approximately halve the output of the last brood and would also have to spend time searching for a new mate and nest site before being able to start again. So monogamy and parental care by both male and female are not hard to understand. In some seabirds like the kittiwake, *Rissa tridactyla* (Coulson 1966) and the Manx shearwater, *Puffinus puffinus* (Brooke 1978), long-term mate fidelity seems to pay

Table 9.1 Frequent types of parental care and mating system in three classes of vertebrates. These are very broad generalizations and many exceptions are discussed later in the chapter

	Parental care	Mating system
Birds	Both male and female	Monogamous
Mammals	Female only	Polygynous
Fish	Male only	Polygamy/ promiscuity

because pairs that have bred together previously have higher reproductive success than new pairs.

When the constraint of having to have two parents feeding the young is lifted, it is usually the male who deserts and the female who is left caring for the brood. Comparative evidence shows that polygyny often occurs in fruit and seed eaters, probably because these food supplies become so seasonally abundant that one parent can feed the young almost as efficiently as two (e.g. weaver birds, Chapter 2). Why is it the male who deserts? There are two factors which may be important. First, the male has the opportunity to desert before the female. With internal fertilization, she is left literally holding the babies inside her. Second, the male can gain more by desertion than the female because his lifetime reproductive success depends more on his number of matings (see Chapter 8).

MAMMALS

In mammals, females usually care for the young

In mammals, females are even more predisposed to care for the young. The offspring often have a prolonged period of gestation inside the female, during which the male can do little direct care (though he can protect and feed the female). Once the young are born they are fed on milk and only the female lactates. Because of these constraints on the opportunity to care for offspring and also because, with internal fertilization, the male can desert first, it is not surprising that most mammals have a polygynous mating system and parental care by the female alone.

Monogamy and biparental care occurs in a few species where the male contributes to feeding (carnivores) or to carrying the young (e.g. marmosets). Perhaps it is surprising that male lactation has not evolved in these cases (see Daly 1979).

FISH

In fish, either sex may care

In the bony fish (teleosts), most families (79 per cent) have no parental care (Gross & Sargent 1985). In those families which do care for the eggs or young, it is usually done by one parent; biparental care occurs in less than 25 per cent of the families which show care. Compared with the elaborate care of offspring by birds, parental care in fish is a simple affair often consisting of just guarding or fanning eggs. These tasks can usually be done effectively by one parent alone. Which parent will provide care? Table 9.2 shows that female care is commonest with internal fertilization (86 per cent female care) and male care with external fertilization (70 per cent male care). The overall predominance of

Table 9.2 Distribution of male and female parental care with respect to mode of fertilization in teleost fishes. The table shows number of families; a single family may appear in more than one category, but is not listed under 'no parental care' unless care is completely unknown in the family. From Gross and Shine (1981)

	Internal fertilization	External fertilization
Male parental care	2	61
Female parental care	14	24
No parental care	5	100

Male care is commonest with external fertilization and female care with internal fertilization: Three hypotheses

male parental care in fish is related to the prevalence of external fertilization.

Three hypotheses can be proposed to explain this result (see Gross & Shine 1981).

Hypothesis 1.   Paternity certainty

Trivers (1972) suggested that reliability of paternity will be affected by mode of fertilization. Since external fertilization occurs at the time of oviposition, reliability of paternity may be greater than with internal fertilization where sperm competition may take place inside the female's reproductive tract (Chapter 8). According to this hypothesis, with internal fertilization, therefore, a male should be less prepared to provide parental care because he is less certain than the female of genetic representation in the offspring. However, paternity certainty alone should not influence a male's decision whether to care or desert because he will presumably be equally uncertain of his paternity if he deserts and mates again. A male will only gain by desertion if this increases his reproductive output above that which he would achieve by staying. Furthermore, it may not necessarily be true that external fertilization leads to greater paternity certainty. In some cases, for example sunfishes *Lepomis*, cuckoldry takes place during oviposition (see Chapter 10).

Hypothesis 2.   Order of gamete release

Dawkins and Carlisle (1976) suggested that, as with birds and mammals, internal fertilization gives the male the chance to desert first and thus leave the female to care. With external fertilization they suggested that the roles may be reversed. Because sperm are lighter than eggs the male must wait until the eggs are laid before he can fertilize them, or else his gametes will float away. Therefore, the female has the opportunity to desert first and swim off while the male is still fertilizing the eggs! This is an ingenious idea but it must be rejected on empirical grounds. In fact, the most common pattern of gamete release is simultaneous release by male and female. In these cases both sexes have an equal chance to desert, but 36 out of 46 species which have simultaneous gamete release and monoparental care, have care provided by the male. Secondly, in some families of fish (Callichthyidae, Belontiidae) the male builds a foam nest and releases sperm before the female lays eggs. In these cases, the 'opportunity for desertion' hypothesis predicts that males can desert first, but nevertheless parental care is provided by the male. Therefore, male parental care remains correlated with external fertilization independently of the order of gamete release or opportunity to desert.

Hypothesis 3.   Association

Williams (1975) suggested simply that association with the embryos preadapts a sex for parental care. For example, with internal fertilization the female is most closely associated with the embryo and this may set the stage for the evolution of embryo retention and live birth, followed by care of the young fry. With external fertilization, on the other hand, the eggs are often laid in a male's territory and it is the male who is most closely associated with the embryos. Defence of the territory in order to attract further females becomes, incidently, defence of the eggs and young, and therefore provides a preadaptation for more elaborate parental care by males. This hypothesis is the best predictor of the data in Table 9.2. Male parental care is more common in territorial species and the prevalence of male parental care with external fertilization results from the fact that male territoriality is particularly common with external fertilization.

Male care in fish is associated with male territoriality

ANCESTRAL ORIGIN OF UNIPARENTAL CARE

Another factor that may influence which sex cares for the zygote is the ancestral origin of uniparental care. In birds, the ancestral state is probably biparental care because in most species both male and female look after the eggs and young. This means that the female cannot use up all her resources in egg laying since she has to care for the young later on. It is well known that female birds can often readily lay replacement eggs if their first clutches fail. Therefore, if the male should desert, the female will still have some resources left to care for the brood.

Plate 4.1 (*right and below*) Brood parasitism in birds. (a) A female cuckoo parasitizing a reed warbler nest. First the female removes a host egg; holding it in her bill, she then lays her own. Her time at the nest is less than 10 seconds. (Photo by Ian Wyllie.) (b) The newly hatched cuckoo chick, still naked and blind, heaves the host eggs and young, one by one, out of the nest. In this case the host is a dunnock. (Courtesy Oxford Scientific Films. Photo by Dr J.A.L. Cooke.) (c) A sedge warbler host feeds the cuckoo chick. (Courtesy FLPA. Photo by Martin B. Withers.)

(a)

(b)

(c)

(a)

(b)

Plate 8.1 Sexual selection and elaborate traits. (a) Male satin bower birds (*Ptilonorhynchus violaceus*) build a bower consisting of two parallel rows of vertical fine twigs on a court of cleared ground. The male displays and attracts females to the bower, where mating takes place. The bower is decorated by the male with flowers, feathers, leaves, snake skins, snail shells and human debris such as pens and toothbrushes. Blue and yellow decorations are especially attractive to females, and males with more decorations get more matings. In part, the number of decorations reflects a male's ability to steal from other bowers (Borgia 1985). (Courtesy FLPA. Photo by L. Robinson.) (b) The fantastically elaborate plumes of birds such as this Emperor of Germany Bird of Paradise (*Paradisaea gulielmi*) may have evolved by sexual selection. Female choice for elaborate plumes may involve Fisher's runaway process and/or selection via the handicap process. (Courtesy FLPA. Photo by M.D. Mackay.)

[facing page 212]

Plate 9.1 Diversity of mammalian mating systems illustrated by ungulates.

Plate 9.1(a) The dik-dik (*Madoqua kirki*) is monogamous, probably because female ranges are sufficiently large for a male to be unable to defend more than one mate. (Courtesy Oxford Scientific Films. Photo by Zig Leszczynski.)

Plate 9.1(b) In species such as the impala (*Aepyceros melampus*) males are territorial but female groups wander over a wide area searching for food. Males defend herds of females temporarily during oestrus, when the group wanders through the male's territory. Here a male is preventing a group of three females from leaving his territory. (Courtesy Peter Jarman, photo by Martha Jarman.)

Plate 9.1(c) In the Uganda kob (*Kobus kob thomasi*) males defend tiny territories (15–30 m diameter) on leks. The male in the centre of this photo is mating with a female who has visited his territory. (Photo by James Deutsch.)

Plate 9.1(d) In the buffalo (*Syncerus caffer*), both males and females wander over wide areas in search of food. Several males associate with a large group of females and compete for matings within the multi-male group. (Courtesy Oxford Scientific Films. Photo by G.I. Barnard.)

Plate 9.2 Sex role reversal in birds. This male jacana, about to incubate a clutch of eggs, performs all the parental duties. Females compete for males by defending large territories. (Photo by N J Demong.)

Plate 12.1 A group of dwarf mongooses. A breeding group, typically of about nine individuals, consists of a breeding pair and helpers. Helpers may defend the den, often in a used termite mound, against predators and bring food to the young. Some helpers are related to the breeding pair, whilst others are not. The latter benefit by eventually gaining breeding status, whilst the former may gain both indirect and direct fitness benefits. (Courtesy Jan Terrl and Peter Waser.)

Plate 14.1(a) A mantis shrimp (*Odontodactylus brevirostris*) in its cavity. Mantis shrimps threaten intruders and may cause them to retreat. Normally the threat is backed up by the potential of a real attack, using the powerful, hard, forelimbs. After moult, when the forelimbs are soft, the mantis shrimp still uses the threat display, sometimes successfully, to deter intruders. (Courtesy R. Caldwell.)

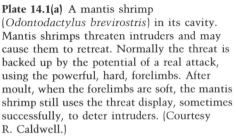

Plate 14.1(b) A stotting springbok (*Antidorcas marsupialis*). Stotting is a stylized leaping movement of some species of ungulate when approached by a predator. It may be a reliable signal to the predator of the body condition of the potential prey item. (Courtesy Oxford Scientific Films. Photo by Rafi Ben-Shahar.)

In fish, the ancestral state is probably no parental care. The female may therefore use up all her resources in egg laying and so cannot afford to guard. This may mean that it is the male who often has to look after the young if care is necessary. Furthermore, Gross and Sargent (1985) suggest that in fish, female fecundity usually increases more with size than does male fecundity. The predominance of male parental care may arise because the loss of growth that results from parental effort has less cost (in terms of decreased future reproductive success) for males than females. However, this may not always be the case (see Fig. 10.10).

An ESS model of parental investment

So far, we have seen that each sex is faced with the decision of whether to stay and care for the young or desert. Ecological factors will determine the costs and benefits of these two options and proximate constraints will also be important (e.g. female mammals lactate, males do not). However, it is also important to realize that the best decision for one sex will depend on the strategy adopted by the other sex. For example, if the female stays, it may pay the male to desert but if the female deserts it may pay the male to stay. Maynard Smith (1977a) has developed a useful model for considering what will happen in such cases of conflict.

The best strategy for one sex depends on the strategy adopted by the other sex

The model looks for a pair of strategies, say I_m for males and I_f for females, such that it would not pay a male to diverge from strategy I_m so long as females adopt I_f, and it would not pay females to diverge from I_f as long as males adopt I_m. In other words, we seek the evolutionarily stable strategies for male and female (see Chapter 7). Suppose that the reproductive success of a pair during the breeding season depends on the amount of parental care and on the number of eggs laid by the female. It is assumed that the more the female invests in eggs, the less she is able to invest in care and vice versa.

Let P_0, P_1 and P_2 be the probabilities of survival of eggs which are not cared for, cared for by one parent and cared for by two parents, respectively; $P_2 > P_1 > P_0$. A male who deserts has a chance p of mating again. A female who deserts lays W eggs and one who cares lays w eggs; $W > w$.

The pay-off matrix for this game is illustrated in Table 9.3. There are four possible ESSs:

ESS 1: female deserts and male deserts. This requires $WP_0 > wP_1$, or the female will care and $P_0 (1 + p) > P_1$, or the male will care.

Table 9.3 An ESS model of parental investment (Maynard Smith 1977a). Each sex has the possibility of caring or deserting. The matrix gives the reproductive success for males and females. See text for details

Male		Female	
		Care	Desert
Care	Female gets	wP_2	WP_1
	Male gets	wP_2	WP_1
Desert	Female gets	wP_1	WP_0
	Male gets	$wP_1 (1+p)$	$WP_0 (1+p)$

ESS 2: female deserts and male cares. This requires $WP_1 > wP_2$, or the female will care, and $P_1 > P_0 (1 + p)$, or the male will desert.

ESS 3: female cares and male deserts. This requires $wP_1 > WP_0$, or the female will desert, and $P_1 (1 + p) > P_2$, or the male will care.

ESS 4: female cares and male cares. This requires $wP_2 > WP_1$, or the female will desert, and $P_2 > P_1 (1 + p)$ or the male will desert.

An ESS model helps us to think about factors that might favour care by one or other sex

For given values of the parameters in the model, ESS 1 and ESS 4 can be alternative possibilities, as can ESS 2 and ESS 3. For example, ESS 2 is favoured if the female can lay many more eggs if she does not invest in caring ($W \gg w$) and if one parent is much better than none ($P_1 \gg P_0$) but two parents are not much better than one ($P_2 \simeq P_1$). This situation probably applies to many fish, as discussed above, where the female tends to desert and the male often cares. However, ESS 3 is an alternative possibility, especially if a male who deserts has a much better chance of mating again; this may apply to some species of birds and mammals (see below). If two parents can raise twice as many young as one ($P_2 \gg P_1$), or if the chance that a deserting parent will remate is small, then ESS 4 is the likely outcome, as in many species of passerine birds.

Maynard Smith's model is useful for identifying the factors which will lead to different patterns of parental investment and mating systems. In the rest of this chapter we will discuss how ecological factors, such as the distribution of resources and predators, and the availability of mates influence the kind of mating system which results.

Mating systems with no male parental care

We begin with cases where the male does not provide any parental

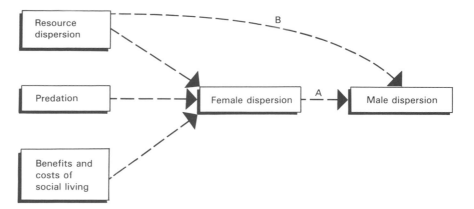

Fig. 9.1 The two-step process influencing mating systems in cases where males do not provide parental care. Because female reproductive success tends to be limited by resources, whereas male reproductive success tends to be limited by access to females, female dispersion is expected to depend primarily on resource dispersion (modified by predation and benefits and costs of social living), while male dispersion is expected to depend primarily on female dispersion. Males may compete for females directly (A) or indirectly (B), by anticipating how resources influence female dispersion and competing for resource-rich sites.

care. In theory the mating system should be easy to understand as the outcome of a two-step process (Fig. 9.1). First, female reproductive success will be limited most by access to resources (e.g. food, breeding sites) so the distribution of females should depend primarily on resource dispersion, modified by predation and the costs and benefits of associating with other individuals (see Chapter 6). On the other hand, male reproductive success will be limited more by access to females, so the second step in the sequence is that males should then distribute themselves in relation to female dispersion. Males could compete for females either directly (A in Fig. 9.1) or indirectly (B) by anticipating how resources will influence female dispersion and competing for resource-rich sites. Two recent studies provide experimental evidence for this scheme.

(a) Grey-sided voles, Clethrionomys rufocanus

Female dispersion is influenced by food, males are influenced by females

Ims (1987) showed that female dispersion was influenced by food; when food was provided in abundance at particular sites female ranges became smaller and overlapped in the resource-rich areas. The males also homed in on these sites. Was the change in male dispersion due to males following females or because they, independently, followed changes in resource distribution? To test this, Ims (1988) introduced a small population of voles on to a little wooded island in south-east Norway. In one experiment, females were kept individually in small cages and their positions were moved each day to simulate movement about a home range. When females were spaced out, free-ranging males (tracked by radiotelemetry) became dispersed, overlapping their ranges with the female ranges. When females were clumped, by placing cages close together, the males aggregated on the female clumps. By contrast, when males were then kept in individual cages the dispersion of free-ranging females was not affected by experimental changes in male dispersion. This study shows that the causal links are from resources to female dispersion and then from female dispersion to male dispersion, as in Fig. 9.1.

(b) Blue-headed wrasse, Thalassoma bifasciatum

Warner (1987, 1990) studied these coral reef fish in the Caribbean. Females spawned in favourite sites on the downcurrent edges of a reef where the pelagic eggs were swept quickly into the open sea and so avoided predation from other reef fish. Individual females returned almost every day to particular sites where they laid a few eggs. Males competed to defend territories at these preferred sites, with the largest males defending the best sites and so gaining most mates.

To assess the roles of the two sexes in determining spawning sites, Warner removed either all the breeding males or all the females from local isolated populations and replaced them with fish from other populations. When males were replaced, most of the spawning sites remained the same as before. By contrast, when females were replaced there were marked changes in sites used, even though some males initially continued to display at and defend the original sites. This neat experiment shows that females choose the spawning sites and males simply compete to defend sites which females prefer.

Experimental changes in female dispersion cause changes in male dispersion, but not vice versa

VARIATION IN MATING SYSTEMS

According to the scheme in Fig. 9.1, the variation in mating systems in cases where males do not provide parental care arises because of differences in resource dispersion, which will influence how females are distributed and hence their economic defendability by males (Emlen & Oring 1977). The economics of female defence will depend on their distribution in both space and time. When mates or resources are more patchily distributed in space there will be greater opportunities for polygyny (Fig. 9.2).

The key factor for determining temporal distribution of mates

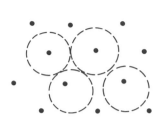

Even distribution.
Little polygamy potential

Patchy distribution.
High polygamy potential

Fig. 9.2 The influence of the spatial distribution of resources (food, nest sites) or mates on the ability of individuals to monopolize more than others. Dots are resources and circles are defended areas. With a patchy distribution of resources or mates there is greater potential for some individuals to 'grab more than their fair share'.

The operational sex ratio influences the intensity of competition for mates

is the 'operational sex ratio' (see Chapter 8) which is the ratio of receptive females to sexually active males at any one time. If all the females bred in synchrony, then with a real sex ratio of 1 : 1 in the population, the operational sex ratio at breeding would also be 1 : 1 and there would be little opportunity for a male to mate with more than one female because by the time he had mated once all the other females would have finished breeding. This applies, for example, to common toads (*Bufo bufo*) which are 'explosive breeders': all the females spawn in a week or so and a male has time to mate with one, or at most two, females before the breeding season is ended. Bullfrogs, *Rana catesbeiana*, by contrast, are 'prolonged breeders' with females arriving at the pond over several weeks. Males which can defend the best spawning sites may mate with up to six females in a season (Wells 1977).

A COMPARATIVE SURVEY OF MAMMALIAN MATING SYSTEMS

The influence of variation in resource and female dispersion on mating systems is well illustrated by mammals, where the economics of female monopolization by males is influenced by three main factors: female group size, female range size, and the seasonality of breeding (Plate 9.1, between pps 212–213). The following comparative survey is based on the review by Clutton-Brock (1989).

1 *Females solitary: range defensible by male.* In over 60 per cent of mammalian species females are solitary and a male defends a territory which overlaps one or more female ranges. If female

Different mating systems
in mammals arise from
variation in female home
range, group size and
movements

ranges are small relative to the area which a male can defend
then the male can be polygynous. If female ranges are larger, then
the male may only be able to defend one female, hence monogamy
(e.g. most rodents and nocturnal prosimians; Kleiman 1977).
Usually the male simply mates with the female then leaves
her to care for the offspring alone. More rarely (3 per cent of
mammalian species) the male may help defend the young against
predators (e.g. klipspringer, *Oreotragus oreotragus*) or carry them
(e.g. siamangs, marmosets, tamarins) or help to feed them (e.g.
jackals, wild dogs). Such obligate monogamy occurs where female
ranges are small enough for a male to defend but where a male is
unable to defend a large enough area to have more than one
female (Rutberg 1983). In these cases a male may then maximize
his reproductive success by providing parental care. Species with
obligate monogamy tend to have large litter sizes. For example
this mating system is common in canids, which have large litters,
and rare in felids, which have smaller litters. Marmosets, which
have male parental care, produce twins whereas most monkeys
produce only single offspring. If a male marmoset dies then the
female often deserts the young so male assistance with parental
care seems to be important.

2 *Females solitary: range not defensible by male.* Where females
wander more widely then males may rove over wide ranges,
associating with females temporarily while they are in oestrus.
This occurs in moose, *Alces alces* and orang-utans, *Pongo
pygmaeus*; in the latter species the females move over large
ranges following the fruiting seasons of different species of plants
(Mackinnon 1974).

3 *Females social: range defensible by male.* Where females
occur in small groups in a small range, then a single male may be
able to defend them as a permanent harem within his territory
(e.g. black and white colobus *Colobus guereza*, Hanuman langurs
Presbytis entellus). When a new male takes over the territory, he
often kills the young offspring fathered by the previous male,
thus bringing the female into oestrus sooner and hastening the
day he has a chance to sire his own young (Hrdy 1977). Where
females occur in larger groups several males (often relatives) may
defend the territory together (e.g. red colobus *Colobus badius*,
chimpanzees, lions). Joint defence by several males may increase
the length of tenure of a harem and may also be necessary for
economic defence of large groups of females wandering over a
large range (Bygott *et al.* 1979).

4 *Females social: range not defensible by male.* Sometimes
groups of females wander over ranges which are uneconomic for
one or more males to defend. The ways in which males compete

for females then depends on how predictable female group movements are in time and space.

(a) Daily female movements predictable

Sometimes the group of females wanders over a large range but they use regular routes to particular water holes or rich sources of food. In these places the males may defend small territories, much smaller than the females' range, and attempt to mate with them as they pass through (e.g. topi *Damaliscus lunatus korrigum*, Grevy's zebra *Equus grevii*). Such defence of mating territories may occur where more direct competition between males, such as fights for harems, would be costly because males are unable to build up the food stores necessary for them to engage in intense male—male interactions (Owen-Smith 1977).

More intense competition for territories may lead to the defence of tiny patches of ground in areas where females pass through. Male waterbuck (*Kobus elypsiprimnus*) defend territories of $0.25-0.5 \, \text{km}^2$ and mate with the females as they pass through the territory in small groups. Puku (*K. vardini*) have smaller territories of $0.1 \, \text{km}^2$ and finally lechwe (*K. lechwe*) and Uganda kob (*K. kob thomasi*) have tiny territories which may be no more than 15 to 30 m across. In these two species the males often aggregate into leks (see later). All four of these species, in the genus *Kobus*, inhabit marshes where females may come to particular areas to feed on rich aquatic vegetation. In the lechwe and Uganda kob, females move in large herds and males may only be able to economically defend very small territories (Leuthold 1966). These examples of ungulate leks may simply be extensions, therefore, of resource defence polygyny, with territories being small because of the high costs of defence.

Sometimes males wait for females (margin note)

(b) Daily female movements not predictable

Here males tend to follow the females, rather than waiting for the females to come to them. Where females live in small groups males may rove and associate with individuals in oestrus (e.g. mountain sheep *Ovis canadensis*, elephants *Loxodonta africana*). Where female groups are larger, the males may attempt to defend harems.

1 *Seasonal harems.* If all the females come into oestrus at a particular season then it may pay a male to put on energy reserves to enable him to have a burst of energy expenditure on harem defence. For example, male red deer (*Cervus elaphus*) stags compete to defend harems during the one month in which all females

Sometimes males follow females (margin note)

come into oestrus. A male's reproductive success depends on his harem size and the length of time for which he can defend the harem and this in turn depends on his body size and fighting ability. After the mating season the males are reduced to very poor body condition and are literally 'rutted out'! (Clutton-Brock et al. 1979).

As another example, female Northern elephant seals (*Mirounga angustirostris*) haul up on beaches to drop their pups and mate again for the production of next year's offspring. Because the females are grouped, due to the localized nature of the breeding grounds, they are a defendable resource and the males fight with each other to monopolize them. The largest and strongest males win the biggest harems and in any one year all the matings are performed by just a few males (Le Boeuf 1972, 1974; Cox & Le Boeuf 1977).

Harem defence

To be a harem master is so exhausting that a male usually only manages to be top ranking for a year or two before he dies. In the process of defending his harem against other males he sometimes tramples on his females' new-born pups. Although this is obviously not in the females' interests, these pups were probably not sired by the male himself because he is unlikely to have been a harem master the previous year. From the male's point of view, therefore, there is little cost in damaging or even killing the pups; his main concern is to protect his paternity.

2 *Permanent harems.* Where females do not all come into oestrus at a particular time, males may defend permanent harems for the whole duration of their reproductive life (e.g. hamadryas baboons, *Papio hamadryas* and gelada baboons, *Theropithecus gelada*, Dunbar 1984; Burchell's zebra, *Equus burchelli*, Rubenstein 1986). Often several groups (male plus harem) go around together forming a large 'super-group'. Where females go around in still larger groups, several males may associate with large groups of females and compete with each other for matings. Such 'multi-male' groups occur in buffalo, *Syncerus caffer* and olive baboons, *Papio anubis* (Altmann 1974).

LEKS AND CHORUSES

Leks are aggregations of males on small mating territories

In the examples discussed so far, males compete for females directly (female defence) or indirectly by defending resources to which females are attracted (resource defence). In some cases, by contrast, males aggregate into groups and each male defends a tiny mating territory containing no resources at all — often the territory is no more than a bare patch of ground just a few metres across. The males put a great deal of effort into defending their territories and advertise themselves to females with elaborate visual, acoustic or olfactory displays. In these mating systems, known as leks, females often visit several males before copulating and appear to be very selective in their choice of mate. Mating success is strongly skewed, with the majority of matings performed by a small proportion of males on the lek (Fig. 9.3).

Leks have been reported for seven species of mammals — the walrus, hammer-headed bat and five ungulates — and some 35 species of birds including three shorebirds, six grouse, four hummingbirds, two cotingas, eight manakins, eight birds of paradise, the kakapo and great bustard (Oring 1982). This breeding system is therefore not common. Similar systems occur in some frogs (Wells 1977) and insects (Thornhill & Alcock 1983), where females visit male choruses, choose a mate and then lay eggs away from the display site.

Leks may occur when neither females nor resources can be defended economically

It has been suggested that leks occur when males are unable to defend economically either the females themselves or the resources they require (Bradbury 1977; Emlen & Oring 1977). This may arise where females exploit widely dispersed resources and so have large, undefendable ranges, or because high population density, and thus high rates of interference between males, precludes economic female or resource defence. Thus in both antelope and grouse, the lekking species are those with the largest female home ranges (Bradbury et al. 1986; Clutton-Brock 1989) and in Uganda kob, topi and fallow deer, males lek at high population density but defend resource-based territories or harems at low density, where defence of females is presumably more economic.

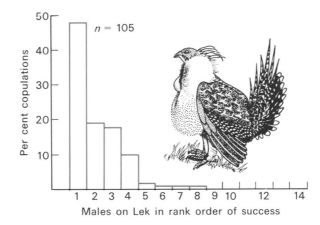

Fig. 9.3 In sage grouse (*Centrocercus urophasianus*) leks almost all of the copulations are performed by just a few of the males. Gibson et al. (1991) have shown that male display characteristics are an important predictor of mating success. From Wiley (1973).

Why do the males aggregate into leks? Four main hypotheses have been proposed (Bradbury & Gibson 1983).

Leks may occur where females are abundant

1 *Males aggregate on 'hotspots'.* Male aggregations may be explained by the familiar scheme in Fig. 9.1, with males settling in areas where female encounter rate is particularly high (hotspots). This seems to apply to the lechwe and Uganda kob discussed earlier. In many cases, however, females only visit lek sites in order to mate and so the males are not simply settling in areas the females would visit in the course of their normal daily routine. Furthermore, males are often aggregated far more closely than would be expected from settlement 'on top' of the female pattern of dispersion.

Avoiding predation

2 *Males aggregate to reduce predation.* In the neotropical frog *Physalaemus pustulosus*, calling males suffer heavy predation by bats, *Trachops* spp., which home in on the male calls. Calling males are safer in larger choruses because of the dilution effect (Fig. 9.4a; Chapter 6). However this is unlikely to be a general explanation of lekking; in many bird leks predation pressure seems to be extremely low.

3 *Males aggregate to increase female attraction.* Males may gain from 'stimulus pooling'; by displaying together they may provide a greater attraction for females and draw in mates from a larger distance. To explain the male aggregation the increase in female attraction would have to be marked, so that pay-offs per individual male increased with lek size. Figure 9.4(b) suggests this is true for *Physalaemus* but the relationship may merely reflect the fact that larger choruses form in areas where females are more abundant anyway. An experimental approach is needed to test whether larger choruses *cause* greater female attraction. Even so, *average* success per male may not be what we really need to measure because individuals may vary in their ability to signal. If some males had particularly effective displays ('hotshots'), it could pay poorer signallers to cluster around them to parasitize their attractiveness (Beehler & Foster 1988).

Synergistic displays

This process certainly seems to explain male aggregation on a small scale, such as that involving calling and satellite male toads (see Chapter 10) but two sources of evidence suggest that it cannot explain the larger scale aggregations of leks. First, when the most successful males are removed from a lek, their territories are quickly taken-over by other males (sharp-tailed grouse, Rippin & Boag 1974; white-bearded manakins, Lill 1974). This suggests that there is something about the site which influences female preference. The 'hotshot model' predicts that the next most preferred male would remain on his territory with the male aggregation rearranging around him, rather than for replacement to

In fallow deer, females choose males and not sites

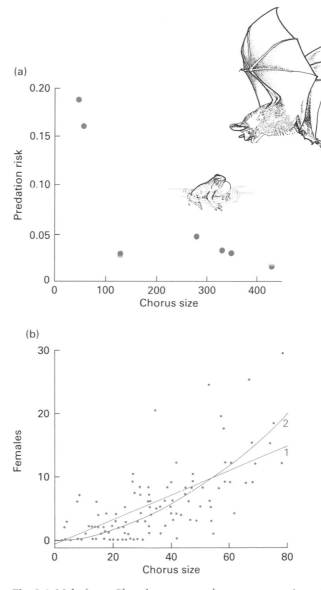

Fig. 9.4 Male frogs, *Physalaemus pustulosus*, aggregate into choruses. In larger choruses individuals are safer from predatory bats (a). The number of females attracted also increases with chorus size (b). The curve (2) gives a better fit to the observed points than a straight line (1) which suggests that the number of females per male increases with chorus size. From Ryan *et al.* (1981).

occur on particular sites. Second, in an experiment on fallow deer (*Dama dama*) leks, Clutton-Brock *et al.* (1989) covered the territories of the most successful males with black polythene, so forcing them to change site. Even though these males set up new territories several hundred metres away, they remained favoured by females. In this case, therefore, females were apparently choosing particular males rather than particular sites. The hotshot model predicts that the other males should have followed the movements of these attractive males to set up a lek at the new site. However, most remained on their old territories. Thus the hotshot model may not explain male aggregation on leks even when females are choosing males rather than particular sites.

4 *Males aggregate because females prefer particular sites or male aggregations themselves for mating.* Although lek mating systems are rare, they have attracted considerable attention because of the fact that females seem to exercise careful choice before they mate, with particular males gaining most of the matings. Because males of lekking species do not provide any parental care, the only possible gains from such choice are either a safe mating (safety from male harassment or predation) or genetic benefits (see Chapter 8). It is still not clear whether one or both of these benefits is important. For example, female black grouse (*Lyrurus tetrix*) prefer the males with the most vigorous displays and these chosen males also have higher survival (Alatalo *et al.* 1991). One possibility, therefore, is that leks provide a testing ground where males reveal their health and viability through the vigour of their displays; if there is any heritability of male viability then females will gain good genes for their offspring by mating with the most vigorous males. The best way to test this idea would be by artificially inseminating females with sperm from different males to see whether offspring sired by males with the best displays had better survival. This experiment certainly needs doing because the alternative possibility is that females simply gain a safer copulation by choosing to mate with the most vigorous male because he is better able to keep competitors at bay during the act of mating.

Whichever type of benefit turns out to be most important, female choice could cause the male aggregation either because choice of a particular site enables females to gain matings with the most vigorous male (the one able to win that site), or because preference for larger male aggregations facilitates comparison among males.

In conclusion, the ecological factors leading to lek mating systems are still not clearly understood. All four of the hypotheses above may be important, with different explanations applying to different species. Recent measures of paternity using DNA fingerprinting (see later) have revealed wide differences in the behaviour of lekking species. For example, in the black grouse, observations show that most females mate with just one male (the dominant

Females may gain genetic benefits by choosing on a lek . . .

. . . or they may just gain a safe mating

DNA fingerprinting reveals male success

male on the lek) and the paternity analysis confirmed that he fathered all the offspring. By contrast, female ruff *Philomachus pugnax* (a shorebird) often mate with several males, including males off the lek, and the fingerprinting revealed many broods with multiple paternity (Terry Burke, in prep.). The reasons for these differences are unknown, but they suggest that it may be unwise to seek a single common explanation for all lek systems.

Mating systems with male parental care

Where males provide parental care the males themselves become a resource which may influence female dispersion, so the simple scheme in Fig. 9.1 no longer applies. As we saw at the beginning of the chapter, male parental care is particularly common in birds, so we shall mainly use examples from birds to illustrate the ideas in this section.

MONOGAMY

Sometimes it pays both male and female to breed as a monogamous pair

David Lack (1968) suggested that monogamy is the predominant mating system in birds (90 per cent of species) because 'each male and each female will, on average, leave most descendants if they share in raising a brood'. This hypothesis certainly explains obligate monogamy in many seabirds and birds of prey, where male and female share incubation or where males feed females on the nest, and where both sexes are essential for chick-feeding. In these species, the death or removal of one partner leads to complete breeding failure.

However Lack's hypothesis does not explain the monogamous mating systems of many songbirds, where both sexes also commonly rear the young together. If males are removed during the nestling period then females are usually able to raise at least some of the young to independence. For example, in song sparrows (*Melospiza melodia*) male removal caused success to decrease to 51 per cent of that of pair-fed broods, and in seaside sparrows (*Ammodramus maritimus*) and dark-eyed juncos (*Junco hyemalis*) the figures were 66 per cent and 38 per cent respectively (Smith *et al*. 1982; Greenlaw & Post 1985; Wolf *et al*. 1990). For some species, removals suggest that male help is more important when food is scarce (Lyon *et al*. 1987; Bart & Tornes 1989). These experiments show that male help can clearly increase reproductive success, but it is not essential. If male desertion reduces productivity to a fraction $1/x$ of a pair-fed brood, then provided a male can gain more than x females, desertion will be the more profitable option from his point of view. Even if success is reduced to a half or less, and a male can gain only two females, polygyny will still pay provided the male helps to provision at least one of the broods. As predicted, male songbirds readily desert to gain extra females if given the chance, for example by removal of a neighbouring male, and they often help to provision either one of their female's broods full-time, or several females' broods part-time. Occasional polygyny has been reported in 39 per cent of 122 well-studied European passerines.

Monogamy in birds maintained by lack of opportunity for males to gain extra mates

These experiments suggest that the predominance of monogamy in many birds arises not, as Lack proposed, because each sex has the greatest success with monogamy but because of the limited opportunities for polygyny. The two most obvious constraints are: (a) strong competition among males may make it difficult for a male to gain a second female; and (b) females are likely to suffer in polygyny through the loss of male help and, as predicted, females are often aggressive to other females which may decrease the chance that their partners are able to gain a second mate (see below).

EXTRA-PAIR MATINGS AND INTRA-SPECIFIC
BROOD PARASITISM

Even where a male is unable to gain a second female, he may be able to increase his reproductive success by more subtle means. Observations have shown that males of many so-called 'monogamous' species of birds adopt a mixed reproductive strategy, not only guarding their own female and helping her to raise a brood but also attempting sneaky copulations with other females, especially those of neighbouring males. Ten years ago the first attempts were made to assess the success of these extra-pair matings in birds by using polymorphic blood proteins as paternity markers. However, in most cases the proteins were not sufficiently variable to assign paternity with precision. In 1985, Alec Jeffreys and his colleagues at Leicester University, England discovered that there was enormous variability in certain regions of the DNA itself which could be used to measure paternity unambiguously. This technique, known as DNA fingerprinting, is explained in Box 9.1 and has begun to revolutionize field studies of mating systems.

Monogamy is not all it seems . . .

Tim Birkhead's study of zebra finches (*Taeniopygia guttata*) provides a good example (Birkhead *et al.* 1988, 1990). Field observations in the wild, in Australia, showed that on average 5 per cent of a female's copulations were with a male other than her mate. The DNA fingerprinting revealed that 8 per cent of broods had one or more chicks sired by these extra-pair matings. Exper-

... extra-pair matings are
common ...

iments in the laboratory, using domesticated strains of zebra finch with genetic plumage markers to assess paternity, showed that there was second male sperm precedence; a single extra-pair copulation occurring as the last mating before the female laid her clutch fertilized 84 per cent of the eggs. This may explain why males of many species perform retaliatory copulations with their partners if they see them involved in extra-pair copulations, and why males often copulate with their females soon after they meet following a period of absence. The old ethological view was that they were 'cementing the pair-bond'; in fact the male's behaviour may be vital if he is to ensure his paternity!

In some species of birds males protect their paternity by following the female closely during her fertile period ('mate guarding', e.g. magpies and swallows). In other species this is not possible because one partner has to defend the nest site while the other goes off to forage (many seabirds and birds of prey). Here males engage in frequent copulations to swamp the sperm of rivals, sometimes copulating several hundred times per clutch, clearly far more than necessary simply to make sure that the eggs are fertilized. Even so, despite these paternity guards the frequency of extra-pair paternity can be very high (25–35 per cent) in some species (Table 9.4). For example, in the red-winged blackbird (*Agelaius phoeniceus*) Lisle Gibbs and his colleagues (1990) found that extra-pair fertilizations accounted for on average 21 per cent of a male's reproductive success (Fig. 9.5).

Studies of sperm competition in wild populations are now a vigorous field of research. The key questions to solve are: Why are there such large differences between species in the frequencies of

Table 9.4 The frequency of extra-pair matings and extra-pair paternity (assessed by DNA fingerprinting) in various species of birds. From Birkhead and Møller (1992)

Species	Per cent extra-pair copulations observed	Per cent broods with extra-pair paternity	Per cent offspring fathered by extra-pair matings
Fulmar	2	0	0
Shag	14	20	18
European bee-eater	4	5	1
Purple martin	–	54	35
Willow warbler	0	0	0
Wood warbler	13	0	0
Pied flycatcher	–	15	4
Blue tit	5	36	18
Indigo bunting	2	48	35
Red-winged blackbird	–	47	28
Zebra finch	5	8	2

Box 9.1 *DNA fingerprinting. From Burke (1989).*

An individual's DNA is isolated (A) from a tissue sample (e.g. bird red blood cells, which have nuclei, or mammalian white blood cells), and cut into pieces with a restriction enzyme (B), which cuts the DNA at positions containing a specific short sequence of nucleotides. The resulting fragments are then spread out according to size by gel electrophoresis (C); smaller fragments move faster and so reach the end of the gel first. The double-stranded DNA fragments are then denatured into their component single strands and transferred to a nylon or nitrocellulose filter membrane (D).

Jeffreys *et al.* (1985) discovered that there are particular sequences of nucleotides ('minisatellites'), highly variable in length depending on the number of tandem repeats of the sequence (which varies from ten to many hundred times). These can be detected with radioactively labelled probes containing complementary DNA sequences, producing dark bands on an autoradiograph. The result is a 'genetic finger-print', unique for each individual.

The example opposite shows the fingerprints of three adult dunnocks, a songbird which often breeds in a polyandrous mating system (one female with two males), together with those of their four offspring. Offspring inherit a random assortment of bands, 50 per cent from each parent. Paternity can be assigned by the presence/absence of diagnostic male-specific bands in offspring (▶, alpha male bands; ▷, beta male bands). In this brood, there is mixed paternity with the beta male siring offspring D, E and F (which, for example, all inherit the unique beta male band marked **) and the alpha male siring offspring G (which inherits, for example, the unique alpha male band *). In this species, with a mean of 10 paternally derived bands per chick and an incidence of band sharing of 0.24, the probability of mis-assigning paternity is less than one in a million (Burke *et al.* 1989).

This is a multi-locus fingerprint, with many minisatellite loci represented. One of the problems of multi-locus fingerprinting is that it is difficult to compare fingerprints from different gels. This is a particular problem in studies of lekking species where offspring may have to be compared with large numbers of potential fathers, too many to fit onto a single gel. Recently, single-locus fingerprinting systems have been developed where huge variability at a single locus can be measured by locus-specific probes to score an individual's genotype at a particular locus.

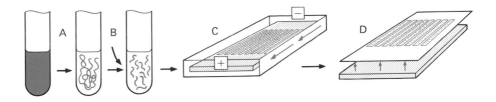

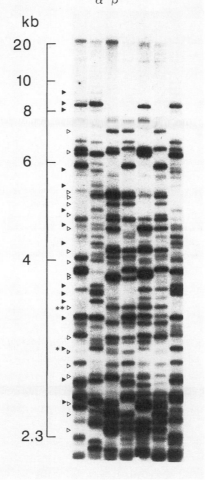

(For explanation, see Box 9.1 opposite.)

extra-pair paternity (Table 9.4)? What role does the female play in extra-pair matings — is she shopping for genetic benefits (Fig. 9.6) or is it simply less costly to accept than resist? What are the physiological mechanisms leading to second male sperm precedence?

While some males are increasing their reproductive success by extra-pair matings, some females are increasing their success

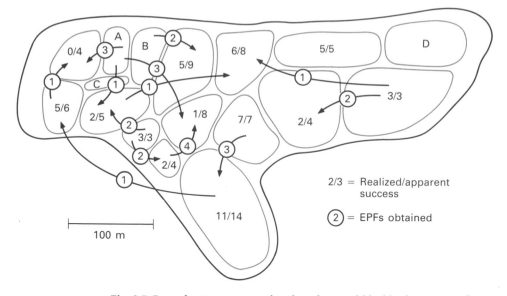

Fig. 9.5 Reproductive success of male red-winged blackbirds on a marsh in Ontario, Canada, assessed by DNA markers. The fractions in each male territory show the number of chicks sired by the resident male over the total chicks raised. Arrows refer to extra-pair fertilizations: the origin of the arrow shows the identity of the cuckolding male; the arrowhead indicates the territory in which he fertilized chicks; the number in the circle indicates the number of extra-pair chicks he sired. The map shows that most, but not all, cuckolders were near neighbours. From Gibbs *et al.* (1990).

... and so is egg dumping in some species by playing at cuckoos, laying eggs in the nests of neighbours in addition to raising a brood in their own nest back home. The proportion of nests parasitized by conspecifics has been estimated as 5–46 per cent for various populations of starlings (*Sturnus vulgaris*) and 3–31 per cent for swallows (*Hirundo* spp.) (Petrie & Møller 1991). In the cliff swallow (*Hirundo pyrrhonota*) some females even move eggs from their own nests to those of other females, so reducing the costs of parental care (Brown & Brown 1988)!

The message for future field studies is clear. Good measures of both maternity and paternity are needed if we are to relate an individual's behaviour to its reproductive success. In many cases the term monogamy may well describe the parental care shared by one male and one female but not their genetic contribution to future generations.

POLYGYNY

If monogamy in birds often occurs because males are unable to gain another female, rather than because it pays males to remain

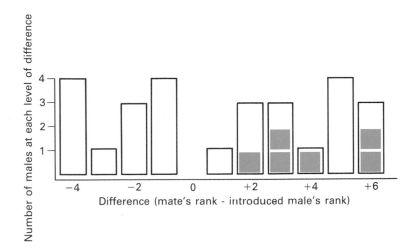

Fig. 9.6 Mated female zebra finches have extra-pair copulations (EPCs) only with males that are more attractive than their mates. The figure shows the distribution of differences in attractiveness (mate – new male) in an experiment in which females were isolated from their partners and introduced to a new male. The black squares show tests in which females had EPCs: these are exclusively with males more attractive than their mates. All males introduced, attractive and unattractive, were equally assiduous in attempting to gain EPCs but unattractive males were rejected. Attractiveness was measured by testing the time spent by unmated females (not the ones used in the EPC trials) sitting near each male. All these females ranked the males in a similar way. Attractiveness is correlated with male song rate and it is heritable, indicating that females, in selecting partners for EPCs, chose males which will yield genetic benefits. From Houtman (1992).

faithful to one mate, what permits regular polygyny in some species? Polygyny in birds usually arises through males monopolizing females indirectly, by controlling scarce resources such as food or nest sites. Where these are patchily distributed, males able to defend the best patches can gain the most mates (Fig. 9.2). It is useful to distinguish various ways in which such 'resource defence polygyny' can arise, bearing in mind that we need to consider the costs and benefits for each sex separately (Searcy & Yasukawa 1989).

Costs and benefits may vary between species

(a) No cost of polygyny to females

In some species the males contribute very little to parental care and so females suffer little, if any, cost from mating polygynously. For example, female yellow-headed blackbirds (*Xanthocephalus xanthocephalus*) build their nests in marshes and feed in fields away from the breeding site. There are apparently no costs or benefits from settling near other females and they settle more or less at random in the marsh (Lightbody & Weatherhead 1988). In yellow-rumped caciques (*Cacicus cela*) females also do not suffer from sharing a male but they benefit from nesting close together in safe sites and by co-operative nest defence against avian predators (Robinson 1986). In both species females may be largely indifferent to the mating system that emerges, which is determined simply by a male's ability to monopolize mates. If a small number of males is able to control the area with the most nesting females then high degrees of polygyny may occur.

(b) Cost of polygyny to females

Sometimes females are forced to accept polygyny costs

In many species, however, females will suffer costs from polygyny through having to share either the resources a male controls (food, nest sites) or his contribution to parental care. Females may be forced to accept these costs if a fraction of the males control all the suitable breeding habitat, their choice being 'accept polygyny' versus 'forego breeding'. For example, in Leonard and Picman's (1987) study of marsh wrens, *Cistothorus palustris*, females settled with mated males only after all the bachelor males had paired. For these later settling females there was no choice but to accept the costs of polygyny.

In other cases, however, most of the males may be able to gain breeding territories. If there is variation in male territory quality then a female's choice may be 'settle on a good territory with an already-mated male, i.e. choose polygyny' versus 'settle on a poor territory with an unmated male, i.e. choose monogamy'.

In other cases they may choose polygyny because the costs are outweighed by benefits

Jared Verner and Mary Willson (1966) suggested that females may choose the polygyny option if the costs of sharing a male's help with parental care were outweighed by the benefits of gaining access to good resources, such as food or nest sites. Gordon Orians (1969) presented this idea in a graphical model, known as the 'polygyny threshold model', explained in Fig. 9.7.

In many species males with the best territories are the ones to attract the most females, just as the model predicts. For example, in red-winged blackbirds experiments have shown that the addition of food to male territories increases the occurrence of polygyny (Ewald & Rohwer 1982), and males with better nest sites are also the first to attract females and attract the most females (Orians 1980). However, showing that females are making the best choice among the breeding options available is difficult unless a great deal is known about the costs of sharing and the choices available to them (Fig. 9.8).

Recent work on the great reed warbler (*Acrocephalus arundinaceus*) comes closest to supporting the polygyny threshold model. This species breeds in reed beds on lake edges, weaving

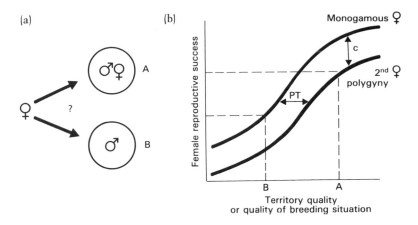

(a)

(b)

Fig. 9.7 The polygyny threshold model. (a) A female has the choice of settling with an unmated male on a poor quality territory B, or with an already-mated male on a good quality territory A. (b) Female reproductive success increases with territory quality. There is a cost C of sharing with another female, so the curve for the second female in polygyny lies below that for a monogamous female. Provided the difference in territory quality exceeds PT (the polygyny threshold), a female does better by choosing to settle with an already-mated male on territory A rather than with an unmated male on territory B. Modified from Orians (1969).

Evidence for the polygyny threshold model: great reed warblers

its nest around the reed stems. In a study in Sweden, Bensch and Hasselquist (1992) captured some newly arrived females in spring and fitted them with radio-transmitters. They then released them onto a study area where male territories had been mapped to see how they sampled territories before pairing. Most females paired up within 24 hours having visited the territories of from three to eleven different males, sometimes going back to pair with a male they had previously sampled. Some females selected already-mated males despite the conspicuous presence of another female on the territory, and even though they had previously sampled the territories of unmated males. These observations show that females sample and choose male territories in exactly the way envisaged in the polygyny threshold model.

Do females make the best choice available? This is a more difficult question to answer. In another study of the same species in Lake Biwa, Japan, Ezaki (1990) found that from 30 to 80 per cent of the males were polygynous each year, some attracting up to four females to their territories, while other males were mono-gamous or remained unpaired. The polygynous males were those who claimed territories containing the best nest sites, namely dense reeds where predation was lowest. Females who settled polygynously as second females did not seem to suffer from their choice because they did at least as well as simultaneously nesting

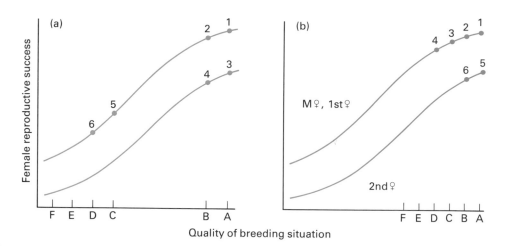

Fig. 9.8 Female settlement patterns predicted by the polygyny threshold model for two distributions of male territory quality (territories A to F). It is assumed that the first female does not suffer from the arrival of a second female, so the top line represents the reproductive success of both monogamous (M) females and the first females in polygyny, while the bottom line refers to second females in polygyny. The sequential settlement patterns of six females (1–6) are shown for the six male territories, assuming that females settle where their expected reproductive success is greatest. In both cases two males become polygynous (A and B), two monogamous (C and D) and two remain unmated (E and F). However, settlement patterns and the reproductive success of monogamous versus polygynous females vary depending on the choices available. From Davies (1989).

monogamous females on poor territories. Thus the differences in male territory quality seem sufficient for females to cross the polygyny threshold (Fig. 9.7).

SEXUAL CONFLICT AND POLYGAMY

Polygamy may arise as an outcome of sexual conflict

The assumptions of the polygyny threshold model are like those of the 'ideal free distribution' which we discussed in Chapter 5. The different resource patches available are male territories of varying quality and females are assumed to be 'free' to settle where they choose. Under 'ideal' conditions, they are expected to settle where their reproductive success is greatest. However, we saw that ideal free conditions rarely hold in nature because dominant individuals attempt to grab more than their fair share of resources, so the assumption that polygyny arises from ideal free female settlement may often be unrealistic. For example, if the first female suffers from the arrival of a second female then it will pay the first female to try to prevent her from settling. Males too may attempt to change the mating system in ways which are detrimental to female success. The following two case studies provide good examples of such sexual conflict.

The pied flycatcher, Ficedula hypoleuca

Rauno Alatalo, Arne Lundberg and their colleagues have studied this bird in the woodlands around Uppsala, southern Sweden. Males defend nest sites, holes in trees or nest boxes, and sing to attract a female. Once a male has attracted one female and she has laid her eggs, he then goes to another nest hole and tries to attract a second female (Fig. 9.9). Males do not simply advertise from the next nearest nest site but go, on average, 200 m away and even up to 3.5 km from their first site! About 10–15 per cent of the males succeed in gaining a second female. They then desert her and go back to help their first female with chick-feeding. Compared to monogamous females the first female suffers little, if at all, from polygyny because she usually gains the male's full-time help, but the second female, who is left to raise

Fig. 9.9 Once a male pied flycatcher has attracted one female, he flies off to another nest site some distance away and tries to attract another. Secondary females suffer because they get little or no help from the male in chick rearing. However females probably are unable to assess whether the male they pair with has another female because of the large distance between a male's two nest sites.

her brood on her own, suffers reduced success, raising on average only 60 per cent of the number of young she would have gained in monogamy (Alatalo *et al.* 1981).

Why, then, do females ever settle polygynously? Three hypotheses have been proposed.

1   *The 'sexy son' hypothesis.* Weatherhead and Robertson (1979) have suggested that although the second females produce fewer offspring, this may be offset if they have sons who inherit their father's ability to be polygynous. The female loses out in the first generation, but then makes up for this in the second generation when her sexy sons, sire lots of grandchildren compared to the sons of monogamous females. According to this hypothesis (which derives from Fisher's sexual selection argument, Chapter 8), secondary females are still making the correct choice when they settle polygynously, but offspring quality is another factor which must be taken into account in the *y* axis of the polygyny threshold model (Fig. 9.7).

For the pied flycatcher, calculations show that the heritability of male mating status would have to be 0.85 for 'sexiness' of sons to offset the loss in offspring numbers. The heritability is not known because young birds disperse and breed away from their natal area. However, the probability that an individual male is polygynous in successive years is only 0.29, which must give an upper limit to the heritability value. Therefore we can reject this hypothesis.

2   *Deception.* Alatalo *et al.* (1981) proposed that second females are deceived into settling polygynously because the male's habit of setting up nesting territories several hundred metres apart (polyterritoriality) prevents females from distinguishing mated from unmated males. By the time the second female has laid her clutch and the male has deserted her to go back to his first female, it is too late in the season for it to be profitable to start another clutch and so she has to make the best of a bad job and rear her offspring alone.

3   *Unmated males hard to find.* An alternative hypothesis is that second females are not deceived but they choose polygyny as their best option simply because unmated males are hard to find. According to this view, male polyterritoriality is not to aid female deception but to decrease the chance that aggression from the first female will prevent the second female from settling (Stenmark *et al.* 1988; Dale *et al.* 1990).

Testing between these last two hypotheses requires detailed observations on how females sample males and territories, and measurements of the profitability of their alternative options. Alatalo *et al.* (1990) performed a clever experiment to test between them. By erecting nest boxes in careful sequence, they arranged for neighbouring boxes, less than 100 m apart, to be occupied by an unmated male and a mated male, whose first female was incubating a clutch in another box 100–300 m away. Boxes were put up at randomly chosen sites, so there was no difference in territory quality between mated and unmated males. In this situation females could clearly sample both males (some were seen to do so) and the songs of both could be heard from either nest site. In 20 such paired choices, nine females settled with the unmated male and 11 with the mated male — clearly no difference. Furthermore, the females who chose the mated males raised significantly fewer young than those who later chose the unmated males they had rejected. This result supports the deception hypothesis; females did not discriminate between mated and unmated males even when they had a simultaneous choice between them, and even though it would have paid them to make a choice.

The dunnock, Prunella modularis

In dunnocks, males prefer polygyny, females prefer polyandry

Conflicts of interest have led to a very variable mating system in another songbird, the dunnock (Fig. 9.10), including simple pairs (monogamy), a male with two females (polygyny) and a female with two (unrelated) males (polyandry). A female has least success in polygyny, where she has to share a male's help with parental care, greater success with monogamy, where she gains a male's full-time help, and greatest success of all with polyandry where she copulates with two males and gains both their help. From a male's point of view, however, reproductive success is greatest in polygyny because despite the cost each female suffers, the total output of two females with part-time help exceeds that from monogamy. Male reproductive success is least in polyandry (the system where a female does best) because although more young are raised the increased production of a trio-fed brood does not compensate a male for shared paternity (Fig. 9.10).

These conflicts of reproductive interests make good sense of male and female conflicts in behaviour. In polygyny, the dominant female attempts to drive the other female away to claim the male to herself, while the male tries to keep between his squabbling females so that both remain with him. On the other hand, monogamous females encourage copulations from other males who approach them in the hope that they will remain and provide parental care, while dominant males attempt to drive subordinate males away to claim full paternity for themselves. The variable mating system can be viewed as the different outcomes of such conflicts of interest. Sometimes the conflict reaches a 'stalemate'

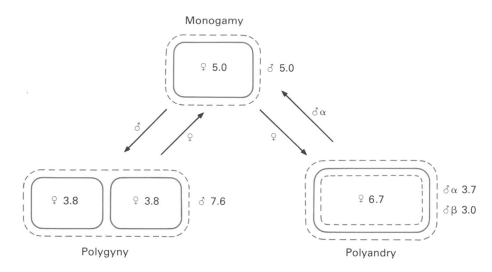

Fig. 9.10 Sexual conflict in dunnocks. Female territories (solid lines) are exclusive and may be defended by one or two unrelated males (dashed lines). The numbers refer to the number of young raised per season by males and females in the different mating combinations (maternity and paternity measured by DNA fingerprinting; Burke *et al.* 1989). Arrows indicate the directions in which alpha male and female behaviour encourage changes in the mating system. The cost of polygyny to females is shared male care. The cost of polyandry to males is shared paternity. From Davies (1992).

in which two males end up sharing two females (polygynandry). Here the dominant male is unable to drive the other male off to claim both females (polygyny) and the dominant female is unable to drive the other female off to claim both males (polyandry). Who is able to win the sexual conflict in particular cases depends on various factors, including individual competitive ability and the population sex ratio (Davies 1992).

FEMALE DESERTION AND SEX ROLE REVERSAL

In most birds, if one sex deserts it is usually the male, because he has the opportunity to desert first. Usually, he also has more to gain from increasing his number of mates because he can potentially fertilize eggs at a faster rate than a female can lay them (see above). Thus polygyny is far commoner than polyandry. Some studies, however, suggest sexual conflict over opportunities to desert first. In the Florida snail kite, *Rostrhamus sociabilis*, either sex may desert leaving the other to care for the brood. Which sex deserts depends on who has the greatest chance of gaining another mate, which varies depending on the operational sex ratio. Desertion is also more frequent when food is abundant so that

In some species either sex may desert

the remaining partner is better able to raise the young unaided (Beissinger & Snyder 1987).

In shorebirds, females often desert ...

In shorebirds (Charadrii), by contrast, although most species are monogamous with biparental care, if one sex deserts it is often the female, leaving the male to care alone. In some species the sex roles are reversed, with the females being larger and more brightly coloured and females competing for males to incubate their clutches for them. In phalaropes (*Phalaropus* spp.) a female defends one male, lays a clutch for him to incubate and then goes off to find a second male (sequential polyandry; Reynolds 1987). In spotted sandpipers (*Actitis macularia*) and jacanas (*Jacana* spp.) (Plate 9.2, between pps 212–213) the females compete to defend large territories in which they may have several males incubating clutches (resource defence polyandry; Jenni & Collier 1972; Lank *et al.* 1985). Female jacanas have been seen to kill other female's chicks in order to gain access to extra males (Emlen *et al.* 1989).

... and females may compete for males

Why should some shorebirds show such sex role reversal? A likely explanation is that shorebirds are characterized by a small clutch size, which never normally exceeds four eggs. The four eggs are large and fit snugly together and experiments suggest that they represent an incubation limit: adding an extra egg reduces hatching success. If shorebirds are indeed 'stuck' with a maximum clutch of four, selection may particularly favour female desertion because, with a fixed clutch size, the only way females can increase their reproductive output if conditions become more favourable is to lay more clutches. In the spotted sandpiper, productivity on the breeding grounds can be so high that the female becomes rather like an egg factory, laying up to five clutches in 40 days, a total of 20 eggs, which represents four times her own body weight. Her reproductive success is no longer limited by her ability to form reserves for the eggs but rather by the number of males she can find to incubate them. This has led to the evolution of sex role reversal with females being 25 per cent larger than males and females competing to gain as many mates as they can.

Ecology and dispersal

As well as influencing the evolution of parental care and mating systems, ecological factors are also an important determinant of another aspect of reproduction, namely dispersal. The movement of a young animal from its place of birth to that of its first breeding attempt is referred to as *natal dispersal*. When birds and mammals are compared, some striking trends emerge (Greenwood 1980; Table 9.5).

Sex differences in
dispersal: birds and
mammals

Table 9.5 The number of species of birds and mammals where natal dispersal is more extensive in males or females. From Greenwood (1980)

	No. species with predominant dispersal		
	By males	By females	No sex difference
Birds	3	21	6
Mammals	45	5	15

1 *In both birds and mammals, one sex usually disperses more than the other.* The result is that close inbreeding is avoided. When an animal mates with a close relative there is a greater chance that any harmful recessive alleles will become homozygous in the offspring and cause low reproductive success.

For example, in the great tit population in Wytham Woods, Oxford, juvenile females disperse more than males so that from a total of 885 pairings only 13 (1.5 per cent) have been inbred pairs. These rare cases were where a male had dispersed more than average or where a female had dispersed less, with the result that a son mated with his mother or a brother with a sister. Such inbred pairs had lower reproductive success than outbred pairs (Greenwood *et al.* 1978). It is known that inbreeding also results in decreased reproductive success in mammals and it is one of the problems associated with trying to breed from small populations in zoos (Ralls *et al.* 1979).

Another way of avoiding inbreeding would be to recognize close relatives. This could come about through early learning because the individuals most closely encountered when young are likely to be kin (see Chapter 11). This mechanism may operate in man: in a study of Israeli kibbutzim, where children are brought up communally, it was found that individuals never married anyone with whom they were associated when young, even when they were not close relatives (Shepher 1971).

2 *In birds, females disperse more than males.* As we saw above, in many species of birds the male helps care for the young. Often he defends a territory and females may choose a male on the basis of his territory quality. It may pay a male to remain near his birth site because it might be easier to set up a territory in the vicinity of relatives, for example by inheriting part of the father's territory (see Chapter 12). Once this happens, it may pay the females to disperse to avoid inbreeding. They may also benefit by moving so that they are able to sample many male territories and choose the best.

3 *In mammals, males disperse more than females.* Male mammals are more often polygynous than birds, and their mating system is usually based on mate defence rather than resource defence, with the male contributing little to the care of the offspring. In mammals, a male will benefit most from gaining access to a large number of a females and so male dispersal may have been favoured.

It is clear that much more data are needed on the costs and benefits of dispersing to breed versus staying at home. For any one individual the pay-offs of the two options must depend on what others in the population are doing and so we really need a theory which will analyse the problem in terms of what will be the evolutionarily stable strategy for a male and a female. We should also remember that in many cases, particularly mammals where the male contributes little to parental care, the cost of inbreeding for a male will be less than that for a female. For example, in the olive baboon (*Papio anubis*) inbreeding is avoided because males transfer from their natal troops whereas females do not. However, before they disperse, young males may still surreptitiously try to mate with females, even close relatives. The females combat this risk of incestuous mating by preferring to associate with transferred (strange) males rather than natal (familiar) males who are likely to be relatives (Packer 1979).

CONSEQUENCES OF SEX DIFFERENCES IN DISPERSAL

A consequence of differential dispersal of the sexes is that members of the sedentary sex will tend to be closely related to their near neighbours, and so we may expect more altruism between them than between members of the dispersing sex where individuals living near by will tend to be unrelated (see Chapter 11).

In mammals, where females are usually more sedentary, the females in an area are often relatives. Altruism between females is common; for example female ground squirrels give alarm calls to warn other females of the approach of a predator and in lionesses, females will suckle other females' cubs (Chapter 11).

In birds it is the other way round. Males disperse less and so in a particular area it is the males who are closely related and who show most altruism to each other. In Chapter 12 we will encounter several species of birds where the males help look after another individual's offspring.

Conclusion

It is clear from our comparative approach to different groups of animals and different habitats that ecological factors are important in shaping parental care and mating system. The distribution of resources, such as suitable egg laying sites and food, and the distribution of the females in space and time all influence the way individuals can behave to maximize their reproductive success. Proximate constraints, like the mode of fertilization, may also determine the mating system by the way they predispose one parent to care for the offspring. We have also seen that the evolution of the behaviour we now see will depend on what came beforehand. Which sex guards the offspring in uniparental care in fish and birds, and the evolution of polyandry in shorebirds are both easier to understand if we take account of the ancestral state of the behaviour.

Categorizing species in terms of their characteristic 'mating system' is fine when we are making broad comparisons across taxa or across habitat types. However we should not expect all individuals within a species to behave in exactly the same way. While some individuals are displaying or defending a resource to attract mates, others may be achieving reproductive success by more devious and sneaky methods. In the next chapter we will take a closer look at the evolution of these alternative breeding strategies.

Summary

Differences between species in parental care and mating systems can be correlated with differences in physiological constraints and ecology. Internal fertilization and specializations such as lactation predispose females to perform parental care, while territory defence, external fertilization and the need for two parents to look after the young may predispose males to show parental care.

Where males do not provide care, mating systems are the outcome of a two-step process in which females distribute themselves in relation to resources and males then distribute in relation to female dispersion (shown by experiments with voles and wrasse). A comparative review of mammalian mating systems shows how male defence of females varies with female group size, range size and the seasonality of breeding. Sometimes males aggregate in leks, where females appear to exercise choice in mating. Four hypotheses for leks are discussed.

Where males provide care (most birds), mating systems vary depending on the costs and benefits of desertion versus caring. DNA fingerprinting has revealed high levels of extra-pair paternity in some birds, and in some species there are high frequencies of intra-specific brood parasitism. Polygyny may occur because there are no other breeding options available to females (marsh wrens), because females choose polygyny on a good territory rather than monogamy on a poor one (great reed warblers), or because females are deceived into polygyny (pied flycatchers). There are often conflicts between the sexes over the mating system which maximizes an individual's success (pied flycatcher, dunnock). Some shorebirds show sex role reversal. Sex differences in dispersal in birds and mammals may be related to ecological factors.

Further reading

Clutton-Brock (1991) reviews parental care and its links with various mating systems. Birkhead and Møller (1992) review sperm competition and mating systems in birds. Kempenaers *et al.* (1992) show that in blue tits extra-pair paternity results from female preference for high-quality males. Good reviews of mating systems in particular animal groups are those by Bradbury and Vehrencamp (1977) on bats, Dunbar (1988) on primates, Wells (1977) on frogs and toads, and Thornhill and Alcock (1983) on insects. Balmford (1991) provides a useful review of mate choice on leks.

Topics for discussion

1 Why are most mammals polygynous whilst most birds are monogamous?
2 What data would you need to collect to test the polygyny threshold model?
3 Could Maynard Smith's model of parental investment be tested experimentally?

Chapter 10. Alternative Breeding Strategies

In the last chapter we used mainly a comparative approach to try and understand differences between species and we characterized each by a 'typical' mating system. During recent years, however, it has become clear that there are marked differences between individuals within a species in the way they compete for scarce resources. A decade ago, if an animal was seen behaving in a different way from the majority of the population it was often thought to be abnormal. Male ducks that engaged in forced copulations instead of courting females by displays were said to be behaving abnormally due to overcrowding. If we observed a male bullfrog sitting silently in the middle of a chorus, while other males were croaking loudly to attract females, we would perhaps have thought that it was ill or having a rest.

Nowadays, whenever we see an individual doing something different we are tempted to label it as a 'strategy'. Silent male frogs may not be tired after all, they may be employing sneaky strategies. There are three main reasons for this change of emphasis. First, the realization that evolutionary arguments must be framed in terms of benefit to the individual, rather than for the good of the spcies, has led us to expect individuals to compete selfishly with each other. If some males are attracting females by calling then we now expect to find others parasitizing their efforts and behaving as sneaks. Once we have seen courtship as a conflict between individuals (Chapter 8), rather than as a co-operative venture, we are not surprised to see some males attempting to force mating with females.

Second, the application of game theory to the study of behaviour (Chapter 7) has shown that it is possible, in theory, to have stable equilibria with individuals in a population behaving in different ways. Finally, an increase in the number of field studies with individually marked animals has shown that indeed there are often several different strategies used within the same species to compete for a mate, a nest site or some other scarce resource.

In this chapter we will examine some examples of individual differences in competitive behaviour and discuss how they might have evolved.

Hypotheses for the occurrence of alternative strategies within a species

Some authors prefer to use the word 'strategy' to give a complete specification of what an animal will do when competing for a scarce resource, and the word 'tactic' for the behavioural components of a strategy. For example, if young male dragonflies searched for females by wandering over large areas while older males defended territories and waited for females to come to them, then the term strategy would describe the whole behaviour pattern 'if young wander, if older defend a territory', and the term tactic would be applied to 'wandering' and 'territorial defence'. Although this distinction is fine in theory, as we shall see it is often difficult to distinguish tactics from strategies and we prefer to use the word strategy somewhat loosely to describe any behaviour pattern or structure used by an individual to compete for a scarce resource.

We consider three main hypotheses for the occurrence of alternative strategies within a species, giving examples under each heading to illustrate the arguments.

CHANGING ENVIRONMENT

The best strategy may depend on the habitat, and if this is patchy in space or changes frequently in time then several strategies may persist. In some cases, individuals may change their behaviour depending on the habitat. When a male speckled wood butterfly searches for females in patches of sunlight on a woodland floor, it defends a territory and sallies out from favourite perches to inspect passing objects. When the same male flies up to the tree canopy, where females are scarcer, it adopts a different searching behaviour and patrols over a large area.

In other cases, different individuals play different strategies. An example is the different colour morphs of the three-spined stickleback (McPhail 1969; Semler 1971; Moodie 1972). Some males have bright red throats while others have dull throats. The red males are more attractive to females; in a choice test with a red male at one end of a tank and a dull male at the other, most females selected the red male. The attractiveness of a dull male can be increased simply by dying his throat red, so it is the colour itself that is attractive to females rather than something special about the behaviour of red males. Why, then, are not all males red?

Different strategies in different habitats: stickleback colours

The answer is that although a red throat brings a benefit in terms of increased attractiveness to females, it also imposes a cost because red males are more susceptible to predation by trout. Experiments showed that red males were particularly preyed upon in bright light, when their colour enabled the predator to spot them more easily. Samples from North American lakes reveal that in deep water most males are of the red type. Presumably in these dark waters the mating advantage of red males outweighs the disadvantage of increased predation. In shallow waters, however, nearly all the males are dull in colour. Here, in the brightest light, dull males do best because the red males are very conspicuous and quickly eaten by predators.

The colour of sticklebacks is therefore selected for by a trade-off between mating and predation. Both red and dull males persist because each does better in a different environment.

MAKING THE BEST OF A BAD JOB

Sometimes, perhaps because of its small size, an individual is unable to compete successfully by fighting or displaying. Instead, it must make the best of its poor circumstances by employing some alternative strategy. Even though it ends up with fewer rewards than others, it will still be doing the best it can given its own competitive ability.

Body size is often age dependent. Unable to compete with larger rivals, young animals may attempt to steal resources by sneaking. We have already seen in Chapter 8 that the largest bullfrogs win the best territories and attract females by croaking. Small, young males are not strong enough to defend a territory and so they behave as satellites, sitting silently near a calling male and attempting to intercept and mate with any females he attracts (Fig. 10.1). They are not very successful; only 2 out of 73

Fig. 10.1 A male bullfrog (left) calls from his territory and attracts a female who is entering the pond on the right. In the middle of the picture is a small male who sits silently in the larger male's territory as a satellite and attempts to intercept the female on her way to the caller.

matings were by satellite males (Howard 1978b). Nevertheless this is probably their best chance of getting a female when they are small, and the satellites make the best of a bad job by parasitizing the largest calling males on the best territories, where most females will be attracted.

A study of natterjack toads, *Bufo calamita*, by Anthony Arak (1988) has shown the various factors that influence a male's decision of whether to call or to be a satellite. The calls are very loud and on still summer nights they can be heard from over a mile away. From a distance of 1 m the call is louder than that legally permitted from a car's engine as heard from the sidewalk! Experiments with loudspeakers showed that females simply moved passively down sound gradients towards the loudest call. Therefore the largest males, who had the loudest calls, attracted the most females. Small males, unlikely to make themselves heard in the chorus, adopted satellite behaviour in an attempt to intercept females on their way to the callers. Callers clearly did better than satellites; on average 60 per cent of the males were callers yet they gained 80 per cent of the matings. Like the bullfrogs, therefore, small males made the best of a bad job until they grew larger and had louder calls. Nevertheless, they varied their behaviour depending on the degree of competition from the larger males. If large males were removed from the chorus, then the small males began to call. If calls were broadcast from a loudspeaker, then large males came over to attack the speaker while small males became satellites next to it!

How do satellites decide which callers to parasitize? Observations showed that when two males were together, one calling and the other a satellite, then they had about an equal chance of capturing the female. (Overall, callers did better because not all callers had satellites.) It was assumed that males had limited knowledge; they could not detect the presence of other satellites and could assess only the calls of their nearest neighbour, not those of more distant callers. Field observations showed that both these assumptions were reasonable. Given that the satellite gains 50 per cent of the caller's females, and that attraction of females depends simply on call intensity, then the male's decision rule should be 'become a satellite on a neighbour if his calls are more than twice as loud as my own'. Figure 10.2 shows that 89 per cent of males adopted the behaviour predicted by this simple rule.

In many cases, small body size is correlated with age and so individuals change behaviour as they get older and so can employ successfully more profitable strategies. In other cases, however, small body size may be fixed throughout an individual's lifetime

Small male natterjack toads are satellites, large males are callers

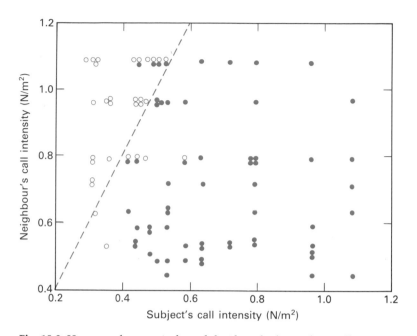

Fig. 10.2 How a male natterjack toad decides whether to be a caller or a satellite. The subject's call intensity is plotted against the call intensity of his nearest neighbour. Males were predicted to become satellites when their neighbours produced calls twice as loud as their own calls (the area to the left of the dashed line). The open circles refer to males who were satellites and the closed circles males who called. From Arak (1988).

and be a consequence of poor feeding conditions when young. An example is the bee, *Centris pallida*, where the largest males are three times the weight of the smallest males (Alcock *et al.* 1977). Large males search for females by patrolling over the ground, searching for buried virgin females about to emerge. When they discover an emerging female they dig her up and copulate. It takes several minutes to dig up a female, during which time other males are attracted to the site by the activity. There are often violent fights and only large males are able to defend their discoveries successfully.

It is not surprising, therefore, that only large males adopt the strategy of patrolling and digging. Small males search for mates by hovering above the emergence areas and pursuing airborne females who have escaped the diggers. Intermediate sized males may adopt both digging and hovering strategies. Observations showed that large males clearly had the greatest mating success and so it is probable that the smaller males are forced to adopt hovering throughout their lives to make the best of a bad job.

These are all examples of strategies that are conditional on an individual's phenotype, for example 'if big, fight; if small, sneak'.

Large male *Centris* bees are diggers, small males are hoverers

The largest and strongest individuals have the greatest success and others are forced by circumstance to adopt less successful, alternative strategies.

ALTERNATIVE STRATEGIES IN EVOLUTIONARY EQUILIBRIUM

Even when there are no constraints from the environment, or phenotypic constraints such as body size, individuals may still differ because the pay-off for one strategy depends on what others in the population are doing. We have seen examples of this in Chapter 7 where different fighting strategies, such as Hawk and Dove, may coexist as an ESS.

If two strategies we observe in nature are an example of an ESS, then at equilibrium we would expect them to enjoy, on average, equal success, just like Hawk and Dove in our hypothetical game in Chapter 7. This is a very different prediction from the previous section where individuals employing a strategy that makes the best of a bad job are expected to have lower success than others. We consider three examples in detail.

(a) Salmon: hooknoses and jacks

Coho salmon (*Oncorhynchus kisutch*) spawn in freshwater streams from November to January on the west coast of North America. After spawning all the adults die. The young fry remain in the streams for a year and then migrate to the ocean as smolts. Females return to breed in the streams at the age of 3 years. For males, however, there are two life history pathways (Fig. 10.3).

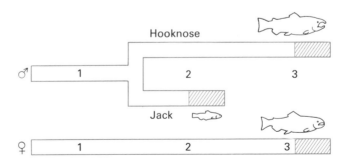

Fig. 10.3 Two life histories exist in male coho salmon: precocious maturity at age 2 years as a small jack, or delayed maturity to age 3 years as a large hooknose. The term 'hooknose' is derived from the exaggerated snout and enlarged teeth which develop at maturity and are used in fighting. By contrast, 'jacks' lack secondary sexual characters and are relatively cryptic on the breeding grounds. Both jack and hooknose males die after breeding, as do the females. From Gross (1985).

Small males, called 'jacks', mature at the age of 2 years while others, called 'hooknoses', do not mature until 3 years of age. Hooknose males may be two or three times the size of the jacks.

In coho salmon, two male strategies coexist in equilibrium

On the breeding grounds females excavate nests in the gravel beds of the streams. The large hooknose males have exaggerated snouts and enlarged teeth and they fight for access to the spawning females. The winners fertilize the eggs externally as they are laid in the nest. By contrast, the small jacks lack secondary sexual characteristics and are relatively cryptic on the breeding grounds. Instead of fighting, they attempt to gain access to females by sneaking. They hide behind rocks or debris to avoid aggression from the hooknoses and surreptitiously gain close proximity to the female during egg laying. Proximity to the female influences the chances of fertilizing the eggs. Since jack density in the streams is too high for all to sneak successfully, some are forced to fight. Likewise sometimes the hooknoses attempt to gain access to females by sneaking. Observation has shown that the smaller males were more successful at sneaking and the larger males at fighting (Fig. 10.4).

The difference between the two male strategies appears to be genetic. Genes coding for an inferior life history in terms of reproductive success should be eliminated from the gene pool by natural selection. How then can we explain the coexistence of the two strategies? The most likely hypothesis is that the two strategies are maintained by frequency-dependent selection such that the average reproductive success of each is the same. For example, an increase in the proportion of jacks in the population would increase competition among them for opportunities to

Frequency-dependent selection maintains the balance

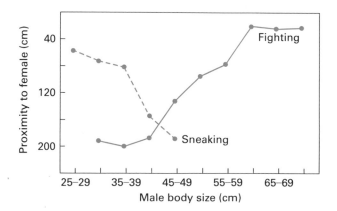

Fig. 10.4 Male proximity to spawning females by either fighting or sneaking in coho salmon. Fighting is most effectively done by large males and sneaking by small males. From Gross (1985).

sneak. More and more jacks would be forced to fight and the average reproductive success of this strategy would decrease. On the other hand, any decrease in the proportion of jacks would increase an individual's chances of sneaking and so the average reproductive success would increase. The same kind of argument holds for hooknoses; the pay-offs for this strategy too would decrease as the proportion of hooknoses increased and the degree of fighting increased.

This may, in theory, give rise to stable alternative life history strategies (Fig. 10.5). If the alternative male strategies represent a stable polymorphism in salmon then the lifetime fitness of jacks and hooknoses should be equal at the frequencies of these strategies observed.

Gross (1985) calculated reproductive success as follows. Measurements of survivorship, from tagged individuals, showed that 13 per cent of jacks survived from leaving the streams as smolts to returning to breed as mature adults. For hooknoses, which spent longer in the ocean before returning, survival was only 6 per cent. On the breeding grounds, jacks were reproductively active for on average 8.4 days and hooknoses 12.7 days. It was assumed that the number of opportunities for spawning was proportional to these times spent on the breeding grounds. The average proximity of jacks to spawning females (averaged over the mix of sneaking and fighting observed) was 124.6 cm, and for hooknoses it was 93.0 cm. It is assumed that relative

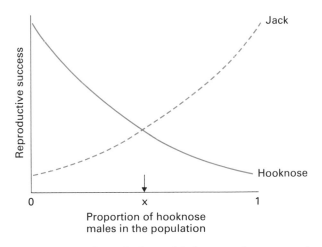

Fig. 10.5 Hypothetical relationship between the success of a strategy (e.g. hooknose versus jack in salmon) and its frequency in the population. The success of each strategy is highest when it is rare. Frequency-dependent selection will result in the proportions of the two strategies being maintained at x, where their reproductive success is equal.

proximity provides a measure of relative fertilization success for males. Jack success was, therefore, 66 per cent $(1 - [124.6 - 93.0]/93.0)$ that of hooknose males.

The relative lifetime reproductive success of jack and hooknose is:

$$\frac{\text{Success jack}}{\text{Success hooknose}} = \left(\begin{array}{c}\text{Survivorship}\\\text{to maturity}\end{array}\right) \times \left(\begin{array}{c}\text{breeding}\\\text{lifespan}\end{array}\right) \times \left(\begin{array}{c}\text{mating}\\\text{success}\end{array}\right)$$

$$= \frac{0.13}{0.06} \times \frac{8.4}{12.7} \times \frac{0.66}{1}$$

$$= 0.95$$

This calculation suggests that the fitnesses are indeed about equal.

Parental and cuckolder males in sunfish

Similar alternative strategies exist in the bluegill sunfish (*Lepomis macrochirus*) which breed in colonies in freshwater lakes in North America (Dominey 1980; Gross 1982). Some males mature at the age of 7 years, defend territories and build nests in which the females spawn. These 'parental' males then look after the eggs and young. Other males mature at 2 years of age and are 'cuckolders'; they do not build nests but instead attempt to sneak fertilizations as the female lays her eggs. Small cuckolders swim along the substrate and make rapid rushes at nests where females are laying eggs, releasing sperm as they swim past. Larger cuckolders, however, mimic female coloration and behaviour; they enter nests slowly and try to insert themselves between the parental male and spawning female and release sperm over the eggs. The different male strategies, parental and cuckolder, are probably (like the salmon) a genetic polymorphism, and calculations again suggest that the reproductive success of the two is equal (Gross & Charnov 1980).

(b) Figwasps: fighters and dispersers

Another dramatic example of male dimorphism with the same species occurs in some figwasps, *Idarnes* spp. (Hamilton 1979). Some males are wingless and put their resources into fighting; they have large heads and mandibles that can chop another figwasp in half. These males remain inside the fig where they were born and fight to mate with newly hatched females which develop from larvae in the fig. Other males are winged and put their resources into dispersal; they have tiny heads and mandibles, are not aggressive and fly off to mate with females that have emerged from the fruits (Fig. 10.6).

It is not known whether the difference between male strategies

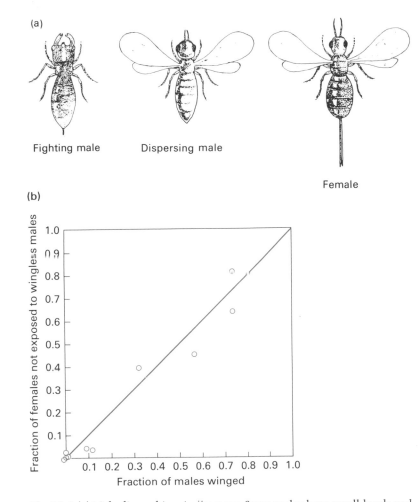

Fig. 10.6 (a) Male dimorphism in figwasps. Some males have small heads and can fly; others are flightless and have enormous mandibles that can chop another male in half. (b) For ten species of figwasps, there is a good relationship between the fraction of males in the population that are winged and the fraction of females leaving their natal fig before mating. Since these females will be mated by winged males, the equality of the two fractions implies equal mating success for the two morphs. From Hamilton (1979).

Fighter and disperser male figwasps coexist in equilibrium

is genetic but it is easy to see that the pay-offs for each strategy must be frequency-dependent. If most males were dispersers, then individuals who remained in their natal figs would have high reproductive success because they would mate with the females before they emerged. If most males were wingless, however, competition for females in the figs would be high and any male which dispersed would have access to all the females which emerged unmated. The pay-offs for the two male strategies will also depend on the dispersion of females among figs. For example, in species where many females are likely to occur in a male's natal fig the advantages will be high for fighters. On the other hand if very few eggs were laid in each fig fruit so that a male may find no females to mate with in his natal fig, it would be more profitable to be a disperser. The relative proportion of fighters and dispersers, therefore, is expected to vary between species depending on how females lay their eggs.

Hamilton (1979) found that the proportions of fighters and dispersers in different species were such that the average reproductive success of the two strategies tended to be equal (Fig. 10.6). This then is another example where alternative male strategies exist as an ESS.

(c) Nesting strategies in female digger wasps: diggers and enterers

Female golden digger wasps, *Sphex ichneumoneus*, lay their eggs in burrows underground which they provision with katydids as food for the offspring (Fig. 10.7). Jane Brockmann discovered that females have two ways of obtaining a burrow: they either dig one for themselves or they enter an already dug burrow. Digging a burrow is hard work and takes on average 100 min so entering may seem a good strategy because the female gets a burrow without having to spend time and energy in digging. However the wasps do not seem to recognize whether the burrow they

Female digger wasps may dig burrows or enter completed burrows

Fig. 10.7 A female digger wasp, *Sphex ichneumoneus*, at a burrow entrance. Photo by Jane Brockmann.

enter is empty (abandoned by its builder) or occupied. If it is empty, then they are able to lay and provision in peace, but if there is another female using the same burrow the two wasps eventually meet and there is always a fight with the result that only one female is successful in breeding in the burrow. Even if a female digs a burrow herself she may be joined by another female who is playing the entering strategy (Brockmann *et al.* 1979).

How can we explain the evolution of these alternative strategies, 'digging' and 'entering'? It can be readily seen how their success could be frequency-dependent. Imagine that all the females in the population are digging all the time. There would be plenty of empty burrows from previous and failed breeding attempts and so it would pay any female that started entering because she would save time not having to dig. Digging is therefore unlikely to be an ESS. However, entering would not spread to take over the whole population because if all the wasps were entering most would end up sharing, there would be lots of fights, and it would certainly pay a female to go off and dig her own burrow, where the chances of sharing would be decreased. Entering is therefore also unlikely to be an ESS. However, because each strategy does best when rare (just like Hawk and Dove), there could be a stable mixture of entering and digging (an ESS) brought about by frequency-dependent selection, where the success of the two strategies was equal.

Brockmann *et al.* (1979) measured the success of the two strategies in terms of the number of eggs laid per unit time. Because individual females employed both strategies, they could not measure success by comparing *individuals* who were enterers with those who were diggers. Instead they measured the success of 'entering' and 'digging' *decisions*. Going back to the Hawk–Dove game in Chapter 7, it will be remembered that an ESS can come about either by a polymorphism in the population, some individuals always playing Hawk and others always playing Dove (as was the case for the figwasps and salmon, above), or by individuals themselves playing Hawk and Dove in the proportion that satisfies the ESS. Therefore, if the two strategies are an example of an ESS we would predict that digging and entering decisions should have equal success.

To test this hypothesis data on the nesting behaviour of 68 female wasps were analysed. This field work was based on over 1500 hours of observation and provided a nearly complete record of the histories of 410 burrows. Calculating the eggs laid per unit time from digging versus entering was a complicated exercise. Brockmann *et al.* (1979) had to work out the possible outcomes of the two decisions in order to calculate the overall

The same individual may be both a 'digger' and an 'enterer'

Decisions   **Outcomes**

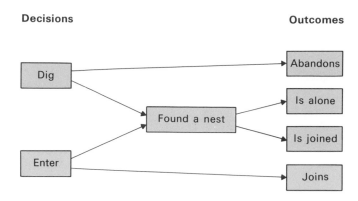

Fig. 10.8 A female digger wasp can either decide to dig a nest burrow herself or enter an already dug burrow. A digging decision may result in one of three outcomes: the wasp may abandon the burrow, she may start a nest and remain alone or she may be joined by another female. An entering decision may also result in one of three outcomes: a wasp may join another female in her nest, or she may have found a nest of her own in which case she may end up nesting alone or in company with another wasp. In order to calculate the overall benefits per unit time for the two decisions, the benefits and times spent in each outcome must be measured. From Brockmann *et al.* (1979).

success (Fig. 10.8). For example, if a female who was entering joined an occupied burrow and was then later evicted before she had the chance to lay any eggs, this had to be included as time spent 'entering' for no rewards.

The two strategies have similar pay-offs

The results showed that a female's decision to enter or dig was not conditional on any obvious phenotypic character, such as body size, nor on the environment, for example time of breeding season. Furthermore, there was no significant difference between the success rates of the two strategies: 59 per cent of the decisions were 'dig' and the average reproductive success of this strategy was 0.96 eggs laid per 100 hours; 41 per cent of the decisions were 'enter' and the success here was 0.84 eggs per 100 hours.

Therefore the hypothesis that best explains the data is that digging and entering are a mixed ESS. Individual females seem to be programmed with a simple rule such as 'dig with probability p, enter with probability $(1-p)$'. The value of p has evidently been fixed by frequency-dependent selection at 0.59, so that the two decisions have the same reproductive success.

Problems of measuring costs and benefits of alternative strategies

Whenever we find individuals within a population employing different strategies it is useful to ask two questions (Caro & Bateson 1986).

1 *What are the causes of the difference in behaviour?* Differences in behaviour may be due to genotypic differences, as is suspected for the salmon and sunfish, or it may simply be due to differences in environment. For example, all male *Centris* bees are likely to be potential 'hoverers' or 'diggers'. Which behaviour they adopt simply depends on their body size, which depends on larval feeding conditions. The interaction between genotype and environment in determining the strategy adopted may be complex. For example, some male field crickets, *Gryllus integer*, call to attract females by rubbing their forewings together. Other males sit silently near the callers and attempt to intercept females as they arrive! Breeding experiments have shown that 'caller' and 'satellite' behaviours are heritable, but although genotype differences influence calling differences, the amount of calling also varies with environment. An individual's calling frequency varies, for example, with the time of day and with degree of isolation from other males (Cade 1981; Cade & Wyatt 1984). Individuals can switch from 'caller' to 'satellite' but their genotype may influence the propensity with which they respond to environmental cues.

2 *Are the pay-offs of the different strategies equal or unequal?* There are two problems here. First, the definition of the alternatives may need care. Continuous variation in behaviour patterns may be the most common situation in nature (e.g. variation in stay times of male dungflies at a cowpat, Chapter 7) and these may be divided up arbitrarily by the observer into discrete categories. Second, there is a difficulty in testing a theory, such as ESS theory, which predicts equality in success. It is impossible to demonstrate statistically that two strategies have exactly equal pay-offs; all we can do is to infer equality if we fail to find a significant difference. The problem is that the smaller the sample size, the less likely we are to find a significant difference, so some supposed examples of ESSs may simply reflect small sample sizes.

Another problem faced by the field worker in measuring pay-offs is that although it is often easy to measure the benefits, such as number of females attracted, it is difficult to measure costs. For example, even though satellite males attract fewer females, their net benefit may be just as great as that of callers because they may incur lower costs. Cade (1979) showed that louder cricket calls attracted not only more females but also more satellites and a parasitic tachinid fly, *Euphasiopteryx ochracea*, which laid larvae into the body of the male cricket or onto a loudspeaker broadcasting cricket calls (Table 10.1)! When Cade sampled crickets he found that 11 out of 14 calling males were

Difficulties in showing equal pay-offs

Table 10.1 When the song of a male cricket, *Gryllus integer*, is broadcast from a loudspeaker, it attracts not only females but also satellite males and a parasitic fly which will kill the cricket. From Cade (1979)

	Attracted to speaker		
Broadcast	No. females	No. satellite males	No. parasitic flies
Silent	0	0	0
80 dB song	7	7	3
90 dB song	21	16	18

Costs must be measured as well as benefits

parasitized by larvae of this fly while only 4 out of 29 satellites were infected. Parasitism must inflict a serious cost on callers because the cricket always dies when the adult flies emerge from its body (Fig. 10.9). It will therefore be difficult to distinguish among our three hypotheses for the occurrence of 'calling' and 'satellite' behaviour in the cricket. Both strategies may persist because each does better in a different environment, calling at times and places where parasitic flies are scarce and satellite behaviour where parasites are abundant. Alternatively, calling may be the best way of maximizing reproductive success and satellites may be individuals in poor body condition making the best of a bad job. Finally, the two strategies may coexist as a frequency-dependent equilibrium each enjoying, on average, equal success with 'calling' bringing high benefits and high costs, and 'satellite' low benefits and low costs.

Where it can be shown that two strategies really do have equal success, it is useful to recognize that this can come about in three different ways.

1 *Polymorphism.* Different individuals may play different strategies and the frequencies of the morphs will be fixed by frequency-

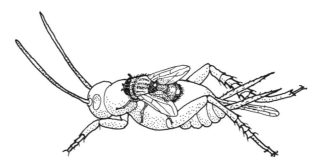

Fig. 10.9 Calling male crickets not only attract females but also a parasitic fly which lays live larvae into the cricket which will eventually kill it.

dependent selection. The polymorphism may arise because of genetic differences between morphs (e.g. salmon and sunfish). However it may arise because of environmental differences. For example, in some species sex is environmentally determined, males (one strategy) being born when the eggs are incubated above a certain temperature and females (the other strategy) when the eggs are incubated below this temperature (e.g. alligators, Ferguson & Joanen 1982; fish, Conover 1984; see Bull 1980 for a review). At equilibrium the reproductive success of males and females must be equal and the temperature threshold for sex determination will presumably have evolved so that males and females are born in the correct proportion to give them equal success.

Three ways of achieving the ESS

2 *Individuals play the different strategies in the proportions that satisfy the mixed ESS.* This occurs, for example, in digger wasps where individuals both dig and enter. Selection could, in theory, have resulted in every individual playing the two strategies with a fixed proportion which results in equal success for 'dig' and 'enter'.

3 *Behavioural assessment.* The third possibility is that instead of having a fixed rule, every individual may have flexible behaviour and base its decision on assessment of what others in the population are doing. For example, imagine that every male frog is free to choose between 'caller' and 'satellite' behaviour. If everyone else was calling it would perhaps pay to be a satellite whereas if all the others were silent it may pay to call. In theory equal success could arise if all individuals simply varied their strategy in relation to the strategies others adopted.

A possible example of an equilibrium arising in this way is the green tree frog, *Hyla cinerea*. There is no difference in body size between callers and satellites, individuals frequently switch between the two tactics and the tactics appear to have equal success in gaining females (Perrill *et al.* 1982).

Of course, assessment may also be involved in cases where satellites make the best of a bad job, as we saw earlier with the natterjack toads.

In conclusion, although it is clear that there are several evolutionary pathways to differences between individuals within a population, there will often be problems in sorting out which hypothesis best applies to a particular example. Measurements of costs as well as benefits, together with data on individual constraints such as age and body size, are needed before we can distinguish 'best of a bad job strategies' from examples of an ESS. Table 10.2 summarizes the examples of alternative strategies which we have discussed.

Table 10.2 Examples of alternative strategies discussed in this chapter

Species	Strategy 1	Strategy 2
Stickleback	Red	Dull
	Perch	Patrol
Bullfrog, green treefrog, natterjack toad, field cricket	Call	Satellite
Bee	Hover	Dig
Salmon	Hooknose	Jack
Sunfish	Parental	Cuckolder
Figwasp	Fight	Disperse
Digger wasp	Dig	Enter

Sex change as an alternative strategy

CHANGING FROM FEMALE TO MALE

Whenever there is intense competition between males for females, it is usually the largest strongest males who will enjoy the greatest reproductive success. We have seen that one way in which young and small males can avoid direct competition with stronger rivals is to adopt sneaky mating strategies. However there is another more startling way of overcoming the disadvantage of small size when young. This is to start reproductive life as a female and then change to a male when large enough to be a successful competitor (Fig. 10.10a). This system of sex change is common in fish and is known as protogynous hermaphroditism. It will be favoured whenever an individual reproduces best as a female when small and as a male when large, and when lifetime reproductive success is greater if an individual changes sex than if it remains as either a male or female throughout its life (Ghiselin 1969; Warner 1975). As mentioned in Chapter 8, the problem of sex change is closely related to that of the sex ratio. Both are part of the general problem of sex allocation.

The blue-headed wrasse, *Thalassoma bifasciatum*, lives on coral reefs in the western Atlantic. The males are brightly coloured and they defend territories on the reef. The females, which are dull in colour, choose the largest and brightest males for mating. The largest male on the reef may spawn 40 times a day at the peak of the breeding season. As we might expect from the fact that only the largest individuals are successful males, this species is a protogynous hermaphrodite; fish start reproductive life as females when small and only change to being males when large (Warner *et al.* 1975). Sex change is socially controlled. If the

In wrasse, females turn into males when they are large . . .

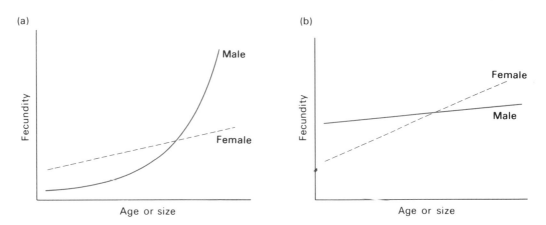

Fig. 10.10 (a) When male—male competition is intense, only the largest individuals will be successful at mating. Although female fecundity also increases with size (larger females are able to lay more eggs) the influence of male size on mating success is much stronger. Under these conditions it may pay an individual to be a female when small, because all females will breed, and a male only when large enough to be a successful competitor. After Warner (1975). An example of this case is the blue-headed wrasse (see text). (b) When male—male competition is less intense, female fecundity may be more dependent on body size than male fecundity. It may then pay an individual to start life as a male when it is small and change to a female when it is older and larger. After Warner (1975). An example of this case is the anemonefish, *Amphiprion* (see text).

largest males on a reef are removed, the next largest individuals (females) will change sex and become bright-coloured males.

The story is a little more complicated than this because there are differences in behaviour on different sized reefs (Warner & Hoffman 1980). On the largest reefs, where populations of wrasse may reach 16 000 individuals, the few largest males have potentially an enormous reproductive success. However, with large numbers of females being attracted to a few males, there are also good opportunities for sneaking by small males. On large reefs some individuals are born as males and remain male throughout their lives (primary males). When small they either go in for sneaky matings, stealing into a large male's territory and attempting to spawn with the females he has attracted, or they may go round in gangs with other small males, chasing females and stimulating them to spawn with the group.

On small reefs, the largest males again attract the most females but the maximum number they can attract is less than on a large reef because the total population is small, perhaps only 20 wrasse. Therefore there is less opportunity for sneaking and small males can be excluded from breeding altogether by the large males. On small reefs there appears to be an advantage for sex-changing

... although some individuals are male throughout their lives

individuals who start life as a female and then become a male when they have grown larger. Warner and Hoffman (1980) suggest that both sex-changers and primary males may be maintained in the blue-headed wrasse because of different advantages on different sized reefs.

CHANGING FROM MALE TO FEMALE

More rarely, an individual may be a male when small and then change to a female when large (protandrous hermaphroditism). This type of sex change may be favoured if male−male competition is not so intense and male size has little effect on breeding success (Fig. 10.10b). An individual may then reproduce best as a male when small because it is able to spawn with some of the large, most fecund females.

An example of a fish that changes from male to female is the anemonefish, or clownfish, *Amphiprion akallopisos*, which lives on coral reefs in the Indian Ocean. It lives in close symbiosis with sea anemones and because there is usually only enough space for two fish to inhabit the same anemone this species lives in pairs (Fig. 10.11). In effect the habitat forces them to be monogamous. The reproductive success of a pair is limited more by the female's ability to produce eggs than by the male's ability to produce sperm and so each individual does better if the larger one is female (Fricke 1979). Like the wrasses, sex change is socially

Anemonefish can change from male to female as they become larger

Fig. 10.11 An anemonefish, *Amphiprion akallopisos*, with its anemone. Photo by Hans Fricke.

controlled. If the female is removed, the male is then joined by a smaller individual, so he changes sex and lays the eggs while the newcomer functions as a male (Fricke & Fricke 1977).

SEX CHANGE VERSUS SNEAKING

Where there is a size advantage for being one sex, the smaller individuals may sneak or may be sequential hermaphrodites, changing sex as they grow larger (as in some fish). Why do not young elephant seals start life as females and then change to males when they are large and strong, like the blue-headed wrasse? The most likely explanation is that in mammals the sexes are more differentiated than in many fish; for example there is internal fertilization and elaborate care of the young by the female in pregnancy and during lactation. It may therefore be too costly to change sex. It is striking that all species of fish known to change sex have relatively simple sex organs and external fertilization. Furthermore for a young mammal, like an elephant seal, experience may be needed to be a successful male. It may be better to forgo reproduction when young and instead put resources into growth and learning the techniques of successful harem defence (Warner 1978).

Summary

Individuals within a species often differ in the way they compete for scarce resources such as food, mates or nest sites. Different strategies may be favoured in different environments (e.g. red and dull male sticklebacks). Sometimes behaviour is conditional on size and strength so that while the largest individuals display and fight to attract mates, smaller individuals employ sneaky strategies which make the 'best of a bad job' (e.g. bullfrog satellites, hoverers in the bee *Centris pallida*). Finally, alternative strategies may exist as an ESS, an evolutionary equilibrium where different strategies enjoy equal success. In these cases there may be a polymorphism in the population, with frequency-dependent selection fixing the frequencies of the different types (e.g. hooknoses and jack salmon), or alternatively each individual may play the various strategies in the proportions that satisfy the ESS (e.g. digging and entering in digger wasps). In many cases there are not enough data on constraints and the success of the different strategies to say whether they are 'best of a bad job', or an example of an ESS.

Some fish change sex as a way of increasing reproductive success when small in size. When male–male competition is intense so that only the largest individuals are successful as males, individuals may be sequential hermaphrodites, changing from female to male as they grow larger (e.g. blue-headed wrasse). More unusually, the reproductive success of a female is more size-dependent than that of a male, in which case individuals may change from male to female as they grow larger (e.g. anemone fish, *Amphiprion*).

Further reading

Dawkins (1980), Dunbar (1983) and Gross (1992) give good reviews of alternative strategies. Thornhill (1981) gives an excellent account of alternative mate searching strategies in scorpionflies. The two papers on nesting strategies in digger wasps (Brockmann & Dawkins 1979; Brockmann *et al.* 1979) give a clear account of the methods and problems of applying ESS ideas to alternative strategies in one species. Charnov (1982) models the optimal timing of sex change and tests his theory with data on shrimps. An especially neat study of an evolutionarily stable mixture of male mating strategies is that of Steve Shuster and Michael Wade (1991) on a marine isopod. There are three male strategies. Large alpha males defend harems within intertidal sponges, small beta males mimic females and tiny gamma males hide inside harems. All three morphs have equal reproductive success. The difference between morphs is due to a single locus.

Topics for discussion

1 In a chorus of male frogs, some are calling and others are silent. What are the hypotheses for the occurrence of these two behaviours? How would you test between them?
2 What explanations, other than the ones in this chapter, can you suggest for variation in reproductive behaviour within a species?

Chapter 11. On Selfishness and Altruism

So far throughout this book we have championed the view that natural selection designs individuals to behave in their own selfish interests and not for the good of their species or for the good of the group in which they live. For example, observed clutch size, foraging behaviour and mating patterns are what would be expected if selection optimized behaviour and life history strategies so as to maximize an individual's reproductive success.

However, it will be obvious to any naturalist that animals do not behave selfishly all the time. Often individuals apparently co-operate with others. Several lions may co-operate to hunt prey; in many species of birds and mammals individuals give alarm calls to warn others of the approach of a predator; sometimes an individual may help others to produce offspring rather than have young itself (see Chapters 12 and 13). Before 1960 such co-operation did not demand special attention. It was seen to be 'good for the species' and therefore adaptive. Since the 1960s, however, people have appreciated again what Darwin said all along, namely that in evolution there is a struggle between individuals to outcompete others in the population. How, then, can we account for the evolution of co-operative behaviour in terms of advantage to individuals? If natural selection favours individuals who do the best and have the most surviving young, how can behaviour evolve which involves helping others to survive and have young at a cost to the helper's own chance of doing so? As E.O. Wilson (1975) put it, the central problem of sociobiology is

Altruism: benefit to others, cost to altruist

how can altruism evolve? *Altruism* is defined as acting to increase another individual's lifetime number of offspring at a cost to one's own survival and reproduction.

In this chapter we will distinguish four hypotheses for the evolution of co-operation. Behaviour which appears to be altruistic at the phenotypic level may turn out to be genetically

Phenotypic versus genotypic altruism

selfish (hypotheses 1 and 4 below). Some forms of co-operation do not involve any altruism (hypothesis 2) and others involve self-sacrifice that is both phenotypically and genotypically altruistic (hypothesis 3).

Kin selection

Theory. The most familiar example of an individual giving aid to another is, of course, parental care. We are not surprised to see a parent bird hard at work feeding its offspring because natural selection favours individuals who maximize their gene contribution to future generations. The young will have copies of their parent's genes and so parental care is genotypically selfish.

We can quantify the probability that a copy of a particular gene in a parent is present in one of its offspring. In diploid species, when an egg and a sperm fuse to form a zygote, each parent contributes exactly 50 per cent of its genes to the offspring. Therefore the probability that a parent and an offspring will share a copy of a particular gene identical by descent (in an outbreeding species) is 0.5. This quantity is called the *coefficient of relatedness*, often denoted by r.

The coefficient of relatedness is a measure of genetic similarity

Now offspring are not the only relatives to share copies of the same genes identical by descent. Again we can calculate the probability that a copy of a gene in one individual will be present, by virtue of descent from a common ancestor, in a brother, sister, cousin and so on. For brothers and sisters r is 0.5, for grandchildren it is 0.25 and for cousins it is 0.125 (Box 11.1). It was W.D. Hamilton (1964) who realized the important implication of this for the evolution of altruism, although the idea was anticipated by Fisher (1930) and Haldane (1953). Just as gene proliferation can occur through parental care so it can through care for siblings, cousins or other relatives. Table 11.1 shows various values of r for descendant kin and non-descendant kin.

Kin selection: benefits to close relatives

The main point is that there is nothing particularly special about offspring as kin; if we saw a bird helping to feed a younger brother or sister this could also be favoured by selection as a means of passing on copies of genes to future generations. Maynard Smith (1964) coined the term *kin selection* to describe the process by which characteristics are favoured due to their beneficial effects on the survival of close relatives, including offspring and non-descendant kin.

As far as evolution is concerned there will be no distinction between gene copies produced by helping offspring as opposed to siblings. Whichever route the genes take, the consequence of selection is a change in relative gene frequencies in the gene pool. However, as we shall see later in this chapter and in Chapters 12 and 13, it is often useful in studies of behaviour to quantify the fitness gain from aiding descendant versus non-descendant kin. Jerram Brown (1980) introduced the useful terms *direct fitness* for the component of fitness gained through personal reproduction (i.e. production of offspring) and *indirect fitness* for the component gained from aiding the survival of non-descendant kin, such as siblings. If we assess the fitness gain through both routes then we will have a measure of an individual's *inclusive fitness* (Hamilton 1964).

Direct and indirect fitness

Box 11.1 *Calculation of* r, *the coefficient of relatedness.*

r is the probability that a gene in one individual is an identical copy, by descent, of a gene in another individual.

General method
Draw a diagram with the individuals concerned and their common ancestors, indicating the generation links by arrows. At each generation link there is a meiosis and so a 0.5 probability that a copy of a particular gene will get passed on. For L generation links the probability is $(0.5)^L$. To calculate r, sum this value for all possible pathways between the two individuals.

$$r = \Sigma(0.5)^L.$$

Specific examples
These diagrams show calculations of r between two individuals represented by solid circles, other relatives are indicated by open circles. The solid lines are the generation links used in the calculations; the dotted lines are the other links in the pedigrees:

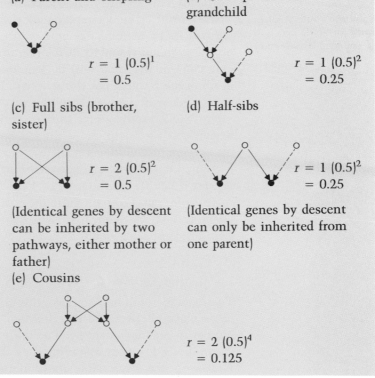

(a) Parent and offspring

$r = 1\,(0.5)^1$
$\quad = 0.5$

(b) Grandparent and grandchild

$r = 1\,(0.5)^2$
$\quad = 0.25$

(c) Full sibs (brother, sister)

$r = 2\,(0.5)^2$
$\quad = 0.5$

(Identical genes by descent can be inherited by two pathways, either mother or father)

(d) Half-sibs

$r = 1\,(0.5)^2$
$\quad = 0.25$

(Identical genes by descent can only be inherited from one parent)

(e) Cousins

$r = 2\,(0.5)^4$
$\quad = 0.125$

Table 11.1 Coefficients of relatedness (r) for descendant and non-descendant kin

r	Descendant kin	Non-descendant kin
0.5	Offspring	Full siblings
0.25	Grandchildren	Half-siblings
		Nephews and nieces
0.125	Great-grandchildren	Cousins

The conditions under which an altruistic act will spread by kin selection are as follows (Hamilton 1964). Imagine an interaction between an altruist (or donor) and a recipient in which the costs and benefits of the interaction can be assessed in terms of survival chances of the donor and recipient. If the donor suffers cost C (through, for example, giving an alarm call) and the recipient gains a benefit B as a result of the altruistic act, then the gene causing the donor to act will increase in frequency if

$$\frac{B}{C} > \frac{1}{r}, \text{ or alternatively, if } rB - C > 0$$

where r is the coefficient of relatedness of the donor to the recipient. This result is known as 'Hamilton's rule'.

Hamilton's rule predicts when altruistic acts will be favoured by selection

An intuitive understanding of this formula can be got as follows. As an extreme example of altruism imagine a gene that programs an individual to die in order to save the lives of relatives. One copy of the gene will be lost from the population in the death of the altruist, but the gene will still increase in frequency in the gene pool if, on average, the altruistic act saves the lives of more than 2 brothers or sisters $(r = 0.5)$, more than 4 nieces or nephews $(r = 0.25)$ or more than 8 cousins $(r = 0.125)$. Having made these calculations on the back of an envelope in a pub one evening, J.B.S. Haldane announced that he would be prepared to lay down his life for the sake of 2 brothers or 8 cousins!

It is often useful to measure the costs and benefits in terms of offspring lost and gained, in which case we use the following form of Hamilton's rule.

$$\frac{B}{C} > \frac{r_{\text{donor to own offspring}}}{r_{\text{donor to recipient's offspring}}}$$

Two examples will help to make this clear. Imagine an individual has a choice between rearing its own offspring and helping its mother to produce offspring. The individual's own offspring and its mother's offspring, assuming they are full siblings, both have $r = 0.5$, so the expression above becomes $B/C > 1$. Therefore helping will be favoured by kin selection if by your help your mother produces more extra offspring than you have 'sacrificed' through providing help (i.e. through forgoing the chance to produce your own offspring). If the individual was faced with the alternative of rearing its own offspring or helping its sister to produce offspring, then the expression becomes $B/C > 2$ (0.5/0.25). In this case helping would evolve only if it resulted in two or more extra offspring produced by the sister for every one offspring lost by the donor. We will discuss some examples of this kind of altruism in Chapter 12.

EXAMPLES OF ALTRUISM BETWEEN RELATIVES

The social insects provide good examples of extreme altruism. Worker bees have barbed stings and attack predators which approach their nests. In the act of stinging the predator the barbs of the sting become embedded in the victim and the worker bee dies as a result. The evolution of such suicidal behaviour poses a problem until we discover that the beneficiaries of the altruistic act are in fact close relatives of the worker. Workers are altruistic in another way because they rarely reproduce themselves, but instead help others in the nest to produce offspring. Darwin regarded this observation as potentially fatal to his theory of natural selection. How can such altruism evolve if the altruists never reproduce? The theory of kin selection immediately suggests a possible answer to this problem because the sterile workers usually help their mother (the queen) to produce offspring (Chapter 13).

Not all acts of altruism are as extreme as suicide or sterility. We now consider two examples where the costs to the altruist are smaller, and where it is likely that kin selection has been a major force in the evolution of the altruism.

Extreme altruism — suicide and sterility

Less extreme altruism — alarm calls

(a) Co-operation and alarm calls in ground squirrels and prairie dogs

Paul Sherman has made an extensive study of Belding's ground squirrel, *Spermophilus beldingi*, a diurnal social rodent which inhabits the subalpine meadows of the far western United States (Fig. 11.1). This species hibernates during the winter, emerging above ground in May. Soon after emergence the females become sexually receptive and mate. After mating the males wander off and leave the females to rear their young alone. A female establishes a territory surrounding its nest burrow and produces a single litter of 3 to 6 young per year. The pups first emerge above

Fig. 11.1 Belding's ground squirrels. (a) Female giving an alarm call. (b) A group of pups at the mouth of a burrow. Photo by © George D. Lepp, Bio-Tech Images.

Unrelated ground squirrels kill each other's offspring, relatives do not

ground at the time of weaning, when they are 3 to 4 weeks old and then soon after this the juvenile males disperse while the juvenile females tend to remain near their natal area. This means that males seldom, if ever, interact with close relatives whereas females spend their whole lives surrounded by close female kin.

Sherman found that closely related females (mother and daughters, sisters) seldom fought for nest burrows and seldom chased each other from their territories. Indeed they co-operated to defend each other's young against infanticidal conspecifics. Eight per cent of all young born were dragged from their burrows and killed by other ground squirrels. The killers were not close relatives of the victims and were either young males wandering in search of an easy meal or immigrant adult females who where searching for new nesting burrows. These females attempted to take over occupied burrows, killing any young they found as a way of clearing the territory of potential competitors (Sherman 1981a,b). Such co-operation among close relatives, in contrast to the conflict among unrelated individuals, is exactly what would be predicted from the theory of kin selection.

Individuals also gave alarm calls whenever a predator, such as a coyote or weasel, approached. Callers probably suffered a cost from giving the alarm because they were more likely to be attacked by the predator, perhaps because the calls made them more conspicuous. Others, however, benefited from the early warning and were more likely to escape. Sherman (1977) found that females were much more likely to give alarm calls than males, and furthermore, females with close relatives nearby were more likely to give calls than females without. Although in most cases the beneficiaries of the alarm calling were likely to be offspring, individuals also gave alarm calls when only parents or non-descendant relatives were nearby. For example, young females who had yet to produce their own young gave alarm calls to warn their mother and sisters of the approach of a predator. In another species of ground squirrel, *Spermophilus tereticaudus*, males were much more likely to give alarm calls before leaving their mother's home area, when they had relatives nearby, but when they dispersed and left their close relatives they were more likely to remain silent when a predator came along (Dunford 1977).

These data on alarm calls show clearly that individuals are more likely to incur the cost of calling when relatives are nearby to gain a benefit. However it could be argued that the major force responsible for the evolution of the calling is an increase in direct fitness because the relatives which usually got the benefit were offspring. Non-descendant kin may simply be common secondary beneficiaries (Shields 1980). Warning offspring (i.e. parental care) and warning sisters are, of course, just different ways of increasing gene propagation to future generations. However it is still interesting to ask whether alarm calling could evolve mainly because of benefit to non-descendant relatives. The best evidence for this comes from a study by John Hoogland (1983) of another colonial rodent, the black-tailed prairie dog, *Cynomys ludovicianus*.

Black-tailed prairie dogs live in social groups called coteries, typically one adult male with 3 to 4 adult females and their offspring. Young females remain in their natal coterie all their lives while young males disperse in their second year. All the females and yearling males within a coterie are therefore usually close genetic relatives. Hoogland studied alarm calling responses by presenting a stuffed specimen of a natural predator, the badger *Taxidea taxus*. This enabled him to get more data than could be obtained by waiting for natural predator attacks and also allowed him to control for proximity of the predator to the prairie dogs. Figure 11.2 summarizes the results of over 700 experiments. The data show that individuals gave alarm calls just as frequently when there were only non-descendant kin in their home coterie as when there were offspring present. Factors other than the warning of relatives must also be involved, however, because immigrants who had no relatives nearby nevertheless sometimes called (Fig. 11.2). There may often be direct benefits to the caller itself in giving the alarm, for example signalling to the predator

Alarms are given when relatives are nearby

Prairie dog alarms are given in the presence of offspring or other relatives

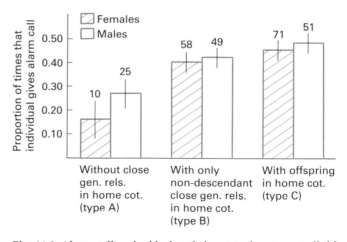

Fig. 11.2 Alarm calling by black-tailed prairie dogs to a stuffed badger. For both males (white histograms) and females (hatched), there are significant differences between type A and type B individuals and also between type A and type C. There was, however, no significant difference for either sex between type B and type C. Data are means ± 1 SE, with number of different individuals observed indicated. From Hoogland (1983).

'I've seen you', which may reduce the likelihood of attack because the predator then does not have the advantage of surprise. Another possibility is that it pays to warn others of the approach of a predator, even if they are not close relatives, because if a neighbour was caught successfully then the predator may be more likely to return to hunt in the same area again. Alarm calling may therefore reduce the likelihood of future attacks by the same predator.

These studies of co-operation among relatives in ground squirrels and prairie dogs are consistent with the kin selection model but they do not constitute a quantitative test of Hamilton's rule. Indeed it would be difficult to assess the costs and benefits of alarm calls in terms of offspring lost and gained. We now turn to an example where these values can be estimated.

(b) Wife sharing in the Tasmanian native hen

A quantitative test of Hamilton's rule

The Tasmanian native hen, *Tribonyx mortierii*, is a flightless rail. There are more males than females in the breeding population and mating combinations include simple pairs (one male with one female) and trios (two males and one female). In trios, the two males are usually, but not always, brothers and both of them copulate with the female and help to raise young. One of the males is usually dominant to the other, yet he apparently allows the other male to mate with the female. Under what conditions could such cooperation evolve? Box 11.2 applies Hamilton's rule

Tasmanian hens: brothers share a wife and the subordinate brother benefits from indirect fitness gains

to derive the conditions under which co-operation would pay the dominant male if he was either unrelated to the other male or was a brother (adapted from Maynard Smith & Ridpath 1972). The conditions for co-operation with a brother are less severe than with an unrelated male, of course, because if a male helps its brother to produce two extra offspring then this is equivalent to producing one offspring itself as far as gene contribution to the next generation is concerned (r to own offspring is 0.5; to brother's offspring is 0.25).

Table 11.2 summarizes the data on reproductive success of pairs and trios. Lifetime success is not well known but it is likely that each bird has at most 5 breeding attempts and at least 2. It is assumed that survival is equal in pairs and trios, and so success for 5 and 2 attempts provide maximum and minimum estimates of lifetime success. If males are unrelated, then co-operation by the dominant males does not make sense because $1/2 N_2$ is always less than N_1 (Box 11.2). If they are brothers, however, then for the maximum estimate of lifetime success $3/4 N_2 = 21.8$ which is only a little less than the value of N_1, 23.1. For the minimum estimate of lifetime success, $3/4 N_2 = 7.2$ which is greater than N_1, 6.6, so the conditions for co-operation derived in Box 11.2 hold. These calculations suggest that co-operation between brothers may satisfy the conditions necessary for kin selection to favour the behaviour. The fact that unrelated males may cooperate is still a problem. Provided $N_2 > N_1$, however, it will pay a female to breed in a trio rather than in a pair so there may be conflicts of interest between the sexes (see also Chapter 9).

It may pay the dominant male to share matings with his brother

Table 11.2 Reproductive success of Tasmanian native hens in pairs and trios. From Maynard Smith and Ridpath (1972)

1 Numbers of surviving young produced per season

	1st year breeders	Experienced breeders
Pair, ♂ ♀, N_1	1.1	5.5
Trio, 2♂ ♀, N_2	3.1	6.5

2 Number of surviving young produced per lifetime

	If 5 breeding seasons	If 2 breeding seasons
Pair, N_1	$1.1 + (4 \times 5.5) = 23.1$	$1.1 + 5.5 = 6.6$
Trio, N_2	$3.1 + (4 \times 6.5) = 29.1$	$3.1 + 6.5 = 9.6$

Box 11.2 *Wife sharing in the Tasmanian native hen.*

The number of offspring raised by a pair is N_1 and by a trio N_2. If the dominant male was selfish and drove the subordinate away we assume that the dominant male then produces N_1 young and the subordinate produces no young (females are scarce and it is unlikely that the evicted male would find another mate). If the two males shared the female in a trio, we assume they share paternity equally and so each produce $\frac{1}{2} N_2$ young. The pay-offs are therefore as follows:

| | Dominant male behaviour | |
Pay-off to	Co-operate	Selfish
Dominant male	$\frac{1}{2} N_2$	N_1
Subordinate male	$\frac{1}{2} N_2$	0

The benefit to the subordinate of remaining in the trio is the number of young he gains through the dominant male's co-operation.

$$B = \frac{1}{2} N_2 - 0$$

The cost of co-operation to the dominant male is the number of young he loses through co-operation with the subordinate compared to what he would get by being selfish.

$$C = N_1 - \frac{1}{2} N_2$$

Applying Hamilton's rule, the dominant male should co-operate if $B \times r$ to subordinate's offspring $> C \times r$ to own offspring.

Consider two cases.

(a) *The males are unrelated* (r to subordinate's offspring $= 0$). The dominant male should co-operate if

$$0 > \frac{1}{2} (N_1 - \frac{1}{2} N_2)$$

i.e. if

$$\boxed{\frac{1}{2} N_2 > N_1}$$

(b) *The males are brothers* (r to subordinate's offspring $= \frac{1}{4}$). The dominant male should co-operate if

$$\frac{1}{4} (\frac{1}{2} N_2) > \frac{1}{2} (N_1 - \frac{1}{2} N_2)$$

i.e. if

$$\boxed{\frac{3}{4} N_2 > N_1}$$

HOW DO INDIVIDUALS RECOGNIZE KIN?

Hamilton's theory of kin selection requires an individual to behave differently towards individuals of different degrees of relatedness, and this could involve assessing relatedness to other individuals. There is now a rapidly growing body of evidence that individuals can indeed recognize kin and even distinguish close kin from distant kin. How do they achieve this? An entertaining but theoretically unlikely possibility was proposed by Hamilton (1964) and coined the 'green beard effect' by Richard Dawkins. The idea is that there may be 'recognition alleles' which express their effects phenotypically so enabling bearers to recognize these alleles in others, and also cause bearers to behave altruistically towards others with the recognizable phenotypic effect. For example, if a gene conferred on its owner both a green beard and a tendency to be nice to others with the same feature, it would be favoured by evolution. This would provide a mechanism of kin recognition without learning.

Simple rules for recognizing kin

A more likely mechanism is that individuals use a simple rule, for example 'treat anyone in my home as kin'. Parent birds, for example, may ignore their own young if they are placed just outside the rim of their nests, yet will readily accept a strange chick placed inside their nest. This can lead to odd results; for example a reed warbler, *Acrocephalus scirpaceus*, may mob an adult cuckoo, *Cuculus canorus*, which approaches its nest and then, a minute later, return to the hard work of feeding a baby cuckoo inside its nest! Usually, however, this simple rule will lead individuals to care for their own offspring.

Another mechanism for recognizing kin is to learn that those you grow up with are kin. Konrad Lorenz gave the name 'imprinting' to the phenomenon observed in young geese, of following the first conspicuous moving object they see after hatching. Usually this will be their mother and so result in them following someone who will keep them warm and protect them. In an experimental situation, however, young birds have been imprinted on humans and even flashing lights. Experiments by Holmes and Sherman (1982) show that sibling recognition in ground squirrels is also in part based on association in the natal nest. They captured pregnant females and used their pups to create four kinds of experimental rearing groups: siblings reared by one mother (their own or a foster mother), siblings reared apart by different mothers, non-siblings reared as a single litter and non-siblings reared apart. When they were older, animals from the four groups were placed in pairs in arenas and their interactions were observed. Holmes and Sherman found that regardless of true genetic relatedness animals that were reared together rarely fought. Figure 11.3 shows that unrelated individuals reared together were no more aggressive to one another than true sibs reared together. This suggests, therefore, that individuals learn who are their kin from association in early life.

Female ground squirrels recognize kin partly by learning ...

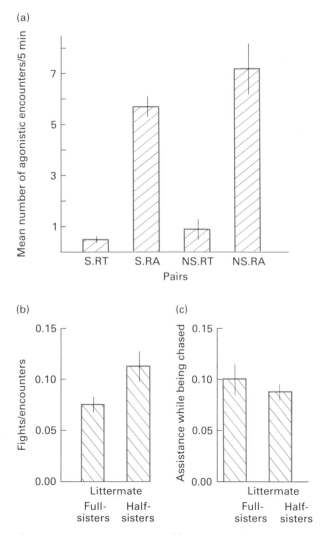

Fig. 11.3 Kin recognition in Belding's ground squirrels. From Holmes and Sherman (1982). (a) Laboratory experiments: mean number (± 1 SE) of agonistic encounters between pairs of yearling Belding's ground squirrels in arena tests. Non-siblings reared together (NS.RT) are no more aggressive than siblings reared together (S.RT). However non-siblings reared apart (NS.RA) are more aggressive than siblings reared apart (S.RA). (b) and (c) Field observations: aggression and co-operation among yearling females which were full or half-sisters (genetic relatedness determined by blood proteins). Full sisters are less aggressive to one another (b), and assist each other more (c).

However, it was also found that among animals reared apart, genetic siblings were less aggressive to one another in an arena test than unrelated individuals (Fig. 11.3a). Interestingly, this effect only occurred among females; thus true sisters reared apart were less aggressive to each other than unrelated females reared apart but genetic relatedness did not affect aggression between male—male or male—female pairs. Only females, the sex that behaves altruistically in the field, show evidence of being able to recognize unfamiliar, but genetically related, kin.

Of course, sisters reared apart may still learn to 'recognize' each other because of prenatal experience in their mother's uterus. Field observations by Paul Sherman, however, suggest that this may not be the whole story. Female Belding's ground squirrels mate with up to 8 different males (mean 3.3 males) on the one afternoon they are sexually receptive in spring. Analysis of polymorphic blood proteins collected from mothers, potential fathers and their offspring, showed that 78 per cent of the litters were sired by more than one male (Hanken & Sherman 1981; this method of assessing probable paternity is rather like the use of blood groups in man to settle paternity disputes in court cases). The exciting discovery was that littermates who were full sisters (same mother, same father) were less aggressive to one another and more co-operative than half-sisters (same mother, different father, i.e. half-sisters arise because of multiple matings). For example, when establishing nest burrows and defending territories full sisters fought and chased less often when they encountered each other than did half-sisters (Fig. 11.3b,c).

Littermates all share the same nest burrow and the same uterus, whether they are full sisters or half-sisters, so some mechanism other than this common experience must be involved. One possible mechanism is known as 'phenotype matching'. A female may be most altruistic to nest mates who are like her phenotypically (e.g. match her own odour). In conclusion, a female Belding's ground squirrel seems to categorize others in two ways. First, she recognizes and co-operates with individuals she shared a burrow with as opposed to those she did not; the former will be full sibs or half-sibs. Second, she may be particularly co-operative with nest mates who are like herself phenotypically, and hence more likely to be full sibs rather than half-sibs.

... and partly by 'phenotype matching'

CONCLUSIONS ABOUT KIN SELECTION

We have dealt at length with kin selection because in many ways it is the most theoretically interesting and possibly the most important mechanism for the evolution of altruism. As predicted by Hamilton's model, much of the altruism seen in nature does occur between closely related individuals and recent work shows that individuals can distinguish close kin from distant kin. However there are many examples of co-operation and altruism among unrelated individuals. How can this evolve? We now consider three other hypotheses.

Mutualism

Individual survival and reproductive success may be greater from co-operative team-work

It may sometimes pay two or more individuals to co-operate simply because each gains a net survival or reproductive benefit from doing so. This is known as mutualism. For example, two pied wagtails, *Motacilla alba*, who join together to defend a winter feeding territory can each enjoy a greater feeding rate than they would by being alone because the benefits of the association in terms of improved territory defence outweigh the costs of having to share the food (Chapter 5).

In other cases, both kin selection and mutualism may be relevant. A single lioness is not very effective at capturing a zebra, but by hunting with another female she improves her capture success so much that the benefits outweigh the costs of having to share the meat once the prey is caught (Caraco & Wolf 1975; Stander 1992). Lionesses in a pride are related (Chapter 1) and so individuals will get kin-selected benefits from the association, but in principle it would also pay two unrelated females to hunt together.

Recent studies of lions, using DNA fingerprinting to assign paternity (Chapter 9), have shown how co-operation among males is influenced by both gains in direct and indirect fitness (Packer *et al.* 1991). Recall from Chapter 1 that prides consist of a resident group of related females defended by a coalition of usually 2–6 males. Larger male coalitions are better able to take-over a pride and can also maintain residence for longer. As a result, the average reproductive success per male increases with coalition size. It would be tempting to conclude that it must therefore pay individual males to co-operate, but measures of *average* reproductive success per male do not tell us what each individual is gaining from the coalition. The fingerprinting showed that the resident males sired all the cubs conceived during their period of tenure of the pride (all 78 young in 7 prides), so they were clearly effective in keeping intruding males at bay. However, paternity was not shared equally among the pride males. Table 11.3 shows that when there are two resident males both gain paternity, but in larger coalitions the subordinate males gain little if any paternity.

Table 11.3 DNA fingerprinting shows how paternity is shared among the males in 7 prides, (a)–(g), studied in Tanzania by Packer *et al.* (1991)

No. males defending pride		No. young fathered by individual males (1–4)			
		1	2	3	4
Two	(a)	8	6	–	–
	(b)	6	5	–	–
	(c)	3	2	–	–
	(d)	8	2	–	–
Three	(e)	8	5	0	–
	(f)	4	3	0	–
Four	(g)	9	8	1	0

Small coalitions of male lions are mutualistic: each individual gains mating success ...

Why, then, do subordinate males join large coalitions if they have little chance of any personal reproduction? Interestingly, Packer *et al.* (1991) found that all the larger coalitions (4 or more males) were of close relatives. Unrelated males only joined to form small coalitions. We can conclude that coalitions involving small groups of unrelated males are a case of *mutualism*, where individuals co-operate because each gains increased reproductive success as a consequence. By contrast, kinship seems essential for the maintenance of larger coalitions where reproduction is highly skewed. Here, co-operation by subordinates depends on gains in indirect fitness, namely through aiding the reproductive success of their dominant kin.

... but in large coalitions subordinates gain only indirect fitness

Manipulation

Our third hypothesis is that what looks like altruism on the part of the donor may in fact be manipulation by the recipient (Dawkins 1982). This is most obvious, for example, when a parent bird feeds the offspring of a brood parasite, such as a cuckoo. The host gains nothing from its altruism; it has simply been tricked by the cuckoo into feeding the wrong species (see Chapter 4). Manipulation also occurs within a species. For example, some female birds lay eggs in the nests of conspecifics, thus avoiding the costs of incubation and parental care (Yom-Tov 1980). Female starlings, *Sturnus vulgaris*, who 'dump' eggs in other female's nests first remove a host egg before laying their own. They then deposit the host egg on the ground nearby. It was at first thought that these eggs which appeared on the ground were laid by females who were unable to get back to their nests in time to lay normally! Then, when eggs were marked in nests as they were laid, it was found that the eggs which appeared on the ground were often

Donors may be tricked into behaving altruistically

marked ones, in other words eggs which had been removed from nests (Feare 1984). In this case, just as with interspecific brood parasitism, the host is being tricked into behaving for another individual's benefit.

Reciprocity

Paying back a favour in the future ...

Whenever the benefit of an altruistic act to the recipient is greater than the cost to the actor, then as long as the help is reciprocated at some later date, both participants will gain (Trivers 1971). For example, A helps B today and then B helps A tomorrow. Reciprocation is common in human society and we use money and laws to regulate its use. The problem for its evolution in animal populations is the possibility of cheating. Because of the time delay between one individual gaining and the other doing so, B may accept help from A today but refuse to repay the favour tomorrow (cf. mutualism, where both individuals get the benefit at the same time). We will explore the conditions under which reciprocity can be evolutionarily stable by using a simple model.

... and the problem of cheating

THE PRISONER'S DILEMMA

The Prisoner's dilemma model was originally developed to help us think about human behaviour, but it provides an elegant model to illustrate the problems of achieving co-operation in animal societies (Axelrod & Hamilton 1981). Imagine two players in a game who have the choice of co-operating or defecting (being selfish). The pay-off matrix is given in Table 11.4 with some illustrative numerical values. These values represent the gain in fitness from the interaction (e.g. number of offspring gained).

Table 11.4 The Prisoner's dilemma game. From Axelrod and Hamilton (1981). The pay-off to player A is shown with illustrative numerical values

		Player B	
		Co-operate	Defect
Player A	Co-operate	$R = 3$ Reward for mutual co-operation	$S = 0$ Sucker's pay-off
	Defect	$T = 5$ Temptation to defect	$P = 1$ Punishment for mutual defection

In the Prisoner's dilemma
game both individuals
would benefit from co-
operation but both are
tempted to cheat

Imagine player A finds another individual B who always co-operates. If A co-operates too it gets a reward of 3, whereas if it defects it gets 5. Therefore if B co-operates, it pays A to defect. Now imagine player A discovers that B always defects. If A co-operates it gains nothing (the sucker's pay-off) whereas if it defects it gets 1. Therefore if B defects, it pays A to defect. The conclusion is that irrespective of the other player's choice, it pays to defect even though with both players defecting they get less (1) than they would have got if they had both co-operated (3). Hence the dilemma!

In other words, co-operation is not an ESS because in a population of 'all co-operators' a mutant who defected would spread. Defect, however, is an ESS; in a population of 'all defect' a mutant co-operator does not gain an advantage. Any population with a mixture of heritable strategies will therefore evolve to 'all defect'. More generally the conditions for this conclusion to hold in the matrix in Table 11.4 are

$$T > R > P > S \text{ and } R > \frac{(S + T)}{2},$$

which define the Prisoner's dilemma game. The problem is essentially that while an individual can benefit from mutual co-operation, it can do even better by exploiting the co-operative efforts of others.

Is there any way in which individuals can escape this dilemma and come to stable co-operation? The answer is no if the two players only meet once; defect is the only stable strategy in Table 11.4. Defect is likewise the stable strategy if the total number of interactions is precisely known in advance because defection will be optimal on the last encounter and therefore also on the next-to-last and so on back to the first encounter. However, if the series of encounters goes on indefinitely or, more realistically, if there is always a finite probability, w, that the two players will meet again, then we have to consider the possibility of more complex strategies involving mixtures of co-operation and defection in various sequences.

Axelrod (1984) ran a computer tournament in which 62 different strategies, submitted by scientists from all over the world, were paired against each other in a round robin tournament. Each entry played every other entry and also played against itself and against 'random', a program that randomly co-operates and defects with equal probability. The pay-offs for each contest were as in Table 11.4 and the game was iterated between the pair of contestants with $w = 0.99654$. This game can be thought of as a contest between intelligent individuals (the scientists who submitted the various strategies) competing to win a prize, or it can be thought of as a model of an evolutionary game, like the Hawk−Dove game (Chapter 7), in which the different strategies represent different genetic mutants competing for representation in the gene pool.

Some of the strategies were very complex, for example 'on each move model the behaviour of the other player as a Markov process and then use a Bayesian inference to select what seems to be the best choice in the long run'. Some were very unforgiving, for example 'never be the first to defect, but once the other player defects, always defect for the rest of the game'. Others were sneaky and tried to get away with occasional exploitation, for example 'if the other defects then defect too, but on 10 per cent of the occasions the other co-operates, defect'. The winning strategy was the simplest one of all those submitted, called 'tit for tat'. This co-operated on the first move and thereafter did whatever its opponent did on the previous move. Thus 'tit for tat' is a strategy of co-operation based on reciprocity. The success of 'tit for tat' against the variety of contenders was due to two features.

1 *It was retaliatory*, which discouraged the other side from persisting whenever a defection was tried.

2 *It was forgiving* after just one act of retaliation, which helped restore mutual co-operation and the greater rewards this gave.

On the other hand, 'tit for tat' did not enjoy the possible advantages of exploiting other rules to any great extent. While such exploitation could sometimes be fruitful, it often led to problems. First, a rule which defects to see what it can get away with risks retaliation from rules which are provocable. Second, once mutual recriminations have set in it can be difficult to escape from long sequences of unprofitable defection. Finally, the attempt to give up on unresponsive rules (e.g. 'random') often leads to giving up on rules which are in fact profitable to co-operate with by a more patient rule such as 'tit for tat'.

Axelrod's computer tournament showed that 'tit for tat' wins in an environment consisting of a variety of strategies. A further analysis examined what would happen if the tournament continued with the frequencies of the different strategies in the next round being proportional to the score they achieved in the previous one. We can imagine, for example, that the scores represent number of offspring, or strategy copies, produced in the previous round. A round of a tournament can therefore be regarded as equivalent to a single generation and a continuation of the game over successive rounds as a simulation of 'survival of the

'Tit for tat' is a rule that can be evolutionarily stable in the Prisoner's dilemma game

fittest' over successive generations. The results showed that as the less successful rules were displaced, 'tit for tat' continued to do well and eventually displaced all the other rules and went to fixation.

Once a strategy has gone to fixation, we have to consider whether it will be evolutionarily stable, in other words resist invasion by a mutant strategy (Chapter 7). Axelrod and Hamilton (1981) showed that 'tit for tat' is an ESS provided that w, the probability that two contestants meet again, is sufficiently large (Box 11.3). Thus, once 'tit for tat' is established, a mutant defector cannot gain an advantage. Therefore, things now look good for the evolution of co-operation based on reciprocity. Unfortunately, however, this is not the end of our analysis because 'all defect' is also an ESS, whatever the value of w. A population of 'all defect' results in pay-off P per move (Table 11.4). A mutant 'tit for tat', which co-operated on the first move would get the sucker's pay-off, S, with no chance of future compensation.

The problems of getting 'tit for tat' started

Our problem, therefore, is that although 'tit for tat' is stable once established, how can it ever get going in the first place in a primeval world full of selfish individuals who always defect? There are two possibilities. One evolutionary scenario is as follows. Co-operation could emerge at first between pairs of relatives, evolving by kin selection. Selection would favour the recognition of cues which gave indications of relatedness. One cue that could be used is simply the fact of reciprocation of co-operation. Then individuals could use rules such as 'co-operate if the other co-operates' (i.e. tit for tat) and co-operation could now emerge even between unrelated individuals (Axelrod & Hamilton 1981).

Another mechanism that can get co-operation started when most of the population is using 'always defect' is clustering. If 'tit for tat' individuals occur in clusters then they can interact with each other more than expected from random encounters in the population at large, and hence enjoy the increased benefits of mutual co-operation. Clustering is often associated with kinship and the two mechanisms can enforce each other in promoting the initial spread of reciprocal co-operation.

Predictions of the model

We have now shown that co-operation based on reciprocity can spread and be stable, provided two conditions hold.
1 Individuals must not be able to get away with defecting without the other individual being able to retaliate effectively. Some animals may be able to recognize individuals and so spot the

cheats. Another possibility is for two individuals to have a fixed place of meeting so they can be more or less certain of interacting again and again with the same individual.

The 'tit for tat' model helps to identify conditions under which reciprocity might be stable

2   For reciprocity to be stable, the probability, w, of the same two individuals meeting again must be high. If w falls, then defection may pay. For example if your partner became ill or old then there would be an incentive to defect and make a one-off gain when the probability of future interactions drops.

We now turn to some examples in nature where individuals do indeed show reciprocity.

Box 11.3 *'Tit for tat' is an ESS provided* w *is sufficiently large. From Axelrod and Hamilton (1981).*

'Tit for tat' has a memory of only one move. An effective challenger will take maximum advantage of it by repeating whichever sequence of co-operate and defect results in greatest pay-off. But because of the short memory of 'tit for tat', we need only consider repeated sequences of two moves. Thus, the most effective challengers will be repeated sequences of 'defect: defect' or 'defect: co-operate'. If neither of these strategies can invade 'tit for tat', then no strategy can. We now examine the iterated Prisoner's dilemma game in Table 11.4.

When 'tit for tat' plays another 'tit for tat' it gets a pay-off of R each move for a total of $R + wR + w^2R\ldots$ moves. The sequence of moves which reiterate with probability w sum to $1/(1 - w)$, so the total pay-off will be $R/(1 - w)$. 'All defect' playing with 'tit for tat' gets T on the first move and then P thereafter, so it cannot invade 'tit for tat' if

$$R/(1 - w) \geq T + wP/(1 - w)$$

The alternation 'defect: co-operate', when playing 'tit for tat' gets a pay-off of

$$T + wS + w^2T + w^3S \ldots = (T + wS)/(1 - w^2)$$

An alteration of 'defect and co-operate' cannot invade 'tit for tat' if

$$R/(1 - w) \geq (T + wS)/(1 - w^2)$$

In conclusion, neither strategy (and hence no strategy at all) can invade 'tit for tat' if both of the following conditions hold: $w \geq (T - R)/(T - P)$ and $w \geq (T - R)/(R - S)$.

EXAMPLES OF RECIPROCITY

(a) Spawning in the black hamlet fish

The black hamlet fish, *Hypoplectrus nigricans*, is a simultaneous hermaphrodite (all individuals have both male and female gonads) which lives in the Caribbean. Fish come together in pairs to spawn in the last two hours of the day before sunset. Fischer (1980) discovered that spawning occurred in bouts, in which each fish assumed alternate male and female roles ('egg trading'). Thus, individual A released a few eggs which B fertilized (fertilization is external), and then B would release some eggs which A fertilized, and so on. Why does this lengthy process occur? Surely it would be more efficient, and simpler, for A to release all its eggs and B to fertilize them and then for B to release all its eggs and A to fertilize them?

Hermaphroditic hamlet fish swap eggs and sperm in bouts

Fischer suggests that egg trading has evolved because it is stable against cheating. The problem is as follows. Eggs are more costly to produce than sperm, so on the first move A is in effect an altruist; B gains the same number of zygotes for less investment. On the second move, B returns the favour. Cheating is a possibility because hamlets do not have mature eggs to release every day, but they can probably produce sperm every day. It would therefore pay an individual to pair up on days when it had no eggs to trade. If A gave up all its eggs at once, B could then get away with cheating. However because A releases just a few eggs at a time it can then easily spot whether B is a cheat and has no eggs to pay back in turn. In fact Fischer discovered that in pairs where the partner failed to reciprocate, the donor refused to give up any more eggs and left. Egg trading is therefore somewhat reminiscent of the 'tit for tat' co-operation envisaged in Axelrod and Hamilton's model.

(b) Regurgitation of blood by vampire bats

Wilkinson (1984) studied a population of individually marked vampire bats, *Desmodus rotundus*, in Costa Rica. Individuals quite often failed to obtain a blood meal during the night and they then begged for blood from other individuals in the daytime roosts. In an experiment, it was found that 5 out of 8 bats captured in the evening before feeding and released into the roosts at dawn were subsequently given blood by a well-fed individual in the roost. By contrast, none of 6 bats captured and released after feeding successfully were given a meal.

Wilkinson discovered that regurgitation occurred only between close relatives or between unrelated individuals who were frequent roost-mates. For reciprocity to evolve, the following conditions are needed.

Unrelated vampire bats feed each other on a reciprocal basis

1 *Donors must be able to recognize cheats, and refuse to feed previous recipients who fail to reciprocate.* Wilkinson performed some clever experiments in the laboratory where he formed a group of bats, some individuals (all unrelated) coming from one roost and others (also unrelated) from another roost. In a series of trials, one bat, chosen at random, was removed and kept hungry while all the others had access to blood. The hungry bat was then reintroduced. It was found that 12 of the 13 regurgitations occurred between individuals from the same roost in the field, in other words individuals which were familiar with each other. Furthermore, the starved bats which received blood later reciprocated the donation significantly more often than expected had the exchanges occurred randomly.

2 *Sufficient repeated pairwise interactions so that there are interchanges of roles and therefore net benefits to all donors.* In the field it was found that some unrelated individuals were constant companions in the roosts, sometimes for several years.

3 *The benefit of receiving aid must outweigh the cost of donating it* ($R > S$ in Table 11.4). Figure 11.4 shows that the bats lost weight with increasing time since their last meal. The decelerating curve means that a donation of a small amount of blood by a well-fed individual results in little cost in terms of time moved along the bottom axis towards the threshold of death. However

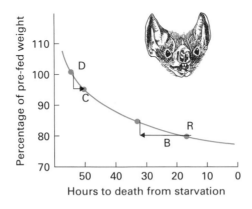

Fig. 11.4 In vampire bats, weight loss after feeding follows a negative exponential decline, with death from starvation occurring at 75 per cent of pre-fed weight at dusk. Therefore a donation of 5 per cent of pre-fed weight when at weight D should cause a donor to lose C hours but will provide B hours to a recipient at weight R. From Wilkinson (1984).

this same amount can bring an enormous benefit to a starving individual, moving its position considerably to the left along the time axis. The blood meal therefore has little cost to the donor and great benefit to the recipient; indeed it may save the recipient's life and enable it to survive until it has the chance to forage again for itself the next night.

(c) Alliances in primates

Grooming (Fig. 11.5) is the most common form of affinitive behaviour in primates and it is also common for two individuals to join together to fight a third. Most grooming and alliances involve close relatives but sometimes friendships form between non-relatives. Seyfarth and Cheney (1984) performed field experiments which showed that grooming between unrelated vervet monkeys, *Cercopithecus aethiops*, increased the probability that they subsequently attended to each other's solicitations for aid.

Fig. 11.5 Two unrelated female vervet monkeys; one is grooming the other. Each female is nursing a ten-week old infant. Unrelated females co-operate in territory defence, alarm calls at the approach of predators and in interactions against other females in the group. Photo by Phyllis Lee.

(a)

(b)

Fig. 11.6 (a) A male olive baboon guarding an oestrus female. (b) Two male olive baboons about to separate a consorting male from an oestrus female. The consorting male is closest to the female in the centre of the photo. The male on the right turning his head, is soliciting aid from the male on the far left. Photo © Craig Packer.

The experiments were done in Amboseli National Park, Kenya, and involved playbacks of vocalizations used by individuals which sought support from others in an alliance. It was found that individual A would look towards the speaker broadcasting B's vocalization for a longer time if B had recently groomed A. Interestingly, this effect worked only between unrelated individuals. Relatives attended to the vocalizations equally whether the caller had recently groomed them or not. These results suggest that vervet monkeys are more willing to aid an unrelated individual if it has behaved affinitively to them in the recent past.

Another example comes from Packer's (1977) study of the olive baboon, *Papio anubis*. When a female comes into oestrus a male forms a consort relationship with her, following her around wherever she goes, awaiting the opportunity to mate (Fig. 11.6a). Sometimes a male who does not have a female enlists the help of another, unrelated male (Fig. 11.6b). This solicited male engages the consort male in a fight and while they are busy doing battle, the male who enlisted help goes off with the female! On a later occasion, the roles are reversed; the male who gave help is now assisted by the one who received help previously and so true reciprocation is involved. Crucial to the conclusion that reciprocity is involved, is the observation that the enlisted male has virtually no chance of mating. In other words, he pays an immediate cost (fighting) with delayed benefits. Bercovitch (1988) suggests that the enlisted male does sometimes mate, in which case alliance may be an example of mutualism.

Unrelated vervet monkeys form alliances on a reciprocal basis

Summary

Altruism is acting to increase the lifetime number of offspring of another individual at a cost to one's own chances of survival and reproduction. Co-operation can evolve in four ways.

1 *By kin selection.* An individual can increase its genetic representation in future generations by helping close relatives, who share copies of its genes identical by descent. The conditions for the spread of an altruistic allele are given by Hamilton's rule. Examples of traits favoured by kin selection may include alarm calling in prairie dogs, wife sharing in Tasmanian native hens and the formation of large male coalitions in lions. Experiments show that individuals can distinguish close kin from others (e.g. Belding's ground squirrel).

2 *Mutualism.* Both individuals gain a net survival or personal reproductive benefit from co-operating.

3 *Manipulation.* One individual is 'tricked' by another into behaving altruistically.

4 *Reciprocity*. The iterated Prisoner's dilemma model explores the conditions under which 'tit for tat' co-operation involving related or unrelated individuals can spread and be stable. Putative examples in nature include blood sharing in vampire bats, egg trading in black hamlet fish and alliances in primates.

Further reading

Grafen (1991) gives an excellent account of the logic and assumptions underlying the idea of kin selection. Sherman (1981a,b) considers whether there is a limit to nepotism (care for relatives) in ground squirrels. Grosberg and Quinn (1986) demonstrate a genetic basis for kin recognition in a colonial tunicate. Grafen (1990c) evaluates the evidence for kin recognition. Dawkins (1982) gives a thought-provoking view of the interaction between genes and phenotypes. The book by Axelrod (1984) explores the lessons of the Prisoner's dilemma model for human society.

Topics for discussion

1 All altruism is genetically selfish. Discuss.
2 Under what conditions would co-operation be expected to evolve (a) by kin selection, (b) by reciprocity?
3 What are the mechanisms that can be used for recognition of kin?
4 Read the papers by Milinski (1987), Reboreda and Kacelnik (1990), Lazarus and Metcalfe (1990) and Milinski *et al.* (1990a,b) and discuss whether or not sticklebacks play 'tit for tat'.

Chapter 12. Co-operation and Helping in Birds, Mammals and Fish

In Chapter 11 we showed that it is possible to estimate the probability of a parent sharing a particular gene with its offspring. The answer for a sexually reproducing species was shown to be 0.5; in other words any gene in a parent's body has a 50 per cent chance of being duplicated in an offspring. Hence to make a genetic profit by sacrificing its life (as for example in certain arthropods such as Cecidomyian gall midges in which a mother offers herself as food for her offspring; (Gould 1978) a parent must produce more than two offspring. But in Chapter 11 we also showed that full siblings have a 50 per cent probability of sharing genes, exactly the same degree of relatedness as for mother and child. Therefore if we view helping and self-sacrifice purely from the point of view of making a profit for selfish genes, our arguments for parental care should apply with equal force to caring for siblings (and also with slightly less force to caring for grandchildren and cousins). The reasons why helping offspring is commoner than helping brothers and sisters are therefore likely to be ecological and practical considerations such as the ease of recognizing offspring and the degree to which offspring or siblings benefit from a given amount of aid rather than genetic factors. Some of these are summarized in Box 12.1.

Helping offspring is genetically equivalent to helping brothers and sisters

Genetic predispositions and ecological constraints

Although helping in animals usually consists of parents caring for offspring, there are many examples of helping directed at other individuals. In more than 200 species of birds and 120 species of mammals, for example, some individuals spend part or all of their lives helping others to reproduce: they help to feed or protect the offspring of others while apparently forgoing the chance to breed themselves. This behaviour appears to be *altruistic*, since the helpers give benefit to others while incurring costs. These costs might be measurable as, for example, energy expended by the helper, but ultimately they must be scored as lost opportunities for the helper to have its own offspring. The natural history of helping varies from species to species, but broadly speaking there are two main categories: *helpers at the nest* (about 80 per cent of species of birds and mammals with helpers (MacDonald and Moehlman 1983, Brown 1987)), in which a group consists of a breeding pair and one or more younger individuals that help with various aspects of parental care; and *plural*

Helping in birds . . .

. . . helpers at the nest

. . . and plural breeders

Box 12.1 *Some hypotheses about the ecological and proximate constraints that favour care of offspring over care of siblings in most species.*

1 Young benefit more from a given amount of aid. One insect fed to a helpless nestling contributes more to its survival than the same insect would to the survival of a healthy sibling. In other words young are 'superbeneficiaries' (West Eberhard 1975). In species with overlap of generations this constraint may not apply since newly born younger siblings might benefit just as much from a helper's aid as would an offspring of the helper.

2 Young are generally easy to identify as relatives of the helper, while siblings may be harder to recognize as kin.

3 While young are often born close to the parents, siblings may disperse from their birth site and therefore not be available for receiving help.

4 Young, especially in slow or continuously growing species (e.g. fish), are very different in size from the helper and are therefore not likely to compete with it for food in the future. In contrast, if the helper increases the survival of one of its siblings from the same brood, it may suffer in the future from increased competition.

5 Young may be more valuable in terms of expected future reproduction. If a youngster is successfully reared to maturity it has a higher expected future reproductive output than does a sibling of similar age to the helper.

6 In a species in which both parents care for the young, suppose that an average individual can raise x youngsters. By helping to raise sibs at home an individual could add up to x offspring-equivalents (full sibs) to the fledged brood, but by mating and having its own offspring it could produce $2x$ offspring, because of the contribution of its mate. In other words the mate's contribution to rearing offspring generally favours parenting over helping (Charnov 1981).

breeders, in which several males and females share a nest and raise a communal brood. In the second category it often transpires that some individuals produce a disproportionately large share of the young, whilst in the first category, helpers occasionally breed, so the two categories can be viewed as two ends of a continuum.

In this chapter we will discuss how helping might be favoured by selection in birds, mammals, and fish. The problem can be considered as having two parts.

1 *The genetic predispositions for helping.* In the last chapter we saw how several evolutionary processes could account for the origin and maintenance of apparent altruism. First, helpers might aid close relatives such as siblings and thereby make a genetic profit in terms of the 'indirect' component of fitness. Second, helpers might make a genetic profit simply by increasing their own chance of survival and reproduction ('direct' fitness gains) by helping unrelated individuals, in which helping could be classified as mutualistic. Thirdly, the other two processes of reciprocity and manipulation might play a role. As a final point, it is worth bearing in mind that a combination of some or all of these processes might be involved.

2 *Ecological constraints which favour helping in some species but not in others.* Even if the potential genetic gains for helping and parenting are similar, practical considerations usually favour parenting (Box 12.1). Perhaps in some species these constraints are reversed in their effects. Sometimes, the option of becoming a parent is simply not available.

In this chapter we will start by describing the Florida scrub jay, the subject of one of the most detailed studies to date of a bird species with helpers at the nest. We then go on to look at how other studies of helpers at the nest test and extend the scrub jay study, before looking at two examples of plural breeders. Finally, we describe the most elaborate helping system of any vertebrate, that of the naked mole rat.

An example of helping in birds — the Florida scrub jay

The Florida scrub jay (*Aphelocoma coerulescens*), as its name implies, lives in oak scrub in Florida. Suitable habitat for birds is patchy and scarce, so there are isolated pockets of breeding jays with totally unoccupied and unsuitable areas between. The breeding birds live as pairs in year-round territories. Just over half the pairs have helpers, the average number being about 1.8 per pair. These birds help the breeding pair to feed their young and to defend the nest (Fig. 12.1); they are generally 1–2-year-old birds and almost without exception are related to the breeders. Out of 165 helpers, 64 per cent helped their parents, 24 per cent helped one parent (the other having died) and only 4 per cent helped breeders that were not their relatives. Both males and females help, but males generally stay at home longer than females. When helpers leave home they disperse and attempt to set up their own territory. Glen Woolfenden and John Fitzpatrick's (1984) long-term study of the scrub jay shows how genetic and ecological factors contribute to the explanation of why birds stay at home and help.

Margin notes:

Ways in which helpers could make a genetic profit

Scrub jays are monogamous and are often helped

Helpers are close relatives

Fig. 12.1 The Florida scrub jay. Helpers provide food for the young in the nest and defend the nest against predators such as snakes. Photo by Glen Woolfenden.

(a) Breeders benefit from the presence of helpers

Breeding groups of scrub jays with helpers rear more young than those without help (Table 12.1). Helpers make two kinds of contribution.

1 They help to defend the nest against predators, such as snakes, by mobbing and giving alarm calls to warn the chicks. This antipredator behaviour is primarily responsible for the increase in survival of chicks when helpers are present in a group.

2 They help to feed the nestlings, providing up to 30 per cent of all food brought to the chicks in some groups. While this apparently does not result in an increase in chick survival, since the

Scrub jay helpers defend the nest and feed young

Table 12.1 Breeding pairs of Florida scrub jays benefit from the presence of helpers. Both experienced and inexperienced (first time) breeders rear more young when they have helpers. From Woolfenden and Fitzpatrick (1984)

	No helpers	With helpers	Average number of helpers in territories with at least one helper
Inexperienced pairs	1.24	2.20	1.7
Experienced pairs	1.80	2.38	1.9

Parents with helpers
survive better

total amount of food brought to the nest is not increased by the efforts of helpers, it does relieve the burden on parents so that they survive better from one season to the next. Breeders with helpers have to put less effort into feeding their chicks and probably as a consequence of this their annual survival rate is 85 per cent, as opposed to 77 per cent for breeders without helpers. Since the helpers and parents are related to one another by about 0.5, this increase in parental survival represents a genetic gain for helpers in addition to that made by rearing younger siblings.

(b) Habitat saturation is an ecological constraint

Nests with helpers raise
more offspring . . .

Do helpers benefit from staying at home and helping? Clearly they make some genetic profit because they help to rear their own younger siblings which, as we said at the beginning of the chapter, is approximately equivalent to rearing offspring. But the crucial question is whether helpers could in theory do better by setting up their own territory and breeding themselves right from the start, or by staying at home and helping. One helper adds on average 0.33 offspring to its parent's production of young, and a novice pair of breeders rears about 1.24 young (a generous estimate of what a helper would do if it set off and bred instead of helping). So a helper could make 0.62 genetic equivalents by going off to breed instead of only 0.14 equivalents by staying at home (Table 12.2). These calculations are greatly oversimplified. They do not, for example, take into account the following: the mortality of

Table 12.2 A scrub jay helper would do better by setting up its own breeding territory if it could find one, than by helping its parents at home. From Woolfenden and Fitzpatrick (1984)

Option		Result
1 Stay at home and help	Young produced by experienced parents with no help	1.80
	Young produced by pair with helpers	2.38
	Extra young due to presence of helpers	0.58
	Average number of helpers	1.78
	Genetic equivalents to a helper (r to nestlings $= 0.43$)	0.14
2 Go off and rear own young	Young reared by first time breeders	1.24
	Genetic equivalents (r to own offspring $= 0.5$)	0.62

young birds is increased by setting off too early in life to seek a territory; by staying at home a youngster can increase the survival of its parents, which as mentioned earlier is part of the genetic pay-off for helping; helping may act as a form of training and increase the youngster's success when it breeds on its own; and the best strategy for a young bird may depend on what all the others do (the more that decide to leave home, the more competition there will be for breeding sites). But these complications are not likely to alter the general conclusion that a young scrub jay would do better to leave home if it could find a territory, in spite of the fact that it rears siblings when it helps at home. The most plausible interpretation is that helpers stay at home because there are no spaces for them to set up a breeding territory: the habitat is saturated. This interpretation is supported by Woolfenden and Fitzpatrick's observation that as soon as vacancies arise helpers usually leave home to set up on their own. In 1979–80, for example, almost half the adults in the study population died of an unidentified disease. The following year most young birds had set up territories in the empty spaces and very few remained at home as helpers.

... but if helpers could establish their own territory they would do better to leave home

Habitat saturation prevents helpers from leaving

(c) Males benefit by inheriting a breeding space

About half of the surviving male helpers eventually acquire a breeding space by inheriting part or all of their parental territory (Woolfenden & Fitzpatrick 1978; Fig. 12.2). The mortality rate of adult males is low and therefore very few vacant territories appear each year; so one way for a young male to acquire a territory is to help his parents to increase the size of their territory until it is large enough for him to split off part of it. By helping his parent the young male increases their production of young and therefore the size of group occupying the territory. Larger groups can expand their territories at the expense of small groups, and eventually the young male buds off to set up his own territory at the edge of the parental area (Fig. 12.2). When there are several male helpers in a territory there is a dominance hierarchy and the oldest dominant male is the first to inherit his own territory.

Males often inherit part of the parental territory

As with most birds, females do not breed in or near their parents' territory, but instead they disperse to find a vacant breeding space elsewhere. As discussed in Chapter 9, this sex difference in dispersal might be a mechanism to avoid inbreeding, since females usually end up mating with an unrelated male as a consequence of dispersal. The female's benefit from helping is therefore less than that of the male: she rears close relatives and has a safe base from which to explore, but she does not inherit

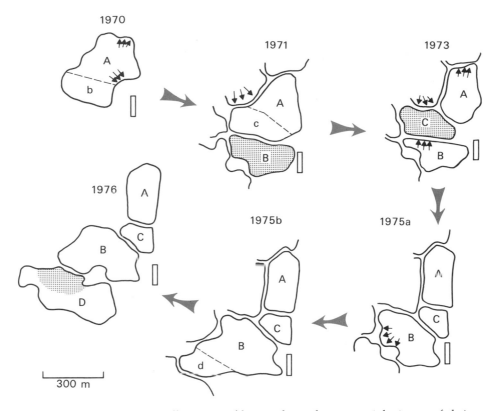

Fig. 12.2 An illustration of how male scrub jays may inherit part of their parents' territory. This is one of the benefits of helping. The diagrams show how three helpers (B,C,D) acquired territories. Shaded areas represent land inherited by two sons (B,C) and one grandson (D) of the pair occupying territory A. Broken lines show the incipient point of budding off and small arrows show how territorial boundaries expanded. From Woolfenden and Fitzpatrick (1978).

Males stay at home longer than females

her own breeding territory as a result of helping. As one might predict from these observations, males tend to stay at home longer than females. In other words help is given to the parents in proportion to the future expected gain for the helper. In addition to the ecological factors predisposing males to help more than females, there may possibly be genetic considerations (Box 12.2).

To summarize, the study of the scrub jay shows that the question 'why do some birds stay at home and help?' has quite a complicated answer. Birds stay at home because (a) they increase their own chance of survival, (b) the chance of successful dispersal to a breeding site is low and (c) if they are males they may inherit all or part of their father's territory. Helping is advantageous because (a) it increases the survival of helpers' parents, (b) it increases the production of young relatives of the helper and (c) it

Two issues: why stay at home? Why help?

may eventually result in an increase in territory size which in turn gives the young male helper a chance to acquire part of the territory for himself.

The scrub jay example illustrates the point that it is difficult to resolve the issue of whether the benefits of rearing siblings (indirect fitness gains) or the various other benefits (direct fitness gains) are the crucial ones that promote helping. A considerable controversy has arisen in the literature about this (Woolfenden & Fitzpatrick 1984; Brown 1987) but the resolution of the debate cannot be in terms of either one or the other. Both factors are important and the emphasis one should place on them depends on the question one asks. If the question is 'why stay at home?', the calculations in Table 12.2 show that benefits from raising siblings are not enough to account for this, but if the question is 'why help?' the fact that helpers are related to the nestlings is important.

Both direct and indirect fitness gains contribute to the evolution of staying at home and helping

Box 12.2 *Genetic predispositions for males to become helpers.*

Ric Charnov (1981) has pointed out that there may be a genetic predisposition for males to become helpers. This arises from uncertainty of paternity. Suppose that a male does not father all his mate's offspring (because he is cuckolded). A daughter of this male is more closely related to her own children than to her siblings. Her relatedness to her children is always 0.5, but if she and her siblings are not all fathered by the same male the average relatedness between her and her siblings is less than 0.5. In the extreme example where each offspring has a different father, the degree of relatedness between brothers and sisters is 0.25.

Now consider the effect of the same situation on a male. If he never fathers his mate's children his relatedness to them is zero, while he is always related to his siblings by at least 0.25 (since all siblings share the same mother). If males father half their mate's children their average relatedness to offspring is $0.5/2 = 0.25$. Their relatedness to siblings in this case is 0.25 (genes shared via the mother) $+ 0.25/2$ (genes shared by siblings through the father). This gives a total of 0.375: again males are more closely related to their siblings than to their own mate's offspring.

These calculations serve to illustrate the point that any degree of uncertainty of paternity might predispose males, more than females, to become helpers and rear siblings instead of offspring.

The scrub jay pattern in other species

Other examples of
helpers that care for
younger siblings

The pattern found by Woolfenden and Fitzpatrick in scrub jays
seems to apply to many bird, mammal, and perhaps fish species
with helpers. Helpers are usually offspring of the breeding pair,
their presence usually increases the breeding success of the group,
and they usually have little opportunity to breed themselves
because of ecological constraints. The black-backed jackal (*Canis
mesomelas*) in the Serengeti fits this pattern. Monogamous breeding pairs have 1–3 young from previous litters that act as helpers
by regurgitating food for the pups and lactating female as well
as grooming, guarding and playing with the pups (Figs 12.3
and 12.4). Similarly, in the Princess of Burundi cichlid fish
(*Lamprologus brichardi*) helpers are young of the breeding pair
from previous broods. These small fish live in the shallows at the
edge of Lake Tanganyika where they defend a territory around a
nesting hole. Young stay on their parents' territory for 2–3 breeding cycles (up to 12 months) where they both guard the nest and
clean the eggs or larvae. As a result of this help the mother is
able to lay more eggs, but at the same time, the energetic expense of
helping causes the young fish to grow more slowly (Taborsky &
Limberger 1981; Taborsky 1984).

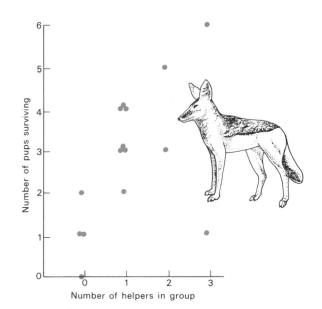

Fig. 12.3 Breeding success of black-backed jackals is positively correlated
with the number of helpers in a group. From Moehlman (1979).

(a)

(b)

(c)

(d)

Fig. 12.4 Black-backed jackals (*Canis mesomelas*) are monogamous and have helpers who are offspring from a previous litter. (a) A mated pair grooming their year-old son (centre) who remains on the parental territory as a helper. Helpers contribute to pup survival by regurgitating food (b), chasing off predators, in this case a spotted hyena (c), and by helping to defend the territory (d), the intruder is on the right. Photos © Patricia D. Moehlman (1980).

DO HELPERS REALLY HELP? EXPERIMENTAL EVIDENCE

A correlation between the presence of helpers and breeding success of the group does not necessarily show that helpers help. The correlation could arise, for example, because good quality parents produce many young each year and therefore have both a high annual success and a large number of helpers from previous seasons. Alternatively, some pairs might live in good territories and have enough food or shelter from predators to rear many young each year and therefore acquire a large number of helpers. When Woolfenden and Fitzpatrick (1984) compared the number of young raised by the same pair in the same territory with and without helpers, the difference was much smaller than in the overall comparisons shown in Table 12.2. In order to disentangle cause and effect it is necessary to do an experiment such as the one carried out by Jerram Brown *et al.* (1978, 1982) on grey-crowned babblers (*Pomatostomus temporalis*). The grey-crowned babbler lives in year-round territorial groups of 2–13 birds in open woodland in Queensland, Australia. As with the scrub jays, each group consists of a breeding pair and a variable number of helpers which are young from previous broods. The breeding success of a group is positively correlated with the number of helpers present, and in order to show that helpers actually cause the increase in success Brown did a removal experiment. He removed all the helpers but one from 10 breeding groups and left another 10 groups as controls with 4–6 helpers per group. The control groups reared more young per season than the experimentals, mainly because the helpers feed the young and relieve the pressure on the breeding female so that she can recover from one breeding attempt and start again sooner. Mumme (1992) has carried out removal experiments on scrub jays and confirmed that the correlation observed by Woolfenden and Fitzpatrick is a cause–effect relationship.

Removal experiments show that helpers cause increased breeding success

EXPERIMENTAL EVIDENCE FOR BREEDING CONSTRAINTS

Observations of young remaining at home until breeding vacancies arise (as in the scrub jays and jackals) provide circumstantial evidence that a full habitat can lead to co-operative breeding. However, it could be argued that some young choose to stay at home because they are unable to breed rather than because they are prevented from doing so by ecological constraints. The best way to test the constraint hypothesis is to experimentally remove the constraint to see whether the helpers then leave home to breed.

Fairy wrens stop helping
and leave home when
vacancies are created

Pruett-Jones and Lewis (1990) did this in their study of superb
fairy wrens (*Malurus cyaneus*) in Canberra, Australia. In this
species most helpers are male. To test whether male helpers
would leave home and breed if vacancies were made available,
breeding males were removed from territories with monogamous
pairs. Of 32 helpers given the opportunity to pair with an unmated
female on a nearby territory, 31 left home to take over the
breeding opportunity, usually filling the vacancy within a few
hours of the removal and beginning a breeding attempt within a
few weeks. By contrast, helpers did not disperse into vacated
territories from which both the male and the female breeder had
been removed, though they promptly filled the vacancy when the
female was reintroduced. Thus in this species vacant territories

Both vacant territories
and unmated females act
as constraints

with available females are a constraint leading to co-operative
breeding.

In some co-operative breeders the young do not leave home
immediately the first breeding opportunity arises, but rather wait
for a high quality vacancy. This applies to the Seychelles warbler,
Acrocephalus sechellensis, studied by Jan Komdeur (1992). Until
recently this species was confined to the tiny island of Cousin
(29 ha) in the Seychelles, Indian Ocean. The island is completely
covered with warbler territories and young birds remained on
their natal territories as helpers until breeding vacancies arose.
There was marked variation in territory quality, with the lusher
insect-rich territories in the centre of the island producing five
times the reproductive success of poor territories on the coast.
When Komdeur created breeding vacancies by removal experiments he found that helpers on high quality territories competed
for vacancies on good territories but preferred to stay at home as
helpers rather than occupy breeding vacancies on poor territories.
These latter vacancies were filled only by helpers from poor
territories. Calculations showed that helpers on good territories
did better by waiting for a high quality vacancy, even if this
meant delaying breeding for several years. By contrast, helpers on
poor territories, where survival chances were much lower, did
better to fill a poor quality vacancy immediately.

When introduced onto a
new island, Seychelles
warblers did not help at
home

Komdeur introduced the removed birds onto an unoccupied
island, Aride. These then began breeding within a few weeks.
When the young became independent they immediately left home
and set up breeding territories in the abundant unoccupied habitat,
so there was no helping. However, as the new population increased
and all the good quality habitat became occupied, young began to
remain at home as helpers rather than breed in poor territories.

These kinds of observations have led Stacey and Ligon (1991)
to propose a 'benefits of philopatry' model for co-operative breeding,

Young on good quality
territories may be less
likely to leave home

which emphasizes that helping can arise because young choose to stay at home, rather than because they are constrained from dispersing by a full habitat. However the two are not so much alternative hypotheses as different sides of the same equation determining the fitness of the young bird's options. The pay-offs from staying at home will depend on quality of the natal territory which will influence survival chances and the indirect fitness that can be gained from helping. The pay-offs from dispersing to breed will depend on the quality of the breeding vacancy available. Different individuals in the population are likely to experience differing degrees of breeding constraints depending on the quality of their natal territory and the quality of the breeding vacancies.

Helpers are not always relatives

(a) The dwarf mongoose: unrelated helpers and pseudopregnancies

Dwarf mongoose groups
have related and
unrelated helpers

The dwarf mongoose (Plate 12.1, between pps 212−213) is a 320-g carnivore that lives in groups of about 9 individuals, consisting of a breeding pair (the alpha male and female) and subordinate helpers. Subordinates become sexually mature in their first year but do not usually breed successfully until they become alphas at the age of 3−4 years: they are presumed to be prevented from breeding before this by the alpha pair. Instead they help by guarding the den (often in the ventilation shaft of a termite mound) against predators, by bringing inverte-brate food to the young, and, in the case of some female sub-ordinates, by lactating and suckling the young. The helpers are either the offspring of the breeding pair, and therefore closely related to the young they help, or immigrants from other groups, in which case they are unrelated to the young they help. Females are more likely than are males to stay at home in their natal group. A 15-year study by Jon Rood (1978, 1990) and Scott Creel (Creel *et al.* 1991; Creel & Waser 1991) in the Serengeti National Park, Tanzania, has revealed how three different strategies — helping as an unrelated immigrant, lactation by subordinates, and breeding by subordinates — reflect different balances of costs and benefits through direct and indirect fitness routes.

1 *Unrelated helpers* gain no genetic profit through the young they help to raise, so the only possible gain they can make is to increase their own success in the future (direct fitness gain). In the dwarf mongoose this direct fitness gain comes about because helpers eventually take over as breeders. The alpha males and

Unrelated helpers benefit
by eventually becoming
breeders

females generally remain as breeders until 5 years of age, when they are nearly always replaced by the oldest subordinates in the group. As with the scrub jay, there is a shortage of breeding opportunities, and helping as an unrelated immigrant is part of the strategy of acquiring a slot as a breeder. But why should the unrelated dwarf mongoose help instead of waiting on the sidelines for a chance to breed? There are three factors which might be important. First, helping may be a form of 'payment' for permission to stay on a breeding territory while waiting to take over. If the resident breeding pair gain nothing from having an unrelated helper close at hand they may simply evict it. Second, helping keeps the group and therefore the territory intact, which is vital to the helper's future success when it becomes a breeder itself. Finally, some of the young reared by the helper will in turn help it when it becomes a breeder, regardless of whether or not they are relatives. These points can be summarized by saying that helping is a long-term investment for selfish future breeding success.

Why do unrelated individuals help?

2 *Subordinate breeders*. Sometimes subordinate females, especially older ones, become pregnant and breed. Creel and Waser (1991) suggest that this may reflect a strategy by alphas of 'allowing' older subordinates to increase their reproductive success. They point out that as subordinates grow older in their natal group they become, on average, less related to the young they are helping (because one or both alphas are likely to have been replaced by new immigrants). Thus the genetic benefit through indirect gains of helping diminishes with age. At the same time, older subordinates are more likely to succeed in taking over another group as an alpha if they emigrate. Hence, as subordinates grow older, staying at home becomes a less attractive option. Perhaps, the alpha pair tip the balance in favour of staying at home (which benefits the alphas because of the help received) by allowing older subordinates to breed.

Helpers sometimes produce young ...

... this may tip the balance in favour of staying at home

3 *Pseudopregnancy*. Young female helpers sometimes mate and become pseudopregnant: they do not actually give birth but they lactate and suckle the young of the alpha pair. These females, unlike the subordinate breeders, are generally very closely related to the offspring of the alpha pair $(r = 0.36)$ and their effect on the production of offspring is sufficiently large to make helping in this way a genetically more profitable option than breeding as a subordinate (Creel *et al.* 1991).

(b) Anemone fish

A similar interpretation may be applicable to helpers in the anemone fish, *Amphiprion akallopisos*. Breeding pairs of fish defend anemones, which are an essential scarce resource for successful reproduction. Sometimes the pair are helped in defence by a non-breeder, which is most unlikely to be a relative of the breeding pair since the young fish disperse in the plankton before settling on a territory. Instead the helpers are probably unrelated and as hypothesized for mongoose helpers, they may be 'hopeful reproductives' waiting for a chance to take over when one of the breeders dies (Fricke 1979). Helping may again be a form of payment for permission to stay in the territory.

(c) Pied kingfisher: primary and secondary helpers

The idea of helping as a payment for permission to remain in a group is clearly illustrated by the pied kingfisher (*Ceryle rudis*) (Fig. 12.5) in which the breeding pair apparently accept unrelated helpers only when the payment is sufficiently advantageous to them. The pied kingfisher, unlike most of the species referred to so far, is a colonial nester and does not live in a year-round territory. Uli Reyer (1980) compared two colonies, one at Lake Victoria and the other at Lake Naivasha in East Africa. At Naivasha, breeding pairs have at most one helper which is nearly

Fig. 12.5 A pied kingfisher bringing food to the nest. Photo by Uli Reyer.

Primary helpers are
relatives, secondary
helpers are not

always a 1−3-year-old male offspring of at least one of the breeding pair (a 'primary helper'). He helps by feeding fish to the female, defending the nest against predators, and bringing food to the nestlings. At Lake Victoria there are sometimes secondary as well as primary helpers; these birds are also males but are unrelated to the breeding pair. They show up at both colonies just after the young hatch out and try to feed the breeding female of a pair; they are persistently chased away by the breeding male at Naivasha, but at Victoria they are eventually tolerated and allowed to stay and help to feed the young. The reason why secondary helpers are accepted at one lake but not the other is probably related to feeding conditions. Victoria is not as good for fishing as Naivasha: the water is rougher so that it takes longer for a kingfisher to catch a fish, the fish are smaller, and the feeding grounds are further away from the colony than at Naivasha. As one might expect, the parents expend more energy per day feeding their chicks at Victoria than at Naivasha (Reyer & Westerterp 1985). This means that parents with only one helper have difficulty in bringing enough food for their chicks and the secondary helpers make a big difference to breeding success (Table 12.3). At Lake Naivasha one helper is enough to rear all the young successfully and a secondary helper could therefore contribute little to the success of a breeding pair. Reyer and Westerterp (1985) confirmed this interpretation by the experiment of adding chicks to broods at Naivasha and taking them away at Victoria. The Naivasha parents, now facing the same degree of energy expenditure as birds would normally have at Victoria, began to accept secondary helpers while the Victoria birds with reduced broods rejected them.

Secondary helpers are
only accepted when the
going is tough

The conclusion is that unrelated helpers are accepted only when their help is effective at increasing the breeder's success or easing their heavy energetic burden. But why do secondary helpers

Table 12.3 Helpers in the pied kingfisher. At Lake Naivasha nests have no helpers or one helper, feeding conditions are good and one helper is therefore enough to ensure that all the chicks get enough to eat. At Lake Victoria there are sometimes second (unrelated) helpers, feeding conditions are poor, and the second helper makes a big difference to chick survival. From Reyer (1980, 1984)

| | Per cent nests with helpers | | | Feeding conditions | | Breeding success | | Young fledged with helpers | | |
	0	1	2	Time to catch a fish (min)	Per cent dives successful	Clutch size	Young hatched	0	1	2
Lake Naivasha	72	28	0	5.9	79	5.0	4.5	3.7	4.3	—
Lake Victoria	37	43	20	13.0	24	4.9	4.6	1.8	3.6	4.6

Secondary helpers
increase their chances of
mating ...

... by taking over the
breeding female

help at all? Adult females are in short supply, probably because they are eaten while on the nest by snakes, mongooses or lizards, and secondary helpers sometimes acquire a wife as a result of helping. Of 17 helpers that survived to breed seven mated with the female they had helped and three of them actually displaced the old breeding male. This is why the male only accepts a helper if there is an immediate large benefit, which has to be weighed against a possible future cost. In addition, secondary helpers have a slightly higher chance of surviving than do birds that simply wait for a year and do nothing.

Primary helpers benefit
through indirect fitness
gains, but would do better
by breeding

Primary helpers obviously make some genetic gain by helping relatives. At Lake Victoria their degree of relatedness to the nestlings is 0.35, and one primary helper on average increases the production of chicks by 2.46 (note that this is not apparent in Table 12.3 where the two kinds of helper are combined). Another advantage is that they increase the survival of their parents (especially the mother) by helping. To be weighed against these genetic gains are the facts that primary helpers, who work harder in feeding the chicks than do secondary helpers, have a lower annual survival than they would by doing nothing, and of course they forgo the possibility of breeding themselves. When Reyer (1984) compared the genetic gains over the first two years of life for four strategies he found that, as in the scrub jay, the highest pay-off would be from breeding in the first year if a mate could be found, while the lowest pay-off of all was for doing nothing in the first year. Helpers achieve an intermediate pay-off. Although the pay-off is similar for the two kinds of helper it comes from different sources: primary helpers gain mainly through helping kin, while secondary helpers gain almost entirely from increasing their own chances of mating.

We can now extend the conclusions of the scrub jay study by saying that helping in birds, mammals and fish may lead to genetic gain through aid to relatives or through future reproduction of the helper (relatedness is not essential for helping to be favoured by selection) and the environmental constraint which leads to helping might be a shortage of territories or mates.

An alternative hypothesis for the evolution of helping

The traditional behavioural ecology view of helping is that 'genes for helping' have spread by kin selection. Recently Ian Jamieson and John Craig (1987; see also Jamieson 1989) have challenged this interpretation. They suggest that 'helping' itself is not a distinct trait, but arises as a by-product of a 'provisioning rule' favoured in the context of parental care. Such a rule might be

Is helping an accidental by-product?

'feed begging chicks in my territory'. When the habitat becomes full, so that juveniles are forced to stay at home, the presence of the next brood of begging chicks produced by their parents elicits provisioning by the juveniles, which can be regarded as a case of misdirected parental care.

Jamieson and Craig's main point is that to show that helping enhances fitness does not necessarily mean that helping itself has been favoured by selection. Selection may simply have favoured crude provisioning rules which occasionally cause helping, bringing fitness benefits, just as they occasionally lead to the feeding of a cuckoo chick, which brings costs (Chapter 4). The way to test their 'unselected' hypothesis for helping is not to show whether helping increases fitness, but to look more closely at the design of helping behaviour to see whether it has been modified by selection. The key question is: is 'not helping' an alternative? Or do individuals blindly follow a crude provisioning rule? Understanding the mechanism will tell us whether we should be measuring the costs and benefits of provisioning or of helping.

No one has yet tested Jamieson and Craig's hypothesis with regard to young birds forced to remain at home by a full habitat. It would be interesting, for example, to test whether non-helping species of jays would automatically show helping behaviour if the young were forced to stay at home or whether selection would need time to favour helping as a new trait. However, in other cases of co-operative breeding we can certainly reject the idea that helping is simply a by-product of a crude provisioning rule. Consider, for example, the contrast in behaviour shown by dunnocks and acorn woodpeckers, two species of birds which often have two males sharing a female. In both species, both males usually help to feed the young provided they each gained a share of the matings. Experiments were performed in which one of the males was removed temporarily during the mating period and then released to see if he would help provision even though he had no chance of paternity. The results showed that male dunnocks did not help if they failed to gain matings (Davies *et al.* 1992) but male acorn woodpeckers often did help (Koenig 1990). (Control removals, done during incubation, after all the mating was over, showed that the experience of the removal itself did not adversely affect chick-feeding.) This difference between the species makes good sense because in the dunnocks the polyandrous males are not related (so there can be no fitness gain in the absence of paternity) whereas in the woodpeckers the males are close relatives (see below) (so a male can gain indirect fitness from helping to raise the other male's offspring). The difference

Helping is not simply a by-product of a provisioning rule, because help is directed to related young

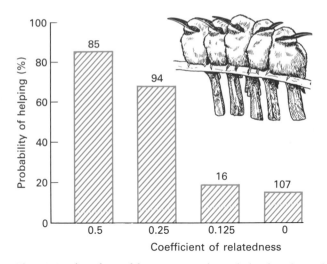

Fig. 12.6 White-fronted bee-eaters prefer to help close kin. The histograms show the probability that a non-breeder with a recipient nest available becomes an actual helper, plotted as a function of its coefficient of relatedness to the recipient nestlings. Numbers refer to number of potential helpers in each category. From Emlen and Wrege (1988).

thus makes good adaptive sense and suggests that birds are not all automatically equipped with inevitable provisioning rules.

Helpers in bee-eaters aid the most closely related potential recipients

Stephen Emlen and Peter Wrege's (1988, 1989) detailed study of helping in white-fronted bee-eaters (*Merops bullockoides*) in Kenya also shows the intricacy of design of helping behaviour, suggesting that it has been moulded by kin selection to enhance the helper's fitness. These bee-eaters breed in colonies, nesting in burrows in cliff faces. Helpers were either non-breeding younger birds or failed breeders and they often had a large assortment of breeders of varying degrees of relatedness available as potential recipients of their aid. Figure 12.6 shows that potential helpers were more likely to help if there were close relatives breeding nearby. Furthermore, of 115 cases in which a helper had two or more broods of differing relatedness simultaneously available as potential recipients, in 108 cases (94 per cent) the helper chose to aid the most closely related brood. These observations provide strong support for the idea of kin-selected helping. They also raise the fascinating question of how the bee-eaters recognize their close kin.

Conflict in breeding groups

We have already seen in the pied kingfisher that helpers and breeders may have conflicting interests. In the colonially nesting

white-fronted bee-eater (*Merops bullockoides*), fathers sometimes
disrupt the breeding attempts of their 1–2-year-old sons, as a
result of which sons may become helpers at the father's nest!
Disruption may consist, for example, of chasing the son and his
mate or blocking their nest entrance (Emlen & Wrege 1992). Sons
may not suffer too seriously as a result of disruption since the
indirect fitness gains through helping are almost the same as the
direct gains through first-time breeding. However, in some plural
breeding species conflicts of interest can lead to substantial
changes in fitness.

(a) Groove-billed anis

The groove-billed ani (*Crotophaga sulcirostris*) lives in a group
territory defended by 1–4 monogamous breeding pairs. The group
builds a communal nest in which all the females lay some eggs
and to which all the group members contribute parental care.
Not all the eggs laid in the nest survive to be incubated and
females compete to make sure that their eggs are the ones which
survive. Competition takes the form of females rolling each other's
eggs out of the nest, so that by the time incubation starts there
are usually broken eggs strewn over the ground below the nest.

Female anis are apparently unable to recognize their own eggs so
each female rolls eggs out of the nest only up to the time when
she herself starts to lay, otherwise she would run the risk of
evicting her own eggs. Sandy Vehrencamp (1977) found that there
is a dominance hierarchy among the females in a group and that
the most dominant female ends up with the largest number of
eggs in the nest at the time of incubation. She achieves this by
starting to lay after the others, who are therefore unable to evict
her eggs. The subordinate females start to lay earlier than the
dominant, but they have several tactics for increasing the chances
of survival of their eggs: they lay more eggs than dominants,
usually wait two or three days between eggs, often produce a late
egg after incubation has started, and initiate incubation earlier

than the dominant. Once incubation starts the embryos begin to
develop and the dominant female is forced to stop laying since
otherwise her young would be late hatchers and at a disadvantage
in competition between nestlings for food. In spite of these tactics,
nearly all the eggs of dominant females survive to be incubated
while less than half those of subordinates survive (Table 12.4).
Although the existence of conflict between females is clear it is
not yet known how dominant females succeed in laying last.

There is also an asymmetry in the extent to which female
anis care for the eggs and young. Surprisingly the dominant female,

Table 12.4 In groups of anis the first female to start laying lays more eggs but has fewer that survive to be incubated than the last female. From Vehrencamp (1977)

	Order of laying		
	1st ♀	2nd ♀	3rd ♀
1 Eggs laid	7.0	6.3	5.8
2 Eggs tossed out	4.0	2.5	0
Eggs incubated (= (1) − (2))	3.0	3.8	5.8

Asymmetry in parental behaviour

with the biggest stake in the nest, does the least care (although the dominant male is very attentive as a parent). Again it is not known how the asymmetry is brought about, although presumably the subordinate females would not care for young if the nest contained none of their own offspring.

The ecological condition which favours plural breeding in anis seems to be predation pressure. In open pasture habitats where the nests are conspicuous to diurnal predators, groups are better able to defend the nest against predators at least in the egg stage. In dense marsh habitats the incubating anis suffer a great deal of predation at night, and when the incubation is shared between the individuals of a group each one has a higher chance of survival (Vehrencamp 1978). However both these advantages are benefits to the average group member and, as we have seen, dominant birds gain much more out of being in a group than do subordinates. Subordinates must still gain more from group nesting in terms of predator defence than they lose from interactions with dominants, or they would leave the group and nest as solitary pairs (see p. 143).

(b) Acorn woodpeckers

In acorn woodpecker groups, females are often relatives

Conflict between breeding females in a plural breeder may occur even if the females are close relatives, as illustrated by the acorn woodpeckers (*Melanerpes formicivorus*) (Fig. 12.8) studied by Walter Koenig, Ron Mumme and Frank Pitelka in central coastal California (Koenig *et al.* 1984). These birds live in groups of up to 15 individuals typically consisting of about two breeders of each sex and their one-year-old offspring. Groups are often formed by a band of brothers and a band of sisters from a different family coming together. All group members help to defend a communal territory and to feed the young. The breeders share the same nest and may even share paternity: genetic analysis of

Fig. 12.7 The acorn woodpecker lives in groups, including several breeders of each sex and non-breeding helpers which are offspring from previous years. Acorns and other hard-shelled nuts are stored in granary trees, some of which have as many as 30 000 holes, drilled by successive generations of the group. The nuts are stored in the autumn and used during the winter. On the right are two males and a female (lower right, with darker head band).

protein polymorphisms has shown that sometimes both males in a breeding group may contribute genes to the same female's clutch of eggs. However the close genetic relationship between females and the sharing of so many aspects of daily life does not mean that there is no conflict within the group. Females sharing a nest sometimes remove each other's eggs and as with Sandy Vehrencamp's anis, the first female to start laying is susceptible to having her eggs tossed out by the other female. Once the second one has started to lay she stops removing eggs, presumably because she is unable to distinguish between her own and the other bird's. Unlike the anis, an acorn woodpecker female does not simply chuck her nestmate's eggs on the floor; instead she carries them to a suitable nook in a branch and proceeds to peck them open. The whole group (including the female that laid the egg in the first place) then joins in an ovicidal feast!

A distinctive feature of acorn woodpecker natural history is that in many places they build enormous storehouses of acorns and pine seeds to keep them going in the winter and through the following spring breeding season. These stores, called granaries, consist of thousands of holes drilled into oak or pine trees (or even occasionally into telegraph poles). In the early autumn the group collects seeds and carefully stores them in the granary

Female acorn woodpeckers eat each other's eggs

Granaries are an ecological constraint

holes. It is the difficulty of constructing and maintaining granaries (combined with the high adult survival that results from the granary) that probably constitutes the crucial ecological constraint that both prevents young woodpeckers from leaving home and leads to the evolution of communal nesting. In south-eastern Arizona, acorns are much less abundant and here the woodpeckers do not live in groups, nor do they construct granaries or live year round in the same territory. The population migrates south each winter and new territories are set up each spring (Stacey & Koenig 1984).

Division of labour and specialized helpers

Naked mole rats have specialized non-breeding workers and a single breeding pair per group

We have seen in this chapter that helping in vertebrates is usually a stepping stone for young individuals to become breeders. Although many helpers must die before they ever have the chance to breed, helping is not a way of life for which individuals are permanently specialized. This, as we shall see in the next chapter, is in contrast to the social insects, where individuals develop specialized morphology and behaviour for a permanent life as a helper. Is there anything analogous to this in the vertebrates? An example which seems to fit the bill has been described by Jenny Jarvis (1981) (Sherman *et al.* 1991, 1992 give a more detailed account). Naked mole rats (*Heterocephalus glaber*) are unprepossessing, hairless, small, blind, pink mammals that feed on roots and tubers, and live in colonies in underground burrows in East Africa (Fig. 12.8). A colony may contain up to 80 individuals but only one pair in the group breeds. The other females have undeveloped ovaries and the males, while they may have active sperm, are apparently non-breeders. The most striking aspect of mole rat life for our present purposes is that non-breeders come in at least two sorts that are different in size and do different jobs. Small ones dig burrows and search for food, while large ones remain close to the reproductive female in the nest, perhaps guarding or brooding the young. The level of activity of workers reflects conflict between queen and workers (Reeve 1992): 'lazy' workers are shoved into activity by the queen. The difference between the two workers' roles may be related to age; young rats start off as small foraging helpers and grow into the larger nest-based helpers. The mechanisms maintaining different roles are yet to be fully elucidated, but it seems probable that chemical cues are involved.

Although mole rats may be more extreme in their development of division of labour than are other vertebrate species with helpers, the ecological and genetic factors that favour helping are probably

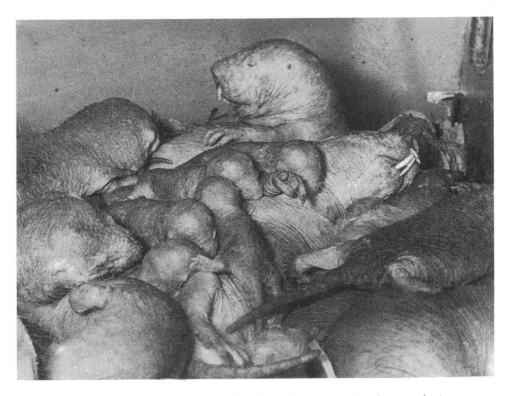

Fig. 12.8 The nest chamber of a colony of naked mole rats. The large individual lying on its back across the middle of the picture is the breeding female suckling some 2-week-old young. Other individuals are sleeping non-workers who feed the breeding female and young with faeces. Photo by Michael Lyster © Zoological Society of London.

quite similar to those in other species. The colony members are almost certainly all closely related to one another, and the difficulty of constructing an elaborate burrow system essential for survival severely limits the chances of successful dispersal away from home.

Conclusions

The main points to emerge from this chapter on helping in birds, mammals and fish are as follows.

1 Most but not all helpers aid close relatives (often they help their parents to rear siblings). Thus helping often results in indirect fitness gains.

2 Individuals may stay at home because the alternative of leaving to become a breeder, although in principle a better option, is prevented by limitations such as a shortage of territories or mates.

3   In addition to the genetic gains from helping relatives, helpers may increase their own chance of survival or future reproductive success by staying at home. This has led to disagreements about whether helping is favoured because of the gains from helping kin or because of the direct fitness gains to the individual helper. These are not, however, mutually exclusive and the relative importance of the two kinds of genetic gain may vary from species to species. Where helpers are unrelated to the individuals they help, the second kind of advantage must be the crucial one. Emlen (1991) concludes that in most species with sufficient data to estimate quantitative values, both direct and indirect fitness gains are essential to maintain 'staying at home and helping' as a strategy. These species would be described as mutualistic (direct fitness gain to helper and breeder) in the classification given in Chapter 11. In two species, the pied kingfisher (primary helpers) and white-fronted bee-eater, the only measurable gains from helping are through indirect fitness. In these species, helping is altruistic in the sense used in Chapter 11.

4   Conflicts of interest may arise between helpers and breeders, for example helpers may be potential sexual rivals of one of the breeders. These conflicts of interest are most marked in species where more than one female uses the same nest.

Summary

In some species of birds, mammals and fish there are individuals that do not breed, but help others to rear offspring. These helpers are usually, but not invariably, close relatives of the breeders that they help, for example they might be young from a previous season. Helping close relatives is a way of promoting genetic representation in the next generation, analogous to having children. However, often helpers could in theory usually do better in terms of genetic representation in the next generation by breeding instead of helping. They are prevented from doing this by environmental limitations such as a shortage of suitable breeding sites. Where helpers are unrelated to the breeders, their benefit from helping is that they increase their chances of surviving and breeding in the future.

Further reading

Emlen (1991) gives a good general account of helping in birds and mammals. Brown's (1987) book is the most extensive review of helping in birds and includes discussions of the relative importance of kinship and individual gains in the evolution of helping. The last chapter of Woolfenden and Fitzpatrick (1984) discusses this issue and reaches a different conclusion. Kerry Rabenold's (1984) work on the stripe-backed wren is an exceptionally good example of a careful field study, including one of the first applications of DNA fingerprinting to a helper species (Rabenold *et al.* 1990). Fingerprinting shows that male helpers sometimes sire offspring. Koenig *et al.* (1992) review the costs and benefits of dispersal versus staying at home.

Topics for discussion

1 Why are there often conflicts of interest in breeding groups and how are they resolved?
2 Are gains in indirect fitness necessary for the evolution of helpers in birds and mammals?
3 Discuss the relative merits of long-term field studies and field experiments for evaluating the costs and benefits of helping.

Chapter 13. Altruism in the Social Insects

The social insects

THE PROBLEM

Co-operation and helping in vertebrates pales into insignificance beside what happens in the social insects. In these insects apparent self-sacrifice reaches the point where large numbers of individuals are completely sterile; they never reproduce themselves but instead spend their whole adult lives devoted to rearing the young of others. As Darwin himself and many other biologists since his time realized, this presents as real paradox, for if natural selection favours maximum genetic contribution to future generations, how can it lead to the development of totally sterile individuals that never reproduce? What is more, these sterile individuals are often specialized for various tasks associated with helping (Fig. 13.1). This raises a further problem: if workers do not reproduce how can their specialized traits evolve? In the last chapter we saw that self-sacrifice and helping others to breed may be favoured by selection in vertebrates because the help is usually given to close relatives. Some individuals are prevented from having their own offspring by ecological constraints such as lack of a territory, so instead they help to rear their younger siblings. We also saw that helping non-relatives or relatives·is sometimes part of a long-term strategy by which an individual eventually gains a mate or a territory. In this chapter we will consider to what extent similar ideas can be used to understand how sterile castes and helping have evolved in the social insects.

Two problems: the evolution of sterility and the evolution of specialized castes

THE DEFINITION OF 'SOCIAL INSECT'

What exactly is meant by the term 'social insect'? To be more precise, this chapter is largely about the *eusocial insects*, which are characterized by three features: they have co-operative care of the young involving more individuals than just the mother; they have sterile castes; they have overlap of generations so that mother, adult offspring and young offspring are all alive at the same time. This is important because it provides the opportunity for the young to rear their younger siblings instead of having their own offspring, as we saw in the scrub jays described in the last chapter. All three of these traits are necessary for a species to qualify as eusocial, but there are many species with intermediate stages,

Eusocial insects have co-operative care, sterile, castes and overlapping generations

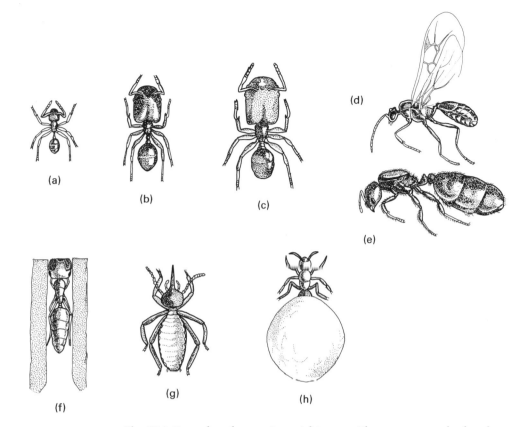

Fig. 13.1 Examples of castes in social insects. The top row are the female castes and male of the myrmicine ant, *Pheidole kingi instabilis*. (a) Minor worker. (b) Media worker. (c) Major worker. (d) Male. (e) Queen. The bottom row are various specialized castes in other species. (f) Soldier of the ant, *Camponotus truncatus*, blocking a nest entrance with its plug-like head which serves as a 'living entrance' to the nest. (g) A sterile caste of the nasute termite, *Nasutitermes exitiosus*, which has a head shaped like a water pistol to spray noxious substances at an approaching enemy. (h) Replete worker of a *Myrmecocystus* ant, which lives permanently in the nest as a 'living storage cask' From Wilson (1971).

such as co-operative nest building without sterile castes. Eusociality occurs in three insect orders: Hymenoptera (ants, bees and wasps), Isoptera (termites) and Homoptera (aphids). The existence of eusociality in the first two orders has been well known for a long time, but in aphids it has been discovered in the past 20 years (Aoki 1977). As we shall see later on, there may be special genetic predispositions towards the evolution of sterile castes in Hymenoptera and aphids.

The social insects are not only important because of their central role in the attempts of evolutionary theorists to understand the origin of altruism, but they are also extremely impressive in terms of their natural history. In a persuasive piece of salesmanship E.O. Wilson (1975) advertises that there are more than 12 000 species of social insects in the world, which is approximately equivalent to all the known species of birds and mammals. The staggering natural history of social insects can be illustrated by the following small sample of facts. In terms of size, a colony of African driver ants (*Dorylus wilverthi*) may contain up to 22 million individuals weighing a total of 20 kg. In terms of communication, the honeybee dance language, in which successful foragers tell other worker bees about the direction and distance to a food source, is one of the only known communication systems in wild animals in which an abstract code (the speed and orientation of the dance) is used to transmit information about remote objects. In terms of feeding ecology, the diets of social insects include seeds, animal prey, fungus grown in special gardens on collections of leaves or caterpillar faeces, and the excreta ('honeydew') of tended herds of aphids. Social insect colonies are often populated by individuals specialized to perform different tasks (so-called castes). Sometimes castes have bizarre morphological modifications to help them carry out their jobs (Fig. 13.1). For example, the head of soldier termites of the species *Nasutitermes exitiosus* is modified into a 'water pistol' used for squirting defensive sticky droplets at enemies, while the head of soldiers in ant species such as *Camponotus truncatus* is shaped like a plug which fits neatly into the nest entrance to keep out intruders, rather like the operculum of a snail.

In the following sections, we will first describe the life history of one example of a eusocial insect to provide a background, then consider two theories which have been put forward to account for the evolutionary origins of sterile castes, before going on to discuss the special features of hymenopteran genetics, which are thought to predispose this group to evolve sterile castes.

The life cycle and natural history of a social insect

Myrmica rubra is a species of ant commonly found in Europe in woodlands, farmland and gardens. It builds a nest, the chambers of which are excavated in the ground underneath flat stones or sometimes in rotten tree stumps or in open soil. The nest is started by a single fertile queen. She is fertilized during a 'nuptial flight' in August or September in which very large numbers of winged reproductive females and males swarm around in the air and copulate (only the sexual forms can fly, and then only at this stage in their life cycle). The queen loses her wings after the

Diversity of social insects

Specialized worker castes

In Myrmica rubra a single queen founds the colony

nuptial flight and spends her first winter sealed inside the nest chamber which she has built by excavating a hole in the ground or in a tree stump. During the following summer, eggs which she has laid develop into larvae and may mature into adult workers (the typical ants one sees scurrying around near ants' nests) before the autumn or in the following spring. Up to the time of maturity of the first worker, the queen feeds herself and her brood entirely off her own reserves of fat and protein, but when the workers mature they start to care for their younger siblings and collect food for them. The workers are female, but they are sterile. They never develop wings, their ovaries do not mature and they never take part in a nuptial flight. In successive years the colony and its nest grow slowly, until after about 9 years it contains roughly 1000 workers and still a single queen who has laid all the eggs. At this stage the colony starts to produce a new generation of reproductives, winged females and males that eventually leave the colony on their nuptial flights. The old colony may then continue for a few more years, but as soon as the old queen dies and stops laying eggs to replenish the worker force, the colony dwindles in size and dies off.

This picture of the life cycle is typical of many temperate zone ants, although of course the details vary greatly from species to species. The details of worker behaviour are also very variable, but the following generalizations are typical of many species. Workers usually spend the first few weeks of their lives inside the colony handling dead prey which have been brought back by foragers, feeding the larvae and the queen with regurgitated food, cleaning the nest, and guarding the entrance. Later in life (this change takes place at an age of about 40 days in *Formica polyctena*, a species which has been studied in detail) workers begin to do jobs outside the colony, mainly foraging and defence against enemies. The total length of life of ant workers is not very well known but probably ranges from a few weeks to a few years. In wasps and bees, workers usually live for about 3−10 weeks. In addition to changes in worker behaviour with age, in some species of ants there are two major castes of worker (both sterile females): soldiers and normal workers. Soldiers are usually larger and have large heads with jaws or glands for producing defensive secretions. As their name implies they are specialists in colony defence.

The females belonging to different castes (queen, worker, soldier) do not usually differ genetically; the determination of caste depends on environmental conditions during larval development. In *Myrmica*, for example, whether a larva develops into a queen or a worker depends on factors such as nutrition, temperature and age of the queen who laid the egg. In honeybees the

In the Hymenoptera workers are female

Typical worker life cycles

Caste differentiation is usually non-genetic

queen can suppress the development of new queens by chemical signals which prevent the workers from feeding larvae the special diet ('Royal Jelly') needed to make them grow into queens.

How eusociality evolved: two pathways

In this section we will reconstruct two possible pathways along which the evolution of sterile castes proceeded. The hypotheses described refer to evolutionary history and therefore cannot be tested by direct experiments. In fact, both pathways are observed in present-day 'primitively eusocial' hymenoptera, so the two hypotheses attempt to generalize from present patterns to evolutionary history. The primitively eusocial hymenoptera have no specialized castes and the queen is often able to raise a brood on her own. They are a good testing ground for ideas about the origins of eusociality because each individual has the potential to play different strategies such as solitary queen, helper and group breeder. Therefore the strategy chosen by an individual should reflect immediate costs and benefits. In contrast, the morphologically specialized workers of advanced eusocial species may lack the flexibility to respond to immediate pay-offs.

In trying to understand how eusociality and in particular sterile castes evolved, it is useful to remember the distinction made in the last chapter between *ecological constraints* and *genetic predispositions*. The ecological constraints are the features of the environment which may either favour group living and co-operative breeding and/or reduce the chances for young individuals to breed themselves. The genetic predispositions refer to the degree of gene sharing between helpers and those they help: the greater the degree of relatedness, the smaller the cost–benefit ratio that is necessary for helping to be favoured by selection (Chapter 11). In the social insects members of a colony are very often part of the same family, so that relatedness is high and workers are genetically predisposed to help. There are two hypotheses about how sterile castes have evolved in insects, both of which postulate certain ecological constraints and genetic predispositions (Fig. 13.2).

HYPOTHESIS 1. STAYING AT HOME TO HELP

(a) Ecological constraints

The ancestors of present-day social bees and wasps were probably parasitoid wasps which laid their eggs inside or on the surface of a host, on which the larvae fed as they grew up (Evans 1977).

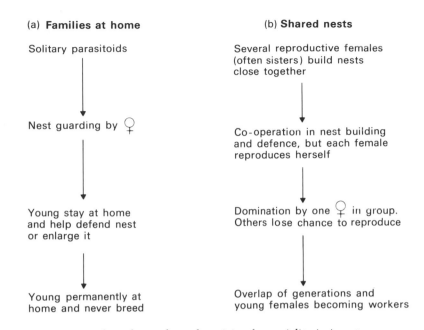

(a) **Families at home**

Solitary parasitoids

↓

Nest guarding by ♀

↓

Young stay at home
and help defend nest
or enlarge it

↓

Young permanently at
home and never breed

(b) **Shared nests**

Several reproductive females
(often sisters) build nests
close together

↓

Co-operation in nest building
and defence, but each female
reproduces herself

↓

Domination by one ♀ in group.
Others lose chance to reproduce

↓

Overlap of generations and
young females becoming workers

Fig. 13.2 Two hypotheses about the origin of eusociality in insects. According to hypothesis (a), sterile castes originated as daughters staying at home to help their mothers (called 'the subsocial route' to eusociality); according to (b), they originated from groups of reproductives nesting together in which one female dominated the rest (the 'parasocial route').

Solitary parasitoid ancestors

A simple form of parental care in present-day parasitoids is to anaesthetize a host by stinging it, and then to carry it back to a safe site such as a hole in the ground or crevice in a tree before laying an egg on it. A slightly more advanced form of parental behaviour, seen for example in the digger wasp (Chapter 10) is to create a safe place by burrowing into the ground. The burrow is then provisioned with one or more paralysed insect prey and an egg laid on each before sealing up the burrow entrance.

Parental care of this type could lead to the formation of larger groups if there was sufficient ecological pressure for the offspring to stay near their birth site after reaching maturity and for mothers to tend their broods. Two main ecological factors are likely to have favoured staying at home.

1 *Defence of the eggs and larvae against parasites.* A major cause of mortality of young in provisioning species such as digger wasps is often parasitism by other insects, and in eusocial species one of the important tasks of workers is to defend the nest against parasites and other enemies.

Defence of brood and the nest as ecological constraints

2 *Nest building.* Although the ancestral parasitoids probably carried prey to naturally occurring refuges such as cracks and holes (or even burrows dug by their hosts), this soon gave rise to the construction of artificial refuges known as nests. In habitats where natural sites were in short supply, there would have been strong selection for the ability to build nests out of mud, chewed vegetation, and so on. Nest building is a laborious time-consuming business, so it is easy to imagine that newly matured adults might have done better to stay at home and help their mother to enlarge or repair her nest rather than go off to start on their own. At first they may have simply used their mother's nest for laying their own eggs, but eventually they may have evolved towards rearing younger siblings instead of their own offspring.

Transmission of protozoans in termites

In the termites there may have been an additional ecological pressure for staying at home. These insects digest cellulose by means of protozoans living in their intestine, and the protozoa have to be passed literally from anus to mouth between generations. Thus ancestral young termites had to stay at home long enough to become infected with cellulose-digesting protozoa. As a thought provoking aside, it is worth mentioning Richard Dawkins' (1979) suggestion that termite eusociality evolved because the protozoans manipulate termites into staying at home to build an ideal environment in which protozoans can grow and replicate!

(b) Genetic predispositions

By helping to rear siblings, the workers pass on their genes to the next generation, so there is an obvious genetic predisposition towards helping. But Ric Charnov (1978) has pointed out that there is more to it than this: a mother makes a genetic gain by 'persuading' her children to stay at home and rear younger siblings while the children themselves make no genetic loss and therefore are 'willing' victims of maternal persuasion. The argument, which

An asymmetry in selection on parents to manipulate offspring and offspring to resist ...

is a development of Alexander's hypothesis of 'parental manipulation' (Alexander 1974), goes like this. Suppose there is enough time to rear two broods in a season and the queen could either rear the first brood, die, and let her daughters lay eggs and rear the second brood, or she could lay both lots of eggs and persuade her first batch of daughters to stay at home and rear the second brood. In the first case, the young emerging at the end of the season will be grandchildren of the original queen, related to her by 0.25 (Chapter 11). In the second case they will be children of the queen, related to her by the usual 0.5. Therefore if the numbers of offspring at the end of the season are the same in the two cases, the queen will have doubled her genetic representation by

... may favour parental manipulation

persuading her daughters to stay at home and help. If the mother has a very large capacity to lay eggs, it is possible that she could

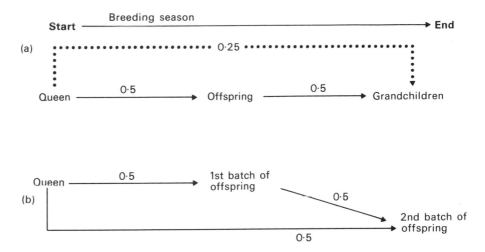

Fig. 13.3 An illustration of Charnov's (1978) idea that offspring may lose nothing by staying at home to help their mother, while the mother makes a large genetic gain. In (a) the queen starts the season by having children and ends with grandchildren (related by 0.25). In (b) she has two successive batches of children and ends with offspring related by 0.5. In (a) the offspring care for their own children and in (b) for their younger siblings, both related by 0.5.

supply each of her daughters with as many eggs as they themselves could lay and rear.

There is therefore strong selection on the queen to 'persuade' her daughters to help. But, and here we come to the crucial part of the argument, the daughters lose nothing by staying at home. They rear younger siblings ($r = 0.5$) instead of children ($r = 0.5$) (assuming the original queen mated once at the beginning of the season). As long as the queen can lay enough eggs, the daughters make the same genetic gain from rearing siblings as from rearing offspring (Fig. 13.3).

HYPOTHESIS 2. SHARING A NEST

Multiple foundresses are often sisters

In many tropical and some temperate zone wasps, nests are founded not by a single queen but by a group of co-operating individuals. Very often the foundresses are sisters (e.g. *Trigonopsis cameronii*) but sometimes they are not sisters and may possibly be unrelated (e.g. *Ceceris hortivaga*) (West Eberhard 1978a). In primitively social wasps each queen lays her own eggs and rears her own young, but if one queen succeeded in dominating the others and preventing their eggs from developing, the stage would be set for the evolution of workers.

(a) Ecological constraints

The ecological pressures favouring nest sharing are probably similar to those suggested earlier as favouring young that stay at home to help: defence against parasites and nest construction. For example in the wasp *Trigonopsis cameronii*, 1–4 sisters build a communal nest out of mud. Although each female builds her own brood cell, lays her own eggs, and provisions her own young, the sisters co-operate in building shared walls of the nest and in driving away ants or other enemies (Eberhard 1972). The way in which this kind of cooperation between sisters might lead to an unequal division of reproductive labour is illustrated by *Metapolybia aztecoides*, another neotropical wasp, this time with a sterile worker caste. Here a nest is built by several queens, all of which lay eggs which develop into workers, that in turn help to enlarge the nest. Co-operation between queens is essential at the start in order to produce enough workers to get the colony and nest going, but once the colony is well established, one queen fights with and evicts the rest (that is, if more than one queen is still surviving) before they have a chance to produce any reproductive offspring. Each queen therefore starts the colony with a certain chance of ending up having reproductive offspring, and the losers are 'hopeful reproductives' whose only chance at the start was to co-operate with the other queens. In *Metapolybia* the founders are probably sisters, so even the losers make a genetic gain from the colony, but if the chances of reproduction without co-operation were small enough, one could imagine co-operation even between unrelated founders (West Eberhard 1978b).

Nest sharing might also arise by accident. In the great golden digger wasp (*Sphex ichneumoneus*) two unrelated females sometimes end up using the same burrow because the second female was attempting to take over an apparently abandoned nest hole which happened to be occupied (Chapter 10). Although in *Sphex* it is disadvantageous for females to share a burrow because they fight with each other and steal each other's prey, it is possible that accidental sharing could be the start of communal nesting if ecological pressures from parasites, for example, were very great.

(b) Genetic predispositions

If, as in *Metapolybia*, the founders are sisters, even those that fail to reproduce will still gain some genetic representation in the next generation. This is nicely illustrated by Bob Metcalf's study of paper wasps (*Polistes metricus*) (Metcalf & Whitt 1977). These wasps sometimes found nests as solitary queens and sometimes

Because foundresses are
usually sisters, indirect
fitness gains are
important

two sisters share a nest. When two queens share, one (the α
female) does nearly all the egg laying while the other (the β
female) passes on her genes largely through her sister's offspring.
Metcalf calculated the degree of relatedness between sisters
sharing a nest (by means of electrophoretically detectable enzyme
polymorphisms) and also estimated the number of offspring
produced by α, β and solitary females. These calculations showed
that the α female does better than the β, while the β does about
as well as a solitary queen in terms of genetic representation in
the next generation, and that the β female's genetic contribution
is made almost entirely through nephews and nieces. Nests of
pairs of females produce more young than those of solitary females
because they are better guarded against predators and parasites.
Metcalf's study shows, therefore, that a *Polistes* female can do as
well by joining her sister as by setting up a nest on her own, even
if she produces hardly any children in the former case
(Table 13.1). Other studies of Hymenoptera, in which nests are
founded by more than one female show that foundresses are
usually full sisters (e.g. Strassmann *et al.* 1991).

The fact that social insect colonies are often founded by sisters
raises the interesting question of how siblings recognize each
other. If dispersal is very limited a simple rule such as 'co-operate
with the first queen you meet' might be adequate, but in at least

Kin recognition in sweat
bees

some species there is a genetically based ability to recognize
relatives. Greenberg (1979) has shown that workers of the sweat
bee, *Lasioglossum zephyrum*, selectively prevent unrelated bees
from entering their nest. There is a linear correlation between
degree of relatedness and the tendency of workers to allow

Table 13.1 Metcalf's calculations of relative gene contribution to future
generations in *Polistes metricus*. The table shows a comparison between the
gene contributions of solitary queens and α and β queens that share a nest.
The shared nests have a higher probability of success than solitary nests. α
Females produce most of the young in a shared nest, while β females make
most of their gene contribution through helping their sister's young. The
measures of success are expressed relative to that of a solitary queen

	Solitary queens	Joint nesters	
		α	β
Relative probability of nest success (±SE)	1	1.38 ± 0.02	1.38 ± 0.02
Average relatedness to offspring	0.47	0.45	0.34
Relative gene contribution	1	1.83 ± 0.57	1.39 ± 0.44

intruders into the colony. Greenberg suggests that the recognition is based on genetically determined family odours and discrimination by workers against bees with an unfamiliar smell (see Chapter 11 for a more detailed discussion of kin recognition).

To summarize the discussion of how sterile castes might have originated in the social insects, the two main hypotheses are that they arose by offspring staying at home to help their mother, or by sisters sharing a communal nest. In both cases the ecological pressures favouring sociality were probably nest defence and nest construction, and the genetic predisposition towards helping was that the recipients of aid were close relatives of the altruists. The two hypotheses are not mutually exclusive. Among present-day primitively eusocial species, sweat bees (Halictidae) have the subsocial pattern and paper wasps (Polistinae) commonly show the parasocial pattern. The two hypotheses are equally applicable to any sexually reproducing diploid organism, but we now turn to a very important special feature of the Hymenoptera which may give them an additional predisposition towards eusociality.

Both hypotheses may be correct, for different groups

Haplodiploidy and altruism

W.D. Hamilton (1964) was the first to fully appreciate the significance of the special genetic predisposition of Hymenoptera to form sterile castes. The special feature is *haplodiploidy*: males develop from unfertilized eggs and are haploid, while females develop from normally fertilized eggs and are therefore diploid.

A haploid male forms gametes without meiosis, so that every one of his sperm is genetically identical. This means that each of his daughters receives an identical set of genes to make up half her total diploid genome. With a diploid father, a female would stand a 50 per cent chance of sharing any particular one of his genes with her sisters, but with a haploid father she is certain to share all of them. The other half of a female hymenopteran's genes come from her diploid mother, so she has a 50 per cent chance of sharing one of her mother's genes with a sister. If we now think about the total degree of relatedness between sisters we come to a remarkable conclusion. Half their genome is always identical, and the other half has a 50 per cent chance of being shared, so the total relatedness is $0.5 + (0.5 \times 0.5) = 0.75$. In other words, because of haplodiploidy full sisters are more closely related to one another than are parents and offspring in a normal diploid species. Hymenopteran queens are diploid and are therefore related to their sons and daughters by the usual 0.5 (Box 13.1, Table 13.2). A sterile female worker can therefore make a greater

Haplodiploidy: a special feature of hymenoptera . . .

genetic profit by rearing a reproductive sister than she could if she suddenly became fertile and produced a daughter! This extraordinary state of affairs also suggests why in the Hymenoptera

---

Box 13.1 *Calculating coefficients of relatedness, r, in haplodiploid species.*

GENERAL POINT. Males develop from unfertilized eggs and so are haploid; all of a male's sperm are genetically identical so the probability of sharing a copy of a gene via the father is 1. Females develop from fertilized eggs and so are diploid; the probability of sharing a copy of a gene via the mother is 0.5, because of meiosis.

METHOD. Draw out a pedigree, linking the two individuals through their recent common ancestors. To determine the coefficient of relatedness between individual A and individual B, draw arrows along the pathways, pointing from A to B. Indicate on each link in the pathway the probability that a copy of a gene will be shared.

EXAMPLES
(a) Sister–sister

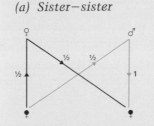

Half of a female's genes come from the father; the probability that a copy of one of these is shared with the sister is 1. The other half come from the mother; the probability that a copy of one of these is shared is 0.5.

Via mother = (0.5×0.5) + Via father = (0.5×1) $r = 0.75$

(b) Sister–brother

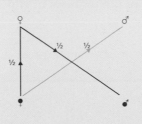

A female is linked to her brother only via her mother, as her brother develops from an unfertilized egg. Half of her genes come from her mother; the probability that a copy of one of these is shared is 0.5. The other half come from her father; the probability that a copy of one of these is shared is zero.

Via mother (0.5×0.5) + Via father (0.5×0) $r = 0.25$

Continued on p. 330

Box 13.1 *Continued*

(c) Brother–brother

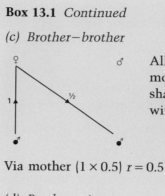

All of a male's genes come from his mother. There is a 0.5 chance of sharing a copy of a particular gene with his brother.

Via mother (1×0.5) $r = 0.5$

(d) Brother–sister

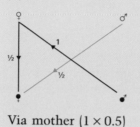

All of a male's genes come from his mother. There is a 0.5 chance of sharing a copy of a particular gene with his sister.

Via mother (1×0.5) $r = 0.5$

Note the asymmetry in relatedness between brothers and sisters [cf. (b) above].

... results in unusual patterns of relatedness which could predispose this group to evolve eusociality

only females help to rear sisters. Males are related to their sisters by 0.5 instead of the 0.75 value calculated above for females. The relatedness of males to their sisters is calculated as follows: a haploid male inherits all his genes from his mother and there is a 50 per cent chance that a given sister will have inherited the same maternal genes. (Note, however, that a female is related to her brother by only 0.25, since the 50 per cent of her genes that come from her father have no chance of being shared with a brother, and the other half of her genes have a 50 per cent chance of being shared: $0.5 \times 0.5 = 0.25$).

By contrast, in the termites in which both males and females

Table 13.2 Degrees of relatedness between close relatives in a haplodiploid species

	Mother	Father	Sister	Brother	Son	Daughter	Niece or nephew (via sister)
Female	0.5	0.5	0.75	0.25	0.5	0.5	0.375
Male	1	0	0.5	0.5	0	1	0.25

are equally related to their siblings, both sexes may become sterile workers.

Haplodiploidy might also help to explain why sterile castes have evolved more often in the Hymenoptera than elsewhere in the insects. Wilson (1975) gives a figure of 11 independent evolutionary origins for sterile castes in the social ants, bees and wasps, which constitute only 6 per cent of all insect species, and only one origin of sterility in all the rest of the insects, namely the termites. Since Wilson wrote this, non-reproductive soldiers have been discovered in a Japanese aphid (Aoki 1977) but aphids are genetically even more prone to sterility than the Hymenoptera. They reproduce asexually (for at least part of the year), which means that the members of an aphid colony are genetically identical just like the cells of a body. So the existence of sterile workers in aphids is perhaps no more of an evolutionary surprise than the fact that our nose cells do not produce sperms or eggs. Recent experiments have shown that soldiers of the gall forming aphid, *Pemphigus spyrothecae*, defend the colony against insect predators, with many soldiers dying as a result, and they also show 'housekeeping' behaviour, cleaning the gall of defecated honeydew and dead aphids. When soldiers were removed, the colony suffered increased predation and reproductive success was also reduced because of the accumulation of waste inside the gall (Foster 1990; Benton & Foster 1992).

Aphid workers

Although haplodiploidy in the Hymenoptera may predispose them to become eusocial, it does not *cause* eusociality to evolve. This is easy to see because not all haplodiploid insects have sterile castes, and in the termites sterile castes have evolved in a normal diploid species. As we have stressed earlier in the chapter, ecological pressures and genetic predispositions work together to determine whether or not sterility will evolve. We will come back to the question of whether or not haplodiploidy paves the way for the evolution of eusociality on p. 338.

Do the simple calculations of relatedness apply to real colonies?

A second cautionary note is that the simple calculations of relatedness in Table 13.2 hold only if the colony is formed by a single queen who has mated once. If two workers do not share the same father, they are related by only 0.25. In honeybees, the queen may mate up to 20 times, but workers are able to discriminate between sisters that are more or less related, and direct their help towards relatives who share the same father (Page *et al.* 1989). If a colony is founded by several sisters as in *Trigona*, helping is still more likely to evolve in a haplodiploid than in a diploid species because in the former a female is related to her sister's children by 0.375 and in the latter by 0.25. Regardless of whether or not the queen is monogamous, the coefficient of relatedness is not always exactly as calculated by working out pedigrees (Craig 1979; Grafen 1984). An hypothetical example will serve to illustrate this point. Suppose that in a social insect there are two strategies for females controlled by a single allele difference: stay at home and help, or set up a nest and accept help. If a female with the helping allele finds a sister with a nest willing to accept help, she can be sure that the sister does not share the helping allele, even though her relatedness by pedigree is 0.5 (Grafen 1986a).

CONFLICT BETWEEN WORKERS AND QUEEN

Selection on present day colonies versus historical origin

Our discussion so far has referred primarily to evolutionary history; we have described the ecological pressures and genetic predispositions that might have been important in leading to the evolution of sterile castes. In this section we are not going to discuss the origin of eusociality, but the selection pressures that act within present-day hymenopteran colonies. Our question will be: 'Given that there are sterile castes, how do workers and queens maximize their genetic profit?'. The answer to this question will probably tell us something about the selective forces *maintaining* eusociality today, but can be used only as indirect evidence for its origin.

Hamilton's theory, which we described in the last section, can be used to analyse how workers and queens might maximize their genetic profit. As we shall show, the theory predicts a conflict of interest between workers and queen over the sex ratio of reproductives in the colony.

CONFLICT OVER THE SEX RATIO

Our account of Hamilton's theory can be summarized as follows. Imagine a young female with the hypothetical choice of going off to rear her own daughters or staying at home to rear a new generation of younger reproductive sisters. Since she is more closely related to sisters than daughters, she would do better to stay at home and have sisters rather than the same number of daughters. In fact the queen seems to be the loser since she is condemned to have offspring!

However there is a further twist to the story that we have not considered. It may well be better for a young female to stay at home and rear sisters, but this presupposes that the queen is going to produce sisters for her to rear. Obviously what the daughters can do will depend on what the queen is doing. As we saw earlier in the book, whenever we encounter situations like this we need to analyse the problem in terms of what will be the evolutionarily stable strategy or ESS.

Let us consider the queen first of all. She is equally related to her sons and daughters ($r = 0.5$ in each case) and so just like diploid females of any sexual species she is expected to produce equal numbers of male and female reproductive offspring. To be more precise she should *invest equally* in the two sexes (Chapter 8). It is important to emphasize that we are referring to equal investment in *reproductive* offspring, not sterile workers. Recall that in Chapter 8 the argument was that a 50:50 sex ratio was stable because the expected *reproductive success* of a male and a female is the same. Hence the discussion of sex ratios is only pertinent to reproductives.

Now for the twist: if the queen produces an equal sex ratio, workers will spend their lives rearing equal numbers of brothers (to which they are related by 0.25) and sisters (related by 0.75). Their average relatedness to reproductive siblings will therefore be only 0.5, exactly the same as they would have to their own progeny if they had decided to leave home and have their own offspring!

In order for female workers to gain the full genetic benefit from staying at home and rearing sisters, they must rear more queens than drones. But how much bias in favour of reproductive sisters should they show? Once again we search for the ESS sex ratio, this time from the workers' point of view. The workers are more closely related to their sisters and so should rear more of them than brothers. But if they rear too many sisters then the sex ratio in the population will become so female-biased that a drone will have very much greater reproductive success than a queen. The stable sex ratio for the workers is 3:1 in favour of reproductive females. When female reproductives are exactly three times as common as males, drones have three times the expected success of queens because on average each drone has three times the chance of finding a mate. From the workers' point of view this would exactly compensate for the fact that brothers are only one third as closely related as are sisters: a worker expects to get three nieces or nephews from her brothers for every one she gets from her sisters. Nieces and nephews on her sister's side are three times as closely related to her, so the total gain per unit investment via brothers and sisters is the same.

To summarize this rather complicated argument, the queen prefers an equal investment in male and female reproductive offspring, but the workers prefer a ratio biased 3:1 in favour of females. There is a direct conflict of interest over the sex ratio between workers and the queen. Who wins?

Queen and workers disagree about the optimal ratio of investment in reproductives

Queens prefer a 1:1 investment ratio, workers prefer a 3:1 ratio female:male

TESTS OF WORKER–QUEEN CONFLICT

Bob Trivers and Hope Hare (1976) attempted to test whether the queen or workers win by analysing the ratio of investment (more accurate than simply looking at numbers) in male and female offspring in 21 species of ants. The ant species were chosen because they were ones in which the conditions for the hypothesis apparently hold (one queen, one mating). Despite a considerable amount of scatter in their data, Trivers and Hare found that on average the ratio of investment was much closer to 3 : 1 than to 1 : 1 (Fig. 13.4). They concluded that the workers win the conflict and successfully manipulate the sex ratio towards their own optimum and away from that of the queen. To put it baldly, the workers are successfully farming the queen as a producer of nieces and nephews: a far cry from the idea of workers as subordinate females making the best of a bad job! Trivers and Hare suggest that the workers win simply because they have practical power; they provision the young and are in a position to selectively kill off males and nurture queens. The queen presumably retaliates

Measuring the ratio of investment in ant nests shows a female bias ...

... workers win the battle over the investment ratio

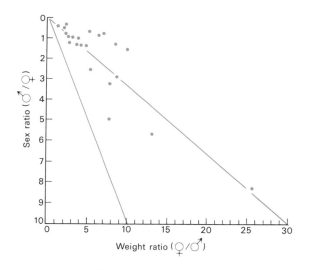

Fig. 13.4 Ratio of investment (measured by weight) in 21 species of ants. The *x*-axis is the ratio of female : male weight and the *y*-axis is the ratio of numbers of males : females in the colony. The lower line is the prediction if the investment ratio is 1 : 1 and the upper line is 3 : 1 in favour of females. The data are closer to the 3 : 1 line, as predicted if workers control the sex ratio. (To understand how the lines are drawn take the example of a ♀ : ♂ weight ratio of 6 : 1. Equal investment would mean six ♂ per ♀, and a 3 : 1 investment ratio in favour of ♀ would mean a ratio of 2 ♂ per ♀.) From Trivers and Hare (1976).

during evolution by attempting to control worker behaviour with pheromones or direct aggression.

Although Trivers and Hare considered two extreme divisions of practical power, total queen control and total worker control of the sex ratio, a more plausible alternative is that power is shared. The queen can choose the sex of the eggs she lays (by whether or not she fertilizes them with stored sperm) and the workers can choose whether or not to rear the larvae. With shared power the problem becomes 'given that the workers control the sex ratio after egg laying, what ratio should the queen lay', and 'given that the queen lays a certain ratio of eggs, how should the workers manipulate it'. In one analysis of the problem, Michael Bulmer and Peter Taylor (1981) suggested that the queen can in general prevent a female-biased investment ratio by limiting the supply of diploid eggs to the workers, although the workers may be able to respond by investing more in each queen egg that is made available to them. The general message of these more complicated analyses of conflict over the sex ratio is that quantitative predictions are not easy to make, although the qualitative predictions of Trivers and Hare generally holds. Indeed it could be argued that the scatter in Trivers and Hare's data reflects the fact that the balance of power varies between species.

Richard Alexander and Paul Sherman (1977) criticized Trivers and Hare's paper on the following grounds. (a) The queen often mates more than once, altering the bias in relatedness from the 3 : 1 assumed by Trivers and Hare. (b) Workers often lay male eggs (in one study of bumblebees, for example, 39 per cent of male eggs were laid by workers (Owen & Plowright 1982)) and with laying workers the 3 : 1 prediction no longer holds. More important, they offer an alternative explanation for the female-biased sex ratio. In Chapter 8 we mentioned that Fisher's theory of equal investment in males and females no longer holds when there is competition between brothers for mates (local mate competition). If a mother 'knows' that her daughters will all be fertilized by her sons, she should produce a brood made up mostly of daughters, with just enough sons to do the fertilizing. Alexander and Sherman suggest that there may be some local mate competition which accounts for the biased sex ratio in the ants studied by Trivers and Hare. This hypothesis is not incompatible with the idea of worker manipulation. If the queen favours a biased sex ratio because of the effects of local mate competition the workers will favour an even more biased ratio because of the additional impact of 0.75 relatedness. Without knowing the exact degree of local mate competition it is not possible for Alexander and Sherman to make a quantitative prediction of the optimal

Practical considerations may determine the outcome of worker–queen conflict

An alternative explanation of female-biased investments

. . . local mate competition

Both local mate
competition and worker–
queen conflict may act
simultaneously

sex ratio for the queen, and the 3 : 1 bias observed by Trivers and
Hare must be taken as suggestive, but not conclusive, evidence
for the idea of worker manipulation in ants.

It will be apparent from the discussion of Trivers and Hare's
paper that in order to test whether workers or queens control the
investment ratio we need to know about the extent of local mate
competition, whether or not workers lay eggs, and the degree of
relatedness of workers and reproductives. A few studies have
succeeded in collecting all this information, and the pattern that
emerges is that when the conditions assumed by Trivers and
Hare are met, sometimes the investment ratio is at the queen's
optimum of 1 : 1, sometimes it is at the workers' optimum and
sometimes it is in between. This probably serves to emphasize
that the outcome of worker–queen conflict depends on how the
life history of the species influences the division of power. In
Polistes metricus, studied by Bob Metcalf (1980) the queen appears
to win. In this species, as we have already described, nests are
founded by one or more mated sisters in the spring. The nests
produce reproductive males relatively early in the summer,
followed by queens at a later stage; workers are produced continually through the season. The young queens mate in the late
summer and overwinter before starting up their own nests the
following spring.

In *Polistes metricus* the
queen appears to win

Genetic analyses of enzyme polymorphisms showed that
workers are related to their reproductive sisters by an average of
0.65 (slightly less than the theoretical maximum of 0.75 because
the queen sometimes mates more than once). Also, workers do
not normally lay eggs. There was little evidence of local mate
competition: this was inferred from the fact that there is little or
no inbreeding and the observation that females do not disperse
far. Males must therefore disperse rather a long way and so
brothers are not likely to compete. Thus, workers should prefer a
female-biased ratio and the queen a 1 : 1 ratio of investment in
reproductives.

Why do the queens win? Metcalf suggests that they win
because of practical considerations. How might the workers
manipulate the sex ratio? The obvious way would be by killing
off male eggs or larvae (given that the queen lays eggs in a 1 : 1
ratio and that workers cannot lay female eggs themselves this
seems to be the only obvious possibility). In *P. metricus* the
queen produces male offspring early in the season. At this stage
few of the workers have emerged and the few that have emerged
can be effectively controlled by the queen. By this ploy it appears
that queens have removed the opportunity for workers to control
the sex ratio. The workers lose, in that they rear reproductives

with an average degree of relatedness of 0.45 (0.25 for brothers and 0.65 for sisters) while the queen is related to her offspring by 0.5. An example of a study in which strongly female-biased sex ratios were found and in which the various conditions of Trivers and Hare's argument were fulfilled (no local mate competition, no worker laying, monogamous queens) is Phil Ward's (1983) of *Rhytidoponera* ants in Australia (see also Fig. 13.5). Evidence

In *Rhytidoponera* ants and numerous other species, the investment ratio is female-biased

Fig. 13.5 *Polistes apachus* workers on the nest. This species has a female-biased sex ratio of reproductives, perhaps indicating that workers benefit more than the queen. Photo by permission of Bob Metcalf.

from numerous studies suggests that Trivers and Hare were essentially correct (Seger 1991).

The evolutionary battle between workers and queen over the sex ratio has taken place in spite of the fact that queen and workers are not genetically differentiated. It is the diet and not the genes of a female that determine whether she becomes sterile or reproductive. If one thought in terms of genes for controlling the sex ratio, these genes must produce a conditional strategy 'if in a worker body favour 3:1, if in a queen favour 1:1'. As Box 13.2 shows, there may also be conflict between workers in an insect colony.

Haplodiploidy and the origin of eusociality

Now let us return to our earlier question (p. 322) of how hymenopteran helpers might have evolved in the first place. Was haplodiploidy important in the origin of helping? It might seem from what we have said so far that the answer should be yes. Let us summarize the argument beginning with 'Hamilton's rule' which we introduced in Chapter 11. The rule is that helping pays whenever the ratio of benefits to costs (K) exceeds $1/r$. Now for our present calculations K is the ratio of (extra siblings raised as a result of helping/offspring lost by helping) and r is ratio (relatedness to siblings reared/relatedness to offspring lost).

Does haplodiploidy help to explain the origin of eusociality?

In diploid animals a full sibling is genetically equivalent to an offspring $(r = \frac{1}{2}$ for both) so it would pay to help whenever K is greater than $(\frac{1}{2}/\frac{1}{2} = 1$. In other words it pays to help if you can replace one lost offspring with just over one sibling. For a female hymenopteran, however, with a bias towards rearing sisters, the average relatedness of a worker to her reproductive siblings is greater than $\frac{1}{2}$; therefore the critical value of K to favour helping is less than 1. For example with the 3:1 bias discussed earlier, the average relatedness to reproductives is $(\frac{3}{4} \times \frac{3}{4}) + (\frac{1}{4} \times \frac{1}{4}) = \frac{5}{8}$ and the critical value of K is $\frac{1}{2}/\frac{5}{8} = \frac{4}{5}$. In other words instead of having to rear just over one sibling for every potential offspring lost, the worker has to raise just over four siblings for every five offspring sacrificed. It would appear that haplodiploidy makes it easier for helping to evolve.

K is the threshold ratio of benefits: costs

This line of reasoning is, however, too simple. To see why we have to introduce the concept of the *value* of males and females, the proportion they will contribute to the gene pool of future generations. Recall that when the sex ratio is 3:1 in favour of females, the expected reproductive success of a male, or in other words his value as a machine for making grandchildren, is three times that of a female. If you include this factor, the pay-off for

Box 13.2 *Worker–worker conflict.*

The conflict in eusocial colonies is not only between workers and queens. Workers may disagree among themselves over the production of eggs. Although workers never mate, in many species of ants, bees and wasps, they are able to lay unfertilized eggs which, being haploid, develop into drones. On the basis of genetic relatedness, it is possible to calculate, all other things being equal, which colony members would benefit from worker egg laying. Let us look at the benefits for the queen, the worker that lays an egg, and the other workers.

If the queen mates only once, the values in Table 13.2 show that:

1 the queen would prefer her sons to her daughters' sons $(r = 0.5 > r = 0.25)$;

2 the laying worker would prefer her sons to her brothers $(r = 0.5 > r = 0.25)$;

3 other workers would prefer their sisters' sons to their brothers $(r = 0.375 > r = 0.25)$.

Thus, workers agree that worker eggs are a good thing. The fact that worker laying occurs at a rather low level in most eusocial colonies probably results from the queen suppressing worker laying by some unknown mechanism. Now consider colonies in which the queen mates many times, as for example in honeybees. This changes the relatedness of sisters from 0.75 to 0.25 (half-sisters) and hence the relatedness of workers to their sisters' eggs from 0.375 to 0.125. In this situation, workers prefer queen-produced males, namely brothers $(r = 0.25)$ to sister-produced males, namely nephews $(r = 0.125)$ (Ratnieks 1988). In other words, worker eggs should be suppressed not only by the queen but also by other workers. This prediction has been tested by Ratnieks and Visscher (1989) who showed that worker honeybees selectively destroy eggs of their half-sisters whilst caring for their younger brothers.

raising siblings and offspring must be calculated as (number raised × value × relatedness). Taking the example where the sex ratio of the population as a whole is 3 : 1 (this applies therefore to the average ratio of offspring and of siblings) we get the following answers:

For helping to raise a sibling:

$$(\tfrac{3}{4} \times 1 \times \tfrac{3}{4}) + (\tfrac{1}{4} \times 3 \times \tfrac{1}{4}) = \tfrac{12}{16}$$

For raising an offspring:

$$(\tfrac{3}{4} \times 1 \times \tfrac{1}{2}) + (\tfrac{1}{4} \times 3 \times \tfrac{1}{2}) = \tfrac{12}{16}$$

With a population sex ratio of 3:1 haplodiploidy does not help to explain eusociality

(In each line the first bracket is for females, the second for males. So on the top line the first bracket means 'three-quarters of sibs are females, with a value of one and a relatedness of three-quarters'.) Therefore with a population sex ratio set at 3 reproductive females for every male, the critical value for helping to pay is $K < \tfrac{12}{16}/\tfrac{12}{16} = 1$, the same as for a diploid species! In other words *when the sex ratio in the population as a whole is 3:1* haplodiploidy gives no advantage to helping: the extra relatedness to females is counter-balanced by the higher value of males.

So can a female-biased sex ratio combined with haplodiploidy ever tip the balance in favour of helping, or is it totally irrelevant? The answer is that a female bias might favour helping if it does not lower the value of females to offset their greater relatedness.

Haplodiploidy might help if the workers raise reproductives with a sex ratio that is female-biased relative to the population as a whole

This could happen if the sex ratio within the nest is female-biased while the overall population sex ratio is not. If the bias in the nest was 3:1 and the population sex ratio was 1:1, the value of males and females would be equal and we would be back to the simple calculation of K based on relatedness which gave $K = \tfrac{4}{5}$. But how could this situation ever arise?

Jon Seger (1983), in an important paper, has pointed out that it could arise as a result of a particular kind of life cycle found in some hymenoptera. In this life cycle, called a 'partially bivoltine' cycle, there are two generations each year, but the generations are not completely distinct; individuals from generation 1 may survive to take part in generation 2. One kind of partially bivoltine cycle is found in halictine bees (Fig. 13.6a). The females overwinter as fertilized adults; in the spring they produce a first generation of offspring. The spring generation then gives rise to the summer generation and at the end of the summer this second generation mates and the fertilized females overwinter. The important fact to add is that some of the spring males survive to mate with second generation females (Fig. 13.6a). These males have the opportunity to mate with two generations of females, so they are more valuable to their mother (as grandchild-makers) than are spring females. The overwintering female should therefore produce a male-biased spring generation of reproductives. Some of these spring males survive to mate with summer females, in effect pre-empting the chances of summer males. It therefore pays a mother in the summer to have a female-biased sex ratio among her offspring, since males, their job having been partly pre-empted by survivors from the spring, are less valuable than females.

The life cycle of halictine bees may produce the necessary female bias

(a) The Halictine life cycle

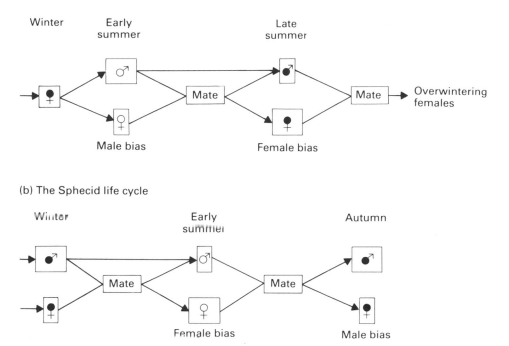

(b) The Sphecid life cycle

Fig. 13.6 Simplified diagrams of two partially bivoltine life cycles. In (a) summer females (open symbols), if they stay at home, would rear a female-biased brood while in (b) they would rear a male-biased brood. The relative size of the male and female boxes indicate a male or female bias in the offspring. After Seger (1983).

Early summer females, by staying at home, may raise a female-biased brood of younger siblings

Now we have just the conditions needed for haplodiploidy to favour helping and it has arisen as a consequence not of selection for eusociality but as a 'preadaptation' resulting from the halictine life history pattern. Imagine an early summer female with the option of staying at home to help her mother rear siblings or going off to breed herself. In either case she will be contributing to the generation that overwinters as fertilized females, a generation that is female-biased relative to the population as a whole (don't forget that the population includes males surviving from the spring).

Let us take a hypothetical numerical example to look at the effect of this. Suppose the population sex ratio of reproductives was $1:1$ and the ratio amongst the overwintering generation was biased $2:1$ in favour of females. If the value of males and females was equal, the pay-off for raising a sibling would be $(3/4 \times 2/3) + (1/4 \times 1/3) = 7/12$, while the pay-off for an offspring would be the usual $1/2$. The critical value of K is therefore $1/2 / 7/12 = 6/7$.

The female halictine has to rear only 6 siblings for every 7 offspring lost, so haplodiploidy combined with a biased sex ratio has shifted the balance slightly in favour of helping. If you try working out examples with different sex ratios, you will find that the general result holds even when the sex ratio in the overwintering generation is not female-biased in absolute terms, as long as it is female-biased *relative* to the population as a whole.

But this is not quite the end of the argument. Remember that summer males are less valuable than females because of pre-emptive matings made by males of the previous (spring) generation. Even if the sex ratio in the summer was not at all female-biased relative to the rest of the population the higher value of females would tip the balance in favour of helping. Suppose that as a result of pre-emptive matings females were twice as valuable as males and the sex ratio was $1:1$, the pay-off for helping is (value × relatedness) = $(\frac{2}{3} \times \frac{3}{4})$ for females + $(\frac{1}{3} \times \frac{1}{4})$ for males = $\frac{7}{12}$, while the pay-off for having offspring would be $(\frac{2}{3} \times \frac{1}{2})$ + $\frac{1}{3} \times \frac{1}{2}) = \frac{1}{2}$ and the value of K would therefore be $\frac{6}{7}$ (note that 'value' has been scaled to add up to one) (Grafen 1986).

To summarize this complicated argument, the halictine life cycle favours helping by females in the summer because (a) the local sex ratio is more female-biased than that of the population and (b) the value of males is decreased because some females are mated by spring males.

There is one final twist to the story. Some partially bivoltine hymenoptera have a life cycle in which the events are shifted around one step; this is the pattern found, for example, in almost all sphecid wasps (Fig. 13.6b). Here the population over-winters as male and female larvae. The larvae complete their development in the spring and mate to give rise to the summer generation. The summer generation then matures and mates to form a second, autumn generation. The autumn generation does not mature but develops into the overwintering last instar larvae of the next year. As in the halictine life cycle, males of the first generation (the overwintering larvae from the previous year) may survive to mate with those of the later generation, so here a male-biased sex ratio would be expected in the autumn and a compensatory female-biased ratio in the summer. Now think of the young summer female with the hypothetical choice of staying at home to help her mother or going off to mate. In either case, the youngsters she rears will contribute to the autumn generation and hence be male-biased. There is no special genetic incentive for becoming a helper arising from haplodiploidy in the sphecid life cycle: haplodiploidy actually works against helping here.

Seger's ingenious idea that the partially bivoltine life cycle of

Spring and summer males compete for matings

The sphecid life cycle does not favour helping

halictines might favour the evolution of sociality is supported by three lines of evidence.

1 There are, as predicted, seasonal changes in the sex ratio: in 10 species of sphecid for which data are available the overwintering generation tended to be male-biased and the summer generation female-biased.

2 As predicted, the early stages of helping are found more often in species with the halictine than with the sphecid life cycle.

3 Both the primitive stages of sociality (representing evolutionary beginnings of helping) and the partially bivoltine life cycle are more common in temperate than tropical regions, suggesting that the former may be facilitated by the latter.

Alan Grafen (1986b) has pointed out that Seger's hypothesis is a special case of 'split sex ratios' in which some nests have male-biased and others female-biased sex ratios. Helping will be favoured in the female-biased nests.

Evidence for Seger's hypothesis

Parental manipulation or daughter advantage?

The threshold value of K for helping to be advantageous for mothers, is below that for daughters

According to Seger's hypothesis, offspring originally stayed at home to help because the sex ratio of the brood was female-biased relative to the population as a whole and this shifted the offspring's critical ratio of benefits to costs for helping from 1 to something less than 1 (e.g. $\frac{4}{5}$ with a $3:1$ bias).

From the mother's point of view, the critical value of K is 0.5. This is because by persuading her offspring to help the mother gets more of her own offspring to whom she is related by $\frac{1}{2}$ instead of grandchildren, to whom she is related by only $\frac{1}{4}$ (Fig. 13.3). It appears, therefore, that ecological conditions (values of K greater than the critical threshold) under which parental manipulation was favoured at the origin of eusociality were likely to be reached before those for which helping was of advantage to the daughter. Does this mean that parental manipulation is more likely as a route for the evolution of eusociality by staying at home, as Alexander (1974) has argued? Robin Craig (1979) has pointed out that if the value of K was below the daughters' threshold, the daughters would have resisted manipulation, so a likely way for the queen to have encouraged workers to help was to alter the value of K to move it above the daughters' critical value. She could have done this, for example, by giving less food to the daughters and thereby stunting their growth and reducing their capacity to lay eggs if they left home (egg-laying capacity in insects is very much size dependent). With a reduced ability to lay eggs, the cost of staying at home would have been lowered, in other words the value of K increased. If we accept this kind of

Parental manipulation and advantage to daughters are not really alternatives

reasoning, although parental manipulation and daughter advantage are theoretically distinct, one would soon have merged into the other in evolutionary time.

Eusociality in hymenoptera arose in groups with parental care

Another line of evidence suggesting that parental manipulation played a role in the origin of eusociality is its taxonomic distribution. All the eusocial hymenoptera belong to the group known as the 'aculeate' (stinging) hymenoptera, and none of the remaining, much larger, group of haplodiploid insects (including the parasitic hymenoptera) have evolved eusociality. The special feature of the aculeates is that they have elaborate parental care, which places the mother in a strong position to exert parental manipulation (Stubblefield & Charnov 1986).

The importance of demography

So far we have considered how ecological factors (nest building and defence) and genetic factors (degree of relatedness) contribute to the evolution of eusociality. Raghavendra Gadagkar (1990), following Queller (1989), has pointed out that a third factor, demography, may also play an important role. His argument goes like this. If a solitary queen dies during the period when she is raising a brood, the offspring, dependent upon her care, will die as well. If, however, the same female was part of a group helping to rear the brood, her death would not condemn the offspring to die, because others would continue to carry out brood care. Thus, as part of a group, a female has 'assured fitness returns'. This advantage could, in theory, make it better to help others breed than to breed as a solitary queen. In an attempt to quantify the importance of this 'demographic effect', Gadagkar studied the social wasp *Ropalidia marginata*. On average, the developmental period from egg to adult in this species lasts 62 days. An individual reproductive female has a probability of only 0.12 of surviving for 62 days. Thus, her expected reproductive success from solitary nesting is not very high. In fact, Gadagkar estimated that a female *Ropalidia marginata* could increase her expected success by 3.6 fold as a result of nesting in a group. All other things being equal, the theoretical maximum extra benefit from kinship in a haplodiploid species is a 1.5-fold increase in fitness (by replacing offspring $r = 0.5$ with sisters $r = 0.75$). Thus Gadagkar concludes that the demographic effect of assured fitness returns could be potentially greater than that of haplodiploidy.

Death of females before the end of brood care favours co-operative breeding

In the wasp *Ropalidia marginata* the demographic effect is larger than that of haplodiploidy

Comparison of vertebrates and insects

With the possible exception of naked mole rats (Chapter 12),

The two routes to
eusociality parallel
the two kinds of helping
in birds

there are no known examples of sterile castes in vertebrates, but
in other respects there are some close parallels between the
conclusions of this chapter and those of Chapter 12. The two
hypotheses about the evolutionary origin of helping in insects are
similar to the two kinds of co-operative nesting described for
birds in Chapter 12. The first resembles scrub jays, in which
young birds stay at home to help their parents rear younger
siblings, while the second is parallel to anis where several females
share the same nest and one female is dominant over the rest and
produces more young.

The ecological pressures thought to favour helping in vertebrates and insects are also quite similar. For example, both
scrub jay helpers and workers in social insects play a similar role
in defence of the nest against predators or parasites. Nest sharing
by reproductive females is also thought to be a response to predator

Similar ecological
factors apply

pressure in both vertebrates and insects. Another major ecological
factor which leads to helping in birds is lack of opportunity for
young birds to find a nesting territory and this again is paralleled
by the importance of co-operative nest building in social insects:
a lone female has little chance of building a nest alone, so she is
forced to co-operate.

In spite of these similarities there must also be important
differences which have led insects but not vertebrates to evolve
sterile castes. The differences could either be ecological or genetic.

First, the ecological and life history constraints are different
in insects and vertebrates. Vertebrate helpers such as young scrub
jays help as part of a long-term strategy of acquiring a breeding
territory or mate in the future. In short-lived insects, such long-
term gains are less (although there are exceptions such as

Different constraints in
insects and vertebrates

Metapolybia), and the emphasis is shifted to gains from helping
siblings. If a young scrub jay, for example, stays at home to help,
it does less well than one that succeeds in setting up on its own
(Chapter 12). This is because the helper is capable of rearing
more young than the mother is able to provide, the constraint on
the mother being her inability to incubate a very large clutch. An
insect queen, in contrast, can often provide her daughters with as
many eggs as they themselves could lay, because the constraint
on brood size is the availability of nest chambers and not the
need for incubation. For both insects and vertebrates one constraint
which prevents helpers from setting up on their own is the
problem of obtaining a nest or territory. In insects nest construction takes relatively longer than in vertebrates, so this constraint
is more severe. In other words, a nest is a scarcer resource for an
insect than is a territory for a vertebrate.

The differences between birds and insects in genetic predispositions towards sterility have already been dealt with in depth. Hymenopteran haplodiploidy and aphid asexuality are features that are not shared by any of the vertebrates discussed in Chapter 12. The termites appear to have no particular genetic mechanisms that could predispose them towards eusociality (but see Box 13.3) and may thus be a closer parallel to the vertebrates.

Box 13.3 *Termites: genetic factors that might favour eusociality.*

1 *Cycles of inbreeding and outbreeding.* Stephen Bartz (1979) has suggested that alternation between inbreeding and outbreeding might favour the evolution of sterile castes in termites. To understand the argument, take an extreme case. Suppose the king and queen of a termite colony are unrelated to one another but both were born in highly inbred colonies and are therefore effectively homozygous. The two parents will be related to their offspring by the usual ½, but the offspring will be related to each other by 1! They will all be heterozygotes like the F1 generation in a Mendelian cross between two homozygotes with alternative alleles. Thus offspring will be genetically predisposed to help their siblings rather than their own offspring. Although this is an extreme example it makes the general point that an alternation between inbreeding and outbreeding could favour helping.

How might these events actually come about? Bartz points out that in termite colonies, winged reproductives are often produced not by the founding king and queen but by secondary reproductives that are offspring of the original founder and foundress. In other words the reproductives that disperse to find a mate and start a new colony are the product of brother–sister matings. Given that the winged dispersers are unlikely to end up mated to a sibling, the alternating pattern of inbreeding and outbreeding seems to hold.

2 *Linked genome.* An additional possible genetic predisposition for eusociality in termites was discovered by Syren and Luykx (1977). In some species an appreciable proportion of the genome is linked to the sex chromosomes. As a result of this, relatedness among siblings of the same sex is higher than 0.5 (Lacy 1980). The general importance of this is hard to assess without more information on its prevalence in termites (Crozier & Luykx 1985).

Summary

In the social insects there are sterile workers that never have offspring but instead help to rear younger siblings. This appears at first sight to go against the idea of natural selection favouring maximum efficiency at passing on genes. However the fact that sterile workers rear close relatives genetically predisposes them to be altruistic.

There are two hypotheses about how sterile helpers arose in evolution: by young staying at home to help their mother and by sisters sharing a nest with one producing most of the offspring. According to both hypotheses the ecological advantages of communal nesting are defence against parasites or predators and construction of an elaborate nest. The genetic predisposition is that help is given to close relatives. In the case of the first theory there may be strong selection for mothers to persuade their daughters to stay at home and little selection for daughters to resist.

In the Hymenoptera there may be an additional genetic predisposition towards helping: haplodiploidy. Sterile female helpers in the Hymenoptera could pass on their genes even more efficiently than the reproductive queen if they bias the sex ratio of the reproductive siblings they help to rear. This is because sisters are more closely related to one another than are mothers and offspring. Haplodiploidy acts as a predisposition for the evolution of helping only when the sex ratio in the nest is more female-biased than in the population at large. Demography may also play a role: by nesting in a group female may increase the certainty that their offspring will survive.

Further reading

Hamilton (1972) summarizes the importance of haplodiploidy in Hymenoptera as a predisposition towards eusociality. Seger (1991) is a very useful overview of social hymenoptera. Wilson (1975) in the chapter on social insects gives an excellent review of the biology of this group. Michener (1974) is the definitive work on social bees.

Dawkins (1979) enumerates an entertaining and salutory list of twelve common misunderstandings in the literature on kin selection. For example, the statement that 'all individuals in a species share a high proportion of their genes' (shown by DNA hybridization studies) and they should therefore be altruistic to one another is fallacious. To see why, look at Dawkins' paper.

Wilson (1980) discusses an aspect of social insect behaviour not covered in this chapter, namely the optimal allocation of workers in a colony to different tasks. He takes the example of leaf cutting ants (*Atta*) and shows that one size of worker is more efficient than the others at cutting and carrying leaves for the colony's store. This is the size at which workers normally change from nest maintenance to foraging duties. A general intro- duction to the economics of caste in social insects is given by Oster and Wilson (1978). Andersson (1984) reviews the origin of eusociality.

Topics for discussion

1 What is the importance of haplodiploidy for the evolution of eusociality?
2 What ecological factors are important for the evolution of helping in vertebrates and insects?
3 What data would you need to collect to evaluate the outcome of worker–queen conflict in hymenopteran colonies?
4 How might parasites influence the evolution of eusociality? (Read Shrykoff and Schmid-Hempel 1991.)

Chapter 14. The Design of Signals: Ecology and Evolution

Most of the interactions between individuals described in this book involve communication. Males attract females or repel rivals, offspring beg from their parents, and poisonous caterpillars warn their predators by using special *signals* or *displays* — behaviours and structures designed for use in communication. This chapter is about how signals are designed by natural selection for effective communication. We will discuss the influence of two kinds of selection pressure on signals: ecological constraints imposed by the environment and the response of reactors at whom the signals are directed. But first let us clarify what we mean by communication.

The most obvious characteristic is usually that a signal or display of one individual (actor) in some way modifies the behaviour of another (reactor). The reactor's response may be immediate and obvious (a male firefly rapidly flies towards a flashing female of the correct species); it may be subtle and difficult to detect (a male antelope slightly alters its direction of walking to avoid crossing a territorial boundary when it detects the scent marks of a resident); it may be delayed in time (the ovaries of a female budgerigar gradually develop as the result of the stimulus of male song); or it may not occur all the time (a territorial male blackbird sings for several hours during which time only one or two intruders hear the song and retreat). If, in spite of these difficulties, we can detect the response of a reactor, we could characterize communication as *the process in which actors use specially designed signals or displays to modify the behaviour of reactors*. The qualification 'specially designed signals' saves our definition of communication from becoming too wide. When we see a drunkard lurching down the street towards us late at night we may cross the road to get out of the way, but since there is no reason to suspect that the drunkard's lurch has been developed by natural selection to signify advanced inebriation we would not include lurching within the definition of communication. In contrast, the grasshopper's incessant buzz, which is created by rubbing the legs together and probably evolved from simple walking movements, is an example of communication. The ancestral male grasshopper may have lurched towards a female accidentally producing a chirrup at the same time, but natural selection has acted on the sound to make it into a loud and conspicuous signal for attracting females.

As an aside it is worth pointing out that our definition of

communication would not satisfy all students of behaviour. Our concern in this chapter is with how signals evolve, and we therefore stress the importance of specially adapted signals in communication. But for someone interested in human or animal social interaction a broader definition 'any aspect of A's presence or behaviour which influences B' might be appropriate. This would encompass observations such as those of Michael Argyle (1972) who found that many subtle aspects of posture (leaning back in a chair, crossing legs, etc.) play a major role in human communication, even though they may have not evolved as special signals.

Ecological constraints and communication

The environment limits the range of possible signals

Different goups of animals rely on different sensory channels for communication. Small mammals live in a world of smells, birds in a world of music, and coral reef fish in a world of brilliant poster colours. Why are there such differences? Part of the answer is that the utility of different channels depends on constraints imposed by the habits and habitats of a species. It is obvious, for example, that sound or scent are more useful than visual signals for nocturnal small mammals and that birds living in dense bushes can more readily hear than see each other. Roe deer living in dense forest mark their territories with loud calls and scents deposited on the vegetation, while in open habitats they use primarily visual signals (Loudon pers. comm.). But differences in the effectiveness of transmission are not the only considerations in assessing the costs and benefits of various communication

Costs and benefits of different modalities

channels (Table 14.1). Sound is very flexible: enormous numbers of signals can be fitted into a short space of time by rapid changes in pitch, loudness and harmonic structure. Scent may be less flexible but is energetically cheap to produce and can last for a

Table 14.1 Advantages of different sensory channels of communication. From Alcock (1984)

Feature of channel	Type of signal			
	Chemical	Auditory	Visual	Tactile
Range	Long	Long	Medium	Short
Rate of change of signal	Slow	Fast	Fast	Fast
Ability to go past obstacles	Good	Good	Poor	Poor
Locatability	Variable	Medium	High	High
Energetic cost	Low	High	Low	Low

very long time, an advantage for an animal such as a fox with a very large territory in which it can announce its presence at any particular site only once every few hours, or even every few days. Brilliant body colours are permanently on display (at least seasonally); while this may be advantageous for attracting females and deterring rivals, it may be a considerable disadvantage in attracting unwelcome predators!

COMMUNICATION IN ANTS

The use of different communication channels in different ecological conditions is nicely illustrated by Bert Hölldobler's (1977) study of recruitment signals in ants. When ant workers return from a foraging trip to the colony they often recruit others to take part in harvesting the supply of food. Hölldobler describes three kinds of recruitment, two of which are illustrated in Fig. 14.1.

1 Species such as *Leptothorax* feed on single, immobile prey items (e.g. dead beetles) which are too big for one worker to carry but which can be brought back to the nest by two individuals pulling together. After finding a prey, *Leptothorax* workers return to the nest, regurgitate some of the food and secrete a chemical signal from their abdomen to attract other workers. One worker is recruited to help bring back the prey and it is 'led by hand' to the foraging site — so-called 'tandem running' in which the recruit follows the leader by keeping contact with its body: the recruit's antennae rest on the leader's abdomen (Fig. 14.1a).

(a)

(b)

Fig. 14.1 Two types of recruitment communication in ants. (a) Tactile. (b) Chemical.

Foraging ants use touch, scent and visual signals to recruit nest mates

2 Fire ants (*Solenopsis*) feed on large mobile prey (large insects and so on) which need several individuals to carry them back to the nest and they use odour trails as a means of recruitment. After finding a suitable prey, a worker returns to the nest laying a trail of scent secreted from a special abdominal gland. The odour trail excites other workers to join in, running along the scent trail to the prey, and returning to the colony laying their own trail if they have found the prey (Fig. 14.1b). The scent trail builds up very rapidly as large numbers of workers add to it, but it also decays rapidly as soon as workers stop renewing it because the scent is very volatile and lasts only for a few minutes. This means that the trail persists only as long as the prey is available, and it can change position to keep up with the moving prey.

3 A third kind of ant foraging trail is characteristic of leaf cutters (*Atta*) and seed eaters such as *Pogonomyrmex*. These ants feed on long-lasting or renewing patches of food, so that workers use the same trails for days, weeks or years on end. There are two ways of marking trails, by long-lasting scents and by cutting a path through the vegetation. Both of these provide durable signals.

These examples illustrate how different kinds of signals using different sensory channels are used according to the feeding ecology of the ants: tactile signals for recruiting a single companion to stationary food, rapidly decaying odour signals for recruiting large numbers of workers to mobile prey, and visual cues or long-lasting scents for large patches of renewing food.

BIRD AND PRIMATE CALLS

The way in which habitat structure and meteorological conditions affect signal transmission has been studied in detail for sound signals, especially bird songs. Differences between species and between populations within a species can sometimes be explained in these terms.

Habitat correlates of birdsong structure

Gene Morton (1975) and Claude Chappuis (1971) were the first to show that the structure of bird songs is correlated with habitat. Morton found that the songs of species living below the canopy in tropical forests in Panama were characterized by lower frequencies, a larger proportion of pure tones and a narrower range of frequencies than songs of grassland birds in the same country (Table 14.2). Typically the songs of tropical forest birds contain many low-pitched pure whistles and those of grassland species sound like buzzy trills.

Several subsequent studies have shown that geographical variation within a species can sometimes be correlated with habitat type. For example, Fernando Nottebohm (1975) found the rufous-

Table 14.2 Differences between the songs of forest and grassland birds in Panama. From Morton (1975)

Habitat	Emphasized frequency (kHz)	Per cent pure tones	Frequency range (kHz)
Forest (below canopy)	2.2	87	1.5
Grassland	4.4	33	3.5

Among birds, songs of forest dwellers have different characteristics from those of open country. This applies both between and within species

collared sparrows (*Zonotrichia capensis*) living in South American forests sing slower trills than their grassland cousins (Fig. 14.2a). The great tit, like the rufous-collared sparrow, lives in many different kinds of habitat over a wide geographical area. Its distribution stretches from Ireland to Japan and from Finnish birch forests to Malaysian mangrove swamps. Mac Hunter and John Krebs (1979) recorded the territorial songs of great tits in two contrasting kinds of habitat, open woodland or parkland and dense forest. Regardless of geographical location (recordings were made in various countries stretching from Norway to Iran) there were consistent differences between songs from the two habitat types (Fig. 14.2b). The birds in more open habitats have songs with higher maximum frequency, more rapidly repeated notes, and a wider range of frequencies than those of forest birds. So striking are the correlations with habitat type that the songs of birds in a park in southern England are more similar to those from the same sort of habitat in Iran (5000 km away) than they are to birds in a dense forest 100 km away in another part of southern England.

What is the survival value of these differences in song structure between habitats? The answer is still not certain. Morton hypothesized that songs are designed to carry the maximum possible distance and suggested that the differences he observed between habitats might be related to differences in the attenuation of sound. He tested this by measuring the attenuation of pure tones of different frequency that he broadcast in the two habitat types from a loudspeaker. He found that in both habitats high frequencies attenuated more rapidly than lower ones. This is not surprising, since on theoretical grounds one would predict that high frequencies would be more readily attenuated by obstacles such as leaves and branches, by air turbulence, and by the viscosity of the air itself. The unexpected aspect of Morton's results was that sounds of about 2 kHz carried particularly well, better than higher or lower pitched tones, in forests but not grassland. Since forest birds sing at about 2 kHz, Morton suggested that their songs are

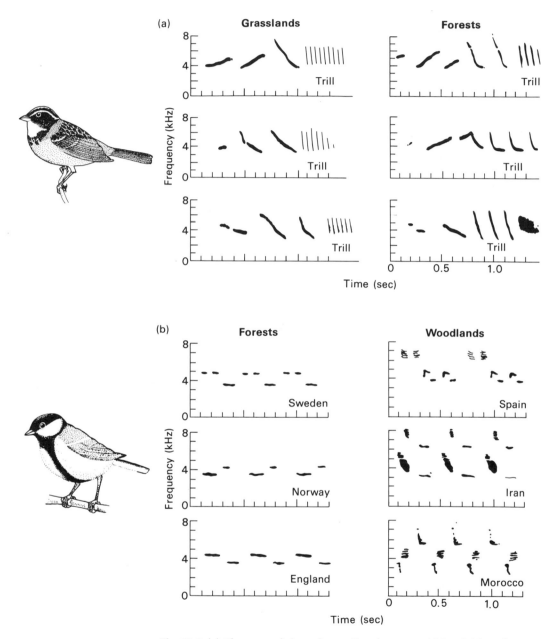

Fig. 14.2 (a) The song of the rufous-collared sparrow (chingolo) has slower trills in forests than in open country. (b) Great tits in dense forests sing songs with a narrower range of frequencies, lower maximum frequency and fewer notes than the songs of open country birds.

tuned into the 'frequency window' of minimum attenuation. The reason for higher pitched, wider frequency band songs in grassland was less clear.

However, other studies of sound attenuation have shown that there are probably only fairly small differences between habitat types (Marten & Marler 1977; Marten *et al.* 1977; Wiley & Richards 1978). While in forests branches and leaves cause attenuation of high frequencies, in more open habitats local air turbulence often has a similar effect. Since open areas are often less sheltered than forests, the overall pattern of attenuation of different frequencies tends to be similar in the two habitats. Furthermore, Morton's frequency window seems to be more related to attenuation of lower (below 2 kHz) frequencies by the ground than to differences between habitats.

Differences between forest and open country birdsongs may be related to differences in acoustic properties of the habitats

Wiley and Richards (1978) propose an alternative explanation for the differences between forest and open habitats. They point out that there are two problems facing a singing bird. One is *attenuation* of sound: if the song attenuates too much the receiver will not be able to detect it because it is lost in the background noise. The other is *degradation*: if the song is degraded or distorted in its passage through the environment it may be confused by the receiver with other sounds. They suggest that there are much greater differences in degradation than in attenuation between habitats. In forests, the major source of degradation is echoes or reverberations from branches and leaves, while in open habitats the major source is irregular fluctuations in amplitude caused by gusts of wind that mask the song. These two kinds of degradation can be reduced by different design features. Reverberations are more severe for high frequency than low frequency sounds (because high frequencies are deflected by small objects such as leaves and branches). They also cause a problem if the song contains rapidly repeated notes as are found in the trills of many bird songs, because the echoes will become confused with the original notes. Therefore songs designed to overcome reverberations in forests should be of low frequency and contain either pure notes as opposed to trills, or trills with widely spaced notes. These are precisely the patterns observed by Morton and in the studies of chingolos and great tits. The irregular amplitude fluctuations of open habitats, in contrast, favour rapid trills. Since the song is masked by wind at irregular intervals it has to contain notes that are short enough and sufficiently rapidly repeated to be detected in a short space of time. The same need to fit the songs into short intervals between gusts may account for the high frequencies of open habitat songs, since a high frequency sound requires a shorter time for a given number of wavelengths. Again the patterns predicted by Wiley and Richards' hypothesis fit well with those observed in the field.

Two hypotheses: attenuation and degradation

An additional complication is that in at least two of the

Differences in bird
density as a confounding
variable

studies referred to (Morton's and the great tit study), forest birds occupied larger territories than those in open habitats. The songs in the two habitats may be adapted not for maximum detection distance, but for an optimum distance which differs between habitats because of the difference in bird density. (An optimum distance could arise if, for example, it was disadvantageous to signal over too great a distance — see below.) Forests songs with their low frequencies and energy concentrated into a narrow band are likely to carry further than the higher pitched, wider frequency band songs of open country birds.

There is a striking parallel between these results of studies of bird song and the interpretation by Busnel and Klasse (1976) of a peculiar kind of human language. In four parts of the world, Andorra, Turkey, Mexico and the Canary Islands, local peasants have developed extraordinary whistling languages with elaborate vocabularies. Although the details of the languages and the methods of sound production vary from place to place, all are designed for long distance communication. The four places are all mountainous with steep-sided, rocky valleys; distances are not very great as the crow flies but are large in terms of the effort needed to cross from one side of the valley to the other. The whistling languages are used for communicating across the steep valleys. They have all their sound energy concentrated into a narrow band of frequencies and are therefore well designed, like forest bird songs, to be detected and interpreted over relatively large distances.

Human whistled
languages carry a long
distance

Long and short distance
calls in mangabey

Not all sounds are used for long-distance signalling, and short-distance sounds may be designed not to carry further than necessary. The grey-cheeked mangabey (*Cercocebus albigena*) (Fig. 14.3) is a monkey which lives in the forests of East Africa; it lives in troops and defends a group territory. Peter and Mary Sue Waser (1977) studied two vocalizations of the mangabey: the 'whoop-gobble' which is used in intergroup signalling, and the 'scream' which is used during agonistic encounters within a group. The two sounds are produced at the same volume by the monkeys, but because the whoop-gobble is lower pitched and has a narrower range of frequencies it carries much further than the scream (Table 14.3). The differences in frequency structure and therefore carrying power of the two sounds reflect their design for communication over different distances.

Probably the reason why the mangabey uses a song which does not carry far for signalling within the troop is to avoid attracting predators. Another way to make sounds less likely to attract a predator is to produce unlocatable sounds, as was first suggested by Peter Marler (1955) in a classic paper about alarm

Fig. 14.3 The grey-cheeked mangabey. Photo by P.M. Waser.

Perhaps hawk alarm calls of small birds are designed to be hard to locate...

calls of small birds. He pointed out that the calls given by many species of small birds when a hawk flies over are remarkably similar. All are thin high-pitched 'seet' sounds which are hard for the human ear to locate. Marler suggested that the calls have evolved a structure which makes them hard for predators to locate and therefore reduces the risk to the caller of attracting the hawk to itself. (Why the caller should warn his flock mates in the first place is a different matter — see Chapter 11.) Elegant

Table 14.3 Comparison between two calls used by the grey-checked mangabey in tropical forests of Uganda. From Waser and Waser (1977)

Call	Function	Distance from which heard by human ear	Sound pressure 5 m from monkey	Frequency (Hz)
Whoop-gobble	Spacing between troops	1000 m	75 dB	300–400
Scream	Within troop agonistic encounters	300 m	78 dB	1000–3000

and appealing though Marler's hypothesis is, attempts to test its assumption that the calls are hard to locate have produced ambiguous results. In one study pygmy owls (*Glaucidium* spp.) and goshawks (*Accipiter gentilis*), both of which prey on birds, were tested to see if they could locate alarm calls. The predators sat on a perch between two loudspeakers and their head movements were used to test if they could locate which speaker played the call. They seemed to be as good at locating alarm calls as they were at locating other, supposedly more locatable calls (Shalter 1978). Brown (1982), on the other hand, found in a somewhat similar experiment that red-tailed hawks (*Buteo jamaicensis*) and great-horned owls (*Bubo virginianus*) had more difficulty in locating alarm calls than other calls.

...although the experimental evidence is ambiguous

Reactors and the design of signals

The constraints of the environment impose broad limits on the design of signals, but within these limits the way in which signals evolve results from selection to increase their effectiveness in altering the behaviour of reactors. Reactors play an important role both in the evolutionary origin of signals and in their subsequent evolution towards increased effectiveness.

HOW SIGNALS ORIGINATE

To the casual observer animal displays often seem to be inexplicably bizarre and absurd. Why should male ducks perform sham drinking and preening movements as part of their courtship ritual? Why do wolves mark their territories by urinating? Why do rhesus monkeys grin as a signal of fear and appeasement?

A great step forward in understanding the answers to these questions came when ethologists such as Lorenz and Tinbergen realized that many signals have evolved from incidental movements or responses of actors which happened to be informative to reactors. Selection favoured reactors who were able to anticipate the future behaviour of actors by responding to slight movements which predicted an important action to follow. If a dog always bares his teeth before biting, reactors who are able to anticipate and escape from an aggressive attack by observing bared teeth will be favoured by selection. Once this happens selection will favour actors who bare their teeth as a means of rapidly deterring reactors, and teeth baring will begin to evolve into a threat display.

We should expect then, that the incidental movements and responses from which signals have evolved are those which were

Many signals evolved from incidental movements that indicated the signaller's future actions

originally most informative about future actions. This is borne out by many studies of the detailed form of displays in birds, fish and mammals. Many signals in these animals have apparently evolved from *intention movements* such as are made by a bird when it crouches and tenses its muscles for take-off: it is not hard to imagine that intention take-off movements were originally good predictors of a lunging attack by a rival, approach by a prospective mate and so on. Other movements performed at moments of transition from one major activity to another are also frequently the raw material from which displays have evolved. Often these movements reflect motivational conflict or indecision as the animal vacillates between, for example, attack and running away (Table 14.4).

Our examples of the duck, wolf and monkey can be interpreted in a similar way. The drinking and preening movements of courting ducks are probably derived from *displacement activities* seemingly irrelevant actions which tend to occur at moments of balance between incompatible motivational states such as aggression and sex. Urinating is like blushing and hair erection, which are also the basis for displays in some species, a consequence of autonomic nervous activity during moments of stress. The ancestral wolf may have urinated uncontrollably when confronted with a rival at its territory boundary (anyone who has nursed a nervous dog will know all about this sort of thing!), and the response has subsequently evolved into a way of signalling 'keep out'. The grin of the terrified monkey is very similar to the reflex response with which a monkey protects the most vulnerable parts of his face such as eyes and mouth from the onslaught of an attacker. As with urinating, an autonomic reaction to stress has become a signal during evolution.

Most of the conclusions described in the above two paragraphs are inferences based on studies of the detailed structure of displays and the contexts in which they occur. For example, the observation that threat displays occur at territory boundaries and in sequences of behaviour which include transitions between attack, threat and retreat suggests that threat signals arise when the animal is in motivational conflict. Similarly the structure of displays can be viewed as reflecting motivational conflict. The 'zig-zag' display of the three-spined stickleback (*Gasterosteus aculeatus*) involves a strange movement in which the courting male approaches the female in a series of short arcs, as if he is in a conflict between approach and avoidance.

There is also some more direct evidence for the conflict hypothesis from experiments in which the tendencies of an animal to attack and flee are manipulated independently. Nick Blurton

... such as intention movements

These ancestral movements often reflected motivational conflict

Experimental manipulation of motivational conflict

Table 14.4 Examples of the kinds of behaviour patterns and other responses from which displays in birds, fish and primates are thought to have evolved. From Hinde (1970)

Behaviour or response from which display evolved	Example of display	
1. Intention movement	Sky pointing in the gannet	
2. Ambivalent behaviour	Forward threat posture of black-headed gull	
3. Protective response	Primate facial expressions	
4. Autonomic response (e.g. sweating, urinating, rapid breathing)	Vocalizations (from rapid breathing). Scent marking	
5. Displacement activities	Preening in duck courtship	
6. Redirected attack	Grass pulling in herring gulls	

Jones (1968) studied the threat displays of captive great tits in this way. He found that the birds would attack a pencil pushed through the bars of their cage and that they would flee from a bright light. When the light and the pencil were presented at the same time the birds tended to perform threat displays. Other threat displays could be elicited by presenting the pencil outside the cage so that direct attack was thwarted. Thus motivational conflict and thwarting of attack seem to generate threat displays.

HOW SIGNALS ARE MODIFIED DURING EVOLUTION: RITUALIZATION

Although signals started off as incidental movements or responses such as those shown in motivational conflict, during their evolution they have become modified by selection to improve their effectiveness as signals. For example, the courtship movement in which male ducks preen their wings has been emphasized in some species by the evolution of conspicuous bright-coloured feathers on the wing, towards which the male points his bill during courtship preening. An extreme case is the mandarin duck (*Aix galericulata*) which has some of its wing feathers modified to form a bright orange 'sail' which is erected during courtship preening. The ancestral preening movement is reduced to a quick turn of the head so that the bill points to the orange sail.

During evolution, signals become exaggerated, repetitive and stereotyped: the process of 'ritualization'

The term *ritualization* is used to refer to the evolutionary modification of movements and structures to improve their signal function. Thus the ancestral preening movement of ducks has become ritualized in species such as the mandarin duck. The changes that occur during ritualization include the following: the movements tend to become highly stereotyped, repetitive and exaggerated; often the movements are emphasized, as in the mandarin duck, by the development of bright colours on the body. It is of course not possible to observe these evolutionary changes taking place and the evidence that they have occurred comes from comparative studies of displays in closely related species. In the example shown in Fig. 14.4 the probable ancestral movement is pecking at pieces of food on the ground during courtship (perhaps originally a displacement activity). This pattern

Ritualization inferred from comparison between species

is seen in present-day jungle fowl (Fig. 14.4). In other species of galliform bird the display is ritualized, for example in pheasants and peacocks it is emphasized by the evolution of a large tail and the original ground pecking movement is reduced to a stereotyped bobbing of the head or pointing of the beak.

We have said that ritualization occurs because it improves the signal function of a display. What exactly does this mean? As

(a)

(b)

(c)

(d)

(e)

Fig. 14.4 Comparative evidence for the origin and ritualization of a display. The ground pecking display of phasianid birds. (a) The least ritualized form is shown in the male domestic fowl (*Gallus*). It scratches the ground with its feet and pecks at food or small stones (perhaps originally a displacement activity). This attracts the female. (b) The male ring-necked pheasant (*Phasianus colchicus*) attracts females by means of a similar display. (c) The impeyan pheasant (*Lophophorus impejanus*) and (d) the peacock pheasant (*Polyplectron bicalcaratum*) both emphasize the pecking display with rhythmic bobbing of the tail or the head. (e) The peacock (*Pavo*) shows little of the ancestral movement. The male spreads his enormous tail and points his beak towards the ground. Redrawn from Cullen (1972).

we saw in Chapter 8 some displays used by males during courtship may be exaggerated and very elaborate as a result of sexual selection: females may exhibit a preference for elaborate displays. While this hypothesis may account for the ritualization of some courtship signals, it cannot be a general explanation of ritualization, since other displays such as those used in threat also appear to have undergone ritualization.

HYPOTHESES FOR THE EVOLUTION OF SIGNAL DESIGN

(a) Reduction of ambiguity

Why do signals become ritualized?

According to the traditional view, ritualization is the result of the selective advantage to signallers in reducing the chance of confusion between their various displays (Cullen 1966). Thus a ritualized threat signal, by its exaggerated, stereotyped and repetitive nature clearly states 'I am about to attack' and not 'I am very frightened' or 'I am a sexy male'. In fact (as Darwin first pointed out) threat and appeasement signals are often extreme opposites: a threatening dog stands erect while a fearful dog crouches near the ground in appeasement. The principle of reduction of ambiguity can also be illustrated with reference to courtship signals. Often the displays of closely related species are clearly distinct so as to minimize the chance of confusion between species (Fig. 14.5).

Ritualized signals may be less ambiguous

Increasing the clarity of signals by increasing their stereotype may at the same time reduce the amount of information they convey about the actor. The exact structure and movement pattern of an ancestral threat signal may have reflected the precise balance between aggression and fear in the actor, but a stereotyped, ritualized signal probably conveys less information about the actor's state. Morris (1957) called this stereotypy 'typical intensity' and he viewed the loss of accuracy of information as the price paid by the actor for reduction of ambiguity. This assumes that it is advantageous for the actor to communicate information about its internal state, but in fact just the opposite may often be the case. As pointed out in Chapter 7, if two animals are contesting a resource by means of ritualized signals, it will not pay either to reveal its exact balance between attack and fleeing until the last possible moment. In fact sterotypy of displays may have evolved precisely *because* it reduces the information available to reactors about the actor's internal state. This leads us to the second hypothesis about ritualization, which emphasizes the use of signals by actors to control the behaviour of reactors.

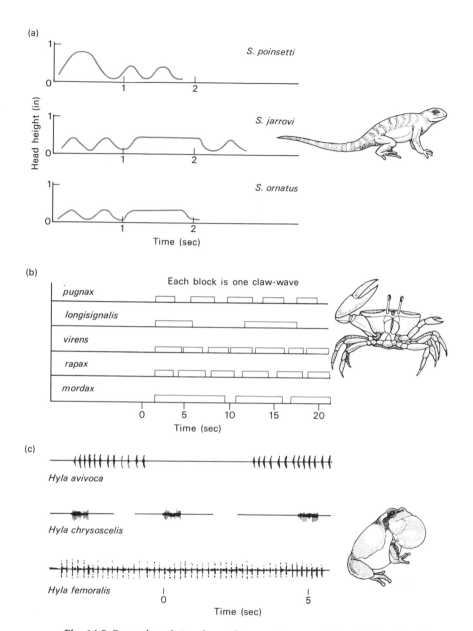

Fig. 14.5 Examples of signals used in species recognition. (a) Lizards of the genus *Scleroporus* have species-specific head bobbing patterns used in courtship and threat. The graph shows the head height as a function of time in the bobbing displays of three species. (b) Male fiddler crabs (*Uca*) attract females by waving their enlarged claw. Each species waves in a different way. The graph to the left shows the waving patterns of four species. A block represents a complete wave of the claw. (c) Tree frogs (*Hyla*) have species-characteristic calls. Females are attracted only to males of their own species. The oscillogram traces to the left show the gross temporal characteristics of each of three species' calls. There are also species-characteristic differences in the fine structure of calls that are important in species recognition.

(b) Manipulation

A rather different view of ritualization emerges if we start to think about costs and benefits of communication to the individual. The basic principles of natural selection that we have used in previous chapters would lead us to expect that animals would only use signals when signalling is to their advantage because of its effect on the behaviour of others. We can put this in more formal and more precise terms as follows. Signals are specially evolved pieces of behaviour with which one individual, the actor, alters or manipulates the behaviour of another, the reactor, to the advantage of the actor. If signals really evolved in this way, how is ritualization to be interpreted? Imagine that individual A manipulated the behaviour of B in a way that was advantageous to A but not to B; it would then have been of advantage to B to resist responding to A's signal, or in other words to develop 'sales-resistance'. This in turn would have generated pressure on A to develop a more effective signal to overcome the resistance of B, for example by increasing the amplitude of the signal, repeating it or exaggerating it. Of course A and B would not have been single individuals, but *roles* played through evolutionary time, and the changes would have taken place over many generations. To summarize this argument, if signals arose as manipulative devices, there would have followed a coevolutionary 'arms race' (Chapter 4) between sales-resistance of reactors and manipulative power of actors. It is the outcome of this coevolutionary race that we see as ritualization. Ritualization, according to this view, did not occur because of selection for more accurate or more unambiguous transfer of information by signalling, but simply to overcome sales-resistance. In overcoming sales-resistance the concepts of information and ambiguity are irrelevant. All that counts is persuasion or manipulation.

Perhaps a useful analogy is with advertising. Most advertisements work because they exploit the psychological susceptibilities of the public. When you see an advertisement with a bronzed, tough, healthy cowboy smoking a cigarette, you do not start to believe that smoking will improve your health or your horse-riding ability, you are (or at least some of you must be) simply persuaded, against your better judgement, by psychological manipulation, that smoking is a 'good thing'. Just as the concept of information is of little relevance to persuasion, so is deceit. Advertisements do not (at least where there is legislation to prevent it) actually tell lies, they simply persuade.

Although the 'manipulation' view offers a different explanation of why signals became ritualized, it takes the same starting point

Ritualization may be the result of an evolutionary 'race' between actors and reactors

Analogy with advertising

as that proposed by Lorenz, Tinbergen and others for the origin of signals. Recall that the probable starting point for many signals was the opposite of manipulation by actors, namely the advantage gained by reactors through responding to subtle cues that allowed them to anticipate the actor's future behaviour. Given this starting point, reactors, were tuned to the signals of actors and primed for actors to turn the situation to their own advantage.

The starting point for the evolutionary race is the benefit to reactors

The term 'manipulation' may seem to be rather emotive, but it is a good description of the effect of ritualized signals on reactors in many kinds of communication. A striking example is the effect of a young cuckoo on its foster parents. By begging in the nest, the cuckoo manipulates the parental behaviour of its foster parents to the advantage of cuckoo genes and the manifest disadvantage of foster parent genes.

So far in our discussion of manipulation we have assumed that communication is not co-operative, that the interests of actor and reactor are not the same. But as we have seen in Chapter 11, co-operative behaviour of many kinds is by no means uncommon, especially where individuals are closely related or where they interact with each other repeatedly and exchange benefits in a reciprocal way. Communication between the members of an ant colony, in a family of jackals, or even between repeatedly interacting territorial neighbours, might well be co-operative. If ritualization of signals is the product of the manipulation — sales-resistance arms race we have described, what about co-operative signals? Should they have become ritualized in the same way? If the ancestral reactor *benefited* from receiving the signal, rather than evolving sales-resistance, it would have evolved heightened sensitivity to the signal. This in turn would have paved the way for actors to decrease the cost of signalling by reducing the amplitude of their signals. All signals have a cost, for example some cicadas spend 20 times their resting energy while they are calling (MacNally & Young 1981), while other animals attract lethal enemies through their conspicuous signals, so that if reactors are highly sensitive to the signals, actors would evolve towards minimizing their cost. The minimum might be set by the sensory limits of the reactors or the noise levels in the environment. In short, co-operative signals should be quite different from the typical ritualized non-cooperative signals which ethologists have studied: they should be quiet or inconspicuous, in fact they may be barely detectable. Unfortunately no one has undertaken a systematic comparison of co-operative and non-cooperative signals, so the predicted difference remains to be tested.

Co-operative signals should be 'hushed whispers'

(c) Honesty

In Chapter 8, we discussed how sexually selected displays may arise as a result of female choice for honest signals of male genetic quality. Elaborate displays such as the elongated tail feathers of swallows may act as diagnostic traits of inherited qualities of disease resistance, foraging ability and so on. An essential property of the display, according to this hypothesis, is that it is costly to develop and maintain (it is a handicap) and can therefore only be expressed by genuinely good quality males. Amotz Zahavi (1979, 1987) has generalized this idea to apply to all animal signals. Females choosing a sexual partner, he argues, are in principle no different from other reactors. All reactors are under evolutionary selection pressure to detect dishonest or manipulative signals. As a result of the evolutionary arms race referred to in the previous paragraphs, only signals that are honest indicators of qualities such as size, strength, motivation to fight, running speed, and so on, will persist. Other signals will be ignored and fall into disuse. Zahavi's hypothesis, which is mainly of relevance to signals that indicate some aspect of the signaller's quality, has three main features: (a) signals are reliable, (b) reliability is maintained by the cost of signalling, and (c) there should be a direct link between signal design and the quality being signalled. Signals of strength should involve demonstrating muscle power, signals of size should directly indicate the outline of the body, and so on. In Chapter 7, we saw how some signals used in fighting are directly related to aspects of fighting ability, for example the pitch of calls used in assessment of size during contests by frogs is physiologically linked to body size.

> Ritualization may be the process by which signals become reliable indicators of quality or intentions

Evidence from present-day signals

The discussion in the last section was about evolutionary history, and the ideas are therefore difficult to test directly. It is possible, however, to look at present-day signals and ask if they are manipulative or honest.

AN EXAMPLE OF BLUFF OR MANIPULATIVE SIGNALLING: MANTIS SHRIMPS

As discussed in Chapter 7, signals used for assessment during contests are often honest and costly, as predicted by Zahavi. However, animals may succeed with bluffing on occasions. Mantis shrimps (*Gonodactylus bredini*) are fearsome animals

(Plate 14.1(a), between pps 212–213). They use their powerful forelimbs, covered with a hard exoskeleton, to smash opponents to pieces (large species are able to smash their way out of an aquarium!). Mantis shrimps do not always fight by hitting each other: they also use a threat display in which the powerful fore-limbs are spread out, perhaps, in Zahavian terms, an honest signal of size. But this threat signal is also used in a dishonest way. Every 2 months, a mantis shrimp moults its hard exoskeleton and has to spend 3 days as a soft, impotent, warrior before the new skeleton hardens up. During the moult, individuals still use the threat to deter intruders, even though it is no longer an honest signalling of fighting ability (Adams & Caldwell 1990). Interestingly, the bluff is more likely to be used against small opponents. It also has a cost: if a soft mantis shrimp threatens and is attacked, it is more likely to be killed than if it flees.

Moulting mantis shrimps bluff

HONEST SIGNALLING: THOMPSON'S GAZELLES

When Thompson's gazelles in East Africa spot an approaching predator, they sometimes, instead of running away as fast as possible, begin to 'stot' (Plate 14.1(b), between pps 212–213). Stotting is a peculiar springing motion, rather as though the gazelle were on a pogo stick. Naturalists first thought that stotting was a signal to other gazelles, warning them of the approach of a predator but, as Amotz Zahavi first suggested and Clare Fitzgibbon has subsequently demonstrated (Fitzgibbon & Fanshaw 1988), it is probably an honest signal, directed to the predator, of a gazelle's ability to escape. Fitzgibbon showed that wild dogs prefer to chase gazelles that stot at a low rate and that if high rate stotters are chased they are more likely to escape. In addition, she found that the gazelles stot more in the wet season, when they are in good condition, than in the dry season.

Stotting gazelles signal their condition

RECEIVER PSYCHOLOGY: TUNGARA FROGS AND SWORDFISH

As Tim Guilford and Marian Dawkins (1991) have pointed out, any hypothesis to account for the evolution of signal design must include not only the selective forces on actors for manipulation and on reactors for honesty, but also constraints on the ability of reactors to detect and recognize signals, which they call 'receiver psychology'. A possible example is the work of Mike Ryan (Ryan *et al.* 1990; Kirkpatrick & Ryan 1991) on two species of Tungara frog, *Physalaemus pustulosus* and *P. coloradorum*. In both species the female ear is most sensitive to sounds of about 2.1 kHz, but only one of the species, *P. pustulosus*, has a call component (the

Sensory bias of receivers may drive the evolution of signals

'chuck') close to this frequency. Ryan *et al.* suggest that the female sensory bias has evolved first (being shown by both species) and that the call of the male *P. pustulosus* has adapted to this bias, whilst *P. coloradorum* has not. An intriguing consequence of this hypothesis is that the female *P. coloradorum* actually prefers the *P. pustulosus* chuck to calls of her own species. A parallel example is found in fish of the genus *Xiphophorus*. Swordtails (*X. helleri*) have elongated tails whilst platyfish (*X. maculatus*) do not. Nevertheless, females of both species prefer to court with long-tailed males (Basolo 1990). The puzzle of why male *P. coloradorum* have not evolved 'chuck' calls and why male platyfish have not evolved long tails, to exploit the respective female preferences for these traits, is unsolved. One possibility is that of evolutionary lag (insufficient time for genetic variation — see Chapter 4). Alternatively, there may be costs associated with these traits (e.g. increased predation) which outweigh the mating advantage for some species but not others.

Frog calls and swordfish tails as examples of sensory bias

Variability of signals and information

We now turn from the question of how signals became ritualized during evolution to that of how they are used in communication on a moment to moment basis. A major difference between animal and human communication is that animals do not, on the whole, communicate about abstract ideas or about remote objects in their environment. Instead they communicate about immediate events connected with the actor and its surrounds; things such as ownership of a territory, readiness to mate, and approach of a predator. Exceptions to this generalization include the well-known dance language of bees, in which information about the direction, distance and quality of remote food sources is communicated. Given that animal signals carry simple messages such as 'keep out' and 'I am hungry' it is surprising how variable they are. Often there is not just one signal for each kind of message, but a repertoire with several threat, mating, and danger signals (Box 14.1). We will illustrate our discussion of this by referring to threat signals.

Why are there many threat signals?

As we saw in Chapter 7, variation *between* individuals in their threat signals is often associated with variation in strength or fighting ability. Therefore when animals engage in threat, one of the things they are doing is gauging their opponent's size or strength from its signals. There is, however, often variation *within* an individual in its threat signals. Magnus Enquist (Enquist *et al.* 1985; Enquist 1985) has studied variable threat signals in the fulmar (*Fulmaris glacialis*), a seabird, competing for pieces of fish

Box 14.1 *Animal signals and information.*

In the literature on animal signals, the word 'information' is used in two distinct ways.

(a) The technical meaning. The reduction in uncertainty of an observer about the actor's future behaviour following a signal. If the actor was equally likely to do one of two activities (say attack and retreat) before giving a signal, but afterwards was certain to do one of them (say attack), then the observer's uncertainty about the future behaviour of the actor has been reduced by one 'bit' as a result of seeing the display. There is no doubt that most animal signals convey information in this sense.

(b) The colloquial meaning. In everyday terms when we talk of information we mean information 'about something'. In animal communication information might be about the species, size, age, intentions, mating status and so on of the actor. In Chapter 7 and again in the present chapter we have referred to the question of whether or not signals convey information in this second sense. The main points are as follows.

1 Animals are not expected to convey information about their intentions, except perhaps in co-operative interactions. Some evidence supports this view.

2 Signals may convey information about size, age and strength, and they tend to do so by variations that are not open to bluff or cheating (so-called reliable cues) (Chapter 7).

3 Signals may convey information about the environment, for example the kind of predator that is in the vicinity (Seyfarth *et al.* 1980) or the whereabouts of food (von Frish 1967).

In fulmars, there are low and high cost signals used in different situations

off the shore in Iceland. The birds use a range of displays which vary both in their effectiveness in causing the rival to retreat and in their cost, measured as the likelihood that the rival will retaliate with physical attack. The central conclusion of Enquist's study is that cost and effectiveness are correlated. The fulmar has a choice of using a low cost display such as wing-raising (the probability of eliciting physical attack is only 0.017) or a high cost display such as rushing from behind (cost = 0.28). The former causes the opponent to retreat on only 12 per cent of occasions while the latter has this effect on 28 per cent of times it is used. Probably an individual chooses which display to use on the basis of how hungry it is, and therefore how much it values the piece of fish.

The interpretation of variable threat displays is not, however, always so straightforward. In another seabird, the Arctic skua (*Stercorarius skua*) there are 11 different kinds of display used as threats in contests, and careful analysis revealed little difference between them with regard either to their association with the actor's future behaviour or in their effect on reactors. Although the different displays are associated with slightly different probabilities of ensuing attack, the differences are not very great and, furthermore, the differences vary between locations and years: there is no consistency about which display gives the best prediction of future attack (Paton & Caryl 1986). This is, of course, exactly what would have been predicted from game theoretic analyses of contests (the animal should conceal its hand, see Chapter 7), but it leaves the puzzle of what the various displays are for. Robert Hinde (1981) has argued that to look simply at the probability of attack (or any other behaviour) following a display may be too simple. His idea is that in contests, and perhaps in other kinds of social interaction as well, signals are often used as a form of interrogation, so that the actor's behaviour following a particular signal may depend on the reactor's response. If this view is correct, there may be very complex relationships between signals and the ensuing behaviours. Contrary to the prediction of this hypothesis, however, Paton (1986) did not find any consistent differences in the response of reactors to the different threat displays. It may in fact be inappropriate to think of this species as having different displays since no differences can be found in their effect. Perhaps the birds simply use a range of postures in contests because they are attacked or start an attack from different physical positions.

In skuas, the range of threat signals may reflect simply the physical position of the actor when giving the signal

Signalling, manipulation, and the animal mind

Earlier in the Chapter, we used the metaphor of 'manipulation' in describing the hypothesis that animal signals may be designed by the process of evolution to optimize the advantage gained by the actor in altering the behaviour of the reactor. It should be clear that we did not use the term 'manipulation' to imply conscious thought or intention by the actor, but merely to describe a hypothesis about evolutionary selection pressures on signals. Although a bee orchid (*Ophrys apifera*) may manipulate the behaviour of bees in such a way that the orchid benefits (by pollen transfer) and the bee does not (it attempts to mate with an inappropriate object), no one would seriously suggest that orchids have conscious intentions!

As a quite separate issue, some studies of primate communi-

The term 'manipulation' does not imply intentionality

cation have asked the question: 'Do primates consciously manipulate others?' One of the most thoughtful and detailed studies of this question is that of Dorothy Cheney and Robert Seyfarth (1990), on communication in the vervet monkey (*Cercopithecus aethiops*) in Amboseli National Park, Kenya. One kind of vocal signal which may be used intentionally to manipulate others is alarm calling. Vervet monkeys live in groups, and when one member of the group spots a potential predator it gives an alarm call, causing the others to take evasive action. The interesting feature of vervet alarm calls is that there are different kinds of calls for different types of predator, the major ones being the snake alarm, which causes other monkeys to stand on tip-toe and peer down at the ground, the hawk alarm call, causing others to scan the sky, and the leopard alarm, which elicits the response by other monkeys of running to the nearest tree and clambering up it. These observations show that vervets have a sophisticated communication system: specific calls refer to particular predators and elicit an appropriate response but, of course, they do not imply any conscious intention. A simpler level of explanation, that the different kinds of predator automatically trigger different kinds of call, would explain the facts.

However, Cheney and Seyfarth report other observations which, they argue, indicate that vervets are more than mere automatons responding in a reflexive way. One such observation is that vervets are much less likely to give alarm calls when they see a predator if they are not in the presence of other group members. This 'audience effect' might be taken to mean that the monkeys intend their calls to be heard by other monkeys, but it could also be explained by saying that the stimulus eliciting an automaton-like response is the presence of both a predator *and* a fellow group member.

A second set of observations which, Cheney and Seyfarth suggest, might imply more than reflex-like behaviour, is the apparently manipulative use of signals. On four occasions, a low-ranking male called Kitui gave a leopard alarm call, even though there was no leopard around, when an interloping male approached the group. The interloper immediately ran to the trees instead of continuing his attempt to enter the group, so Kitui succeeded in averting an incipient invasion by deceptive use of an alarm call. Does this imply conscious intention? Again, a simpler explanation could in principle do: perhaps Kitui, in a state of high arousal, had a lowered threshold for giving alarm calls in response to sudden movements which could signal a leopard attack (just as you might jump at the slightest rustle in the undergrowth if you are walking through a forest in the middle of the night). However, a

Do vervet monkeys intentionally manipulate others?

Different calls for different predators

The audience effect

Anecdotes suggest intentional manipulation, but 'simpler' explanations are possible

further set of observations suggests that this explanation is not the whole story. On two occasions, after giving the alarm call, Kitui left his tree and wandered across to the interloper's tree still giving the alarm call. In response to a real leopard attack (or perceived leopard attack) the alarm caller would be expected to stay in the tree until danger had passed. The fact that Kitui climbed down suggests (a) that he did not simply give the call 'by mistake' in response to a sudden movement, and (b) that his deceptive use of the call, if that is what it was, was inept, since he more or less gave the game away by climbing down the tree and showing that he was not in danger!

Vervet monkeys learn not to respond to inappropriate calls

One noteworthy feature of Cheney and Seyfarth's observations is that the deceptive use of alarm calls was rare (2 per cent of intergroup encounters). In experiments in which a particular individual's alarm calls were repeatedly played over a loudspeaker, the group rapidly learned not to respond to the call which was, in effect, 'crying wolf'. Interestingly, learning to ignore one type of call (say the leopard call) of one individual had no effect on the response of the group to playback of either another individual's call of the same type, or the same individual's call of another

The learning is specific to a call type and one individual

type. These experiments show that vervets are good at remembering both individuals and specific calls that are used inappropriately. This may limit the possibility of deceptive signalling.

This brief analysis of deceptive manipulation and intentionality in vervet monkeys illustrates both the difficulties and the tantalizing glimpses of the possibility of using animal communication to gain insight into the animal consciousness.

Summary

Communication in animals occurs when one individual uses specially designed signals or displays to modify the behaviour of others. The design of signals is influenced both by ecological constraints and by their effectiveness in modifying the behaviour of reactors. The habitat can influence the effectiveness of different sensory channels of communication (e.g. scent versus visual signals) and the exact form of signals within a sensory channel. The latter point is illustrated by differences between the songs of birds living in different kinds of vegetation.

As signals evolve selection improves their effectiveness by making them stereotyped, repetitive and exaggerated. This evolutionary process of ritualization may have come about through a coevolutionary race between signallers and reactors.

The end-point of this coevolution may be either honest or manipulative signalling.

Further reading

Catchpole (1979) is a good general review of bird song including discussions of song repertoires and song mimicry. Cullen (1972) is a good summary of the older ethological literature on animal signals, and Wiley (1983) covers more recent literature.

Dawkins and Krebs (1978) and Krebs and Dawkins (1984) develop the idea that communication is a matter of manipulation by actors of reactors. They contrast this view with the 'classical ethological view' that signals evolve for efficient transfer of information. Hinde (1981) criticizes Dawkins and Krebs, and questions whether their view is very different from that of earlier ethologists. See also Caryl's (1982) reply.

Anderson (1980) discusses the idea of threat displays falling into disuse because they are mimicked by bluff.

Wiley and Richards (1978) provide a technical review of ecological constraints on sound signals and Bowman (1979) describes how the songs of Darwin's finches are adapted to their habitats.

Byrne and Whiten (1988) and De Waal (1986) discuss deceptive signalling in primates.

Endler (1992) reviews the various forces that shape signals.

Topics for discussion

1 Do animals intend to tell lies?
2 How would you test the prediction that co-operative signals should be less conspicuous than non-cooperative signals?
3 How useful is the analogy between animal signs and human advertising?

Chapter 15. Conclusion

The story of behaviour and adaptation that we have told in the last fourteen chapters is inevitably too simple. All the 'ifs' and 'buts' of an impeccably cautious and impregnable account would have made the book twice as long and half as easy to understand. However we do not want to leave the impression that the ideas we have discussed are completely accepted by all evolutionary biologists. Far from it, even our basic assumptions are still very much disputed in the literature.

How plausible are our main premises?

SELFISH GENES

Our discussions of natural selection have always been along the following lines: 'Imagine a gene for such and such behaviour; when would it tend to spread in a population?'. As we saw in Chapter 1 this approach does not imply that there are genes 'for' altruism, spite, long tails or whatever, but merely there are some genetic differences between individuals which are correlated with the behaviour or structure in question.

Genes versus individuals

But how plausible is the view that natural selection is a struggle between selfish genes rather than between individuals or groups? Obviously the field biologist sees *individuals* dying, surviving and reproducing; but the evolutionary consequence is that the frequencies of *genes* in the population change. Therefore the field biologist tends to think in terms of individual selection whilst the theorist thinks in terms of selfish genes. It can be very valuable, however, for the field worker to use selfish gene thinking in formulating ideas. This was apparent, for example, in Chapter 12 where we saw that from the selfish gene viewpoint there is often no difference between parental care and sibling care, so hypotheses about the reasons for one and not the other occurring in a particular species must be framed in ecological rather than genetic terms. An exception to this rule, also clarified by selfish gene thinking, was in the social hymenoptera described in Chapter 13 where females may be genetically predisposed to help their sisters.

Sometimes an analysis in terms of selfish genes may unravel a problem that appears puzzling when considered in the context of individuals maximizing inclusive fitness. An example is the phenomenon of *segregation distortion*, in which an allele gets itself into a disproportionately large number of gametes. For a heterozygous parent, each allele is expected to occur in 50 per cent of gametes, but segregation distorters somehow interfere with the process of gamete formation and increase their proportionate representation. In male *Drosophila* it is thought that chromosomes with segregation distorter alleles somehow cause sperm containing the homologous chromosome to become abnormal (e.g. they may have broken tails) (Dawkins 1982). This phenomenon can be understood within the framework of selfish genes competing for representation in future generations but is hard to explain in terms of maximizing inclusive fitness. In fact since the effect of segregation distorter genes is to decrease fecundity (by killing off gametes), the interests of the selfish gene conflict with those of the individual.

GROUP SELECTION

At the beginning of the book we more or less dismissed group selection as a viable alternative to selection acting on individuals or selfish genes. We acknowledged that it could in principle work, but suggested that the conditions for group selection to be a powerful evolutionary force were not likely to be met in nature very often. However this is not a universally accepted point of view. In particular D.S. Wilson (1980) has published a book in which he claims that group selection is a very major evolutionary force indeed and Leigh van Valen (1980), a distinguished evolutionary theorist, has heralded Wilson's book as a major breakthrough and a change of paradigm for evolutionary biologists. How should we treat this claim?

Differential extinction of groups versus trait group selection

One point to bear in mind from the start is that Wilson's model is more subtle than the simple 'differential extinction of groups' model discussed in Chapter 1. The essential feature of Wilson's hypothesis is that populations are divided into groups (so-called 'trait groups') within which selection for or against altruistic traits (or any other trait for that matter) occurs. After selection has operated on the groups, the whole population mixes together before splitting up again into new groups for the next round of selection. There could be several ways in which individuals sort themselves into trait groups, but Wilson takes as the simplest case random assortment.

In Wilson's model the altruists are at a disadvantage *within* a trait group (because of their self-sacrifice) but trait groups with altruists are more likely to contribute to the next generation than are trait groups with no altruists (Fig. 15.1). If the population consisted of just one trait group, the altruistic gene would only

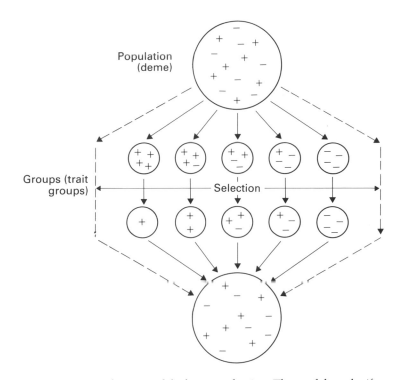

Fig. 15.1 D.S. Wilson's model of group selection. The model works if an allele (− sign) is selected against in mixed groups but increases in frequency across the population. After Harvey (1985), following Wade (1978).

spread if the fitness change for the actor (d) is greater than that for every other member of the population (p). But, and this is the crux of Wilson's argument, if the population is divided into many trait groups the effect of the altruistic act on the small proportion of non-altruists that happen to be in the same trait group as the altruist can be more or less ignored. For example, if there were 100 bird flocks in a forest and one bird in one flock gave a warning call, the effect on the average fitness of all other birds in the forest would be quite small. Wilson concludes therefore that with trait groups, the conditions for an altruistic gene to spread are $d > 0$, instead of $d > p$ if there are no trait groups. (This means that any slight advantage to the altruist will cause the gene to spread regardless of the advantage obtained by recipients of the act.) Thus with selection acting on trait groups the altruistic gene is more likely to spread.

The issue of group selection is still under active discussion by theoreticians. In models such as Wilson's, the gene spreads because of its advantage to individuals but this advantage only comes about because of the way in which the population is divided into groups. It is not yet clear how important this process will be for understanding the evolution of behaviour.

OPTIMALITY MODELS AND ESSs

In nearly every chapter we have used the ideas of optimality and ESSs. An ESS is the equivalent of an optimal solution when the pay-offs are frequency dependent, so the advantages and limitations of the two kinds of model can be discussed together. In Chapter 3 we encountered some of the criticisms which have been levelled at optimality arguments and some of the limitations in putting them into practice. To recap briefly there were three main points.

Criticisms of optimality models

1 *The idea that animals are optimal cannot be tested.* As we saw in Chapter 3 this criticism is based on a mistaken notion. The aim of using an optimality model is not to test whether animals are optimal, but to test whether the particular optimality criterion and constraints used in the model give a good description of the animal's behaviour.

2 *It is hard to tell why the animal's behaviour does not fit the predictions exactly.* Very often the simple models give an approximate but not exact description of the animal's behaviour. This could be because the model makes incorrect assumptions about constraints or goals, or because some component of cost has not been measured. There is no simple way of distinguishing between these possibilities.

3 *Animals are not well enough adapted to optimize.* The main rationale for using optimality and ESS models is the assumption that natural selection produces well-adapted animals, the aim of the models being to find out how they are adapted. There are, however, at least three reasons why animals might not be well adapted.

Reasons for lack of adaptation . . .

(a) The physical or biological environment may fluctuate too rapidly for the animals to 'catch up' in their adaptations (see Chapter 4). For example, the Atlantic gannet (*Sula bassana*) lays a clutch of one egg (Fig. 15.2), but when a second chick is experimentally added to the brood, both are reared without difficulty (Nelson 1964). This result appears to show that gannets are not well adapted to their environment, since natural selection should favour individuals that maximize their lifetime reproductive success (Chapter 1). The reason that gannets are badly adapted seems to be that the food supply has recently changed. Gannets now feed some of the time on fish offal discarded by fishing boats and this extra food allows them to rear more young than in former times, but selection has not

. . . evolutionary lag . . .

Fig. 15.2 The Atlantic gannet, here shown at the nest with a 5-week-old chick, rears only one young per season, although brood manipulations suggest that it could raise two. Photo by J.B. Nelson.

yet had time to change the clutch size from one to two. This view is supported by the observation that the southern hemisphere gannet (*Sula capensis*), which feeds in waters relatively unaffected by fishermen, can still raise only one chick to full fledging weight (Jarvis 1974).

(b) There may be insufficient genetic variation for new strategies to evolve. If the environment changes or if for some other reasons the optimal phenotype changes, animals can adapt to the new conditions only if there is genetic variation in the population. Although the issue is by no means resolved, it seems likely that in small populations the rate of evolution might well be limited by the rate at which new mutations arise (Maynard Smith 1977b).

... lack of genetic variation ...

(c) There may be coevolutionary arms races (Chapter 4) as for example between predator and prey; if one side is ahead in the arms race, the other will appear to be poorly adapted to its environment, for example hosts that are killed or debilitated by pathogens and parasites (see also Rothstein 1986).

... and arms races

Two more criticisms can be added to this list. One is that optimality and ESS models assume that genes exist, but contain no specific genetic mechanisms. For example, the ESS models of fighting strategies in Chapter 7 did not allow for sexual reproduction and the mixing of genes that this entails. The attitude of the optimality or ESS theorist is often 'think of the strategies and let the genes look after themselves'. The population geneticist on the other hand would like to know whether the models can really be couched in terms of genes and whether the rules of inheritance will allow the equilibria suggested by the models to actually develop. The answer to this question is largely unexplored at the moment (Grafen 1984).

Genetic and 'strategy' models

The second criticism is more specifically directed towards the optimality models used and tested in this book. The critical reader will have noticed that although we stressed the value of making quantitative predictions from optimality and ESS models, most of the tests of these predictions were qualitative. The animals were usually seen to do 'approximately the right thing': the dungflies in Chapter 3, for example, copulated for 36 min instead of the predicted 41 min. Some might ask whether it is worth developing quantitative arguments if the tests are only qualitative. The answer is that tests are often qualitative simply because of limitations in the techniques used to carry out the tests. The value of quantitative predictions is still potentially just as great and what is needed is comparable technical developments in ways of testing the models. Once the quantitative predictions can be tested accurately, discrepancies between observed and predicted results help to tell us what is wrong with the models.

The value of quantitative predictions and tests

It is possible to carry on discussing the pros and cons of optimality models for a long time, but the strongest argument in their favour is that over and over again optimality arguments have helped us to understand adaptations. We have illustrated this point in the preceding chapters with behavioural examples — foraging, flock size, territory size and so on — but optimality arguments can equally well be used to understand adaptations at the physiological and biochemical level. For example, the familiar 'herring-bone' arrangement of the swimming muscles of many fish is not merely an incidental design feature. This arrangement allows the muscles to contract at a rate which maximizes their power output (Alexander 1975). At the biochemical level the energy for muscle contraction is generated by oxidation of carbohydrates or fats via the Krebs cycle. It would be chemically feasible to carry out the oxidation by a more direct route, but the advantage of the cycle is that it maximizes the net energy gain per molecule oxidized (Baldwin & Krebs 1981).

Optimality models help us to understand physiological, as well as behavioural, design

The comments discussed so far could apply equally well to simple optimality and to ESS models. Now let us briefly consider a case where the two kinds of model may lead to different interpretations of field data. Richard Dawkins and Jane Brockmann (1980) demonstrated that digger wasps adopt what appears at first sight to be a suboptimal strategy, but further analysis in terms of an ESS revealed a possible explanation as to why the particular strategy might be used. The behaviour in question was the persistence of fighting by female wasps disputing the ownership of a burrow (Chapter 10). The question asked by Dawkins and Brockmann was 'how long should a female persist in a fight?'. They made the assumption that females are designed to maximize the benefit they obtain from the contest, which leads to the prediction that a female's persistence should be related to the value of the burrow. This in turn depends on the number of paralysed insects stored in the burrow. A nest containing four insects is ready for egg laying and the winner of this prize saves herself days of digging and provisioning, while a burrow with only one insect still needs a lot of work before it is ready.

Apparently suboptimal strategies may turn out to be an ESS

Contrary to expectation, Dawkins and Brockmann found that the persistence of females in a fight was not correlated with the total number of insects in the burrow but with the number put in by the loser. Since the loser is the one that determines the length of the fight (because the fight ends when the loser goes away), Dawkins and Brockmann concluded that wasps fight in proportion to their own past contribution rather than in proportion to the total value of the burrow. It is easy to see that this could lead to females giving up very quickly in a fight over a valuable burrow, just because the other wasp has done more of the provisioning, even though the pay-off goes to the winner regardless of who provisioned the burrow.

An example: fighting in digger wasps

Dawkins and Brockmann's first reaction was that the wasps probably cannot 'tell' how many insects are in the burrow, and that the rule 'fight in proportion to own contribution' is a reasonable rule of thumb that approximates the optimal strategy, since there is usually some correlation between the total number and the number put in by each female. This is an example of how the optimal policy depends on assumptions about the constraints. A perfectly knowledgeable wasp should fight in proportion to total burrow contents, but when constrained by ignorance of the total number, the policy 'fight in proportion to own contribution' might be the best option.

However, now consider what would happen if all wasps were endowed with perfect knowledge. If both wasps had the same assessment of the value of the burrow and fought for a time proportional to this value, both would fight equally hard and give up at the same moment! Presumably chance factors would cause one wasp to persist fractionally longer than the other, and if these factors were truly random each wasp would have a 50:50 chance of winning after the long struggle. Imagine now a wasp that decided how long to fight by the toss of a coin and adopted the rule 'if heads give up at once, if tails persist indefinitely' (this is equivalent to the 'bourgeois' strategy in Chapter 7). In a population of perfectly knowledgeable wasps this strategy would win half its fights and waste no time in lost contests. Therefore its net pay-off (subtracting total time wasted from the pay-off for winning) would be higher than that of the omniscient wasp. The message of this example is that 'perfect knowledge', which appeared at first sight to be an optimal strategy, is not an ESS. In general terms, whenever the pay-off depends on what others do, the question should be analysed as an ESS rather than a simple optimality problem.

Causal and functional explanations

Behavioural ecology is about functional explanations (the answers to 'why?' questions) of behaviour. As we emphasized in Chapter 1 a great deal of misunderstanding can arise if functional and causal ('how?') explanations are confused. A simple illustration of this is an objection that is often raised to labels such as 'selfish', 'spiteful', 'sneak' and 'transvestite', used by behavioural ecologists to describe the behaviours they observe. The objection is that the labels are too anthropomorphic and are loaded with the implication that animals are endowed with human-like motives. The answer to this objection is that the labels are used not to describe the causal mechanisms underlying the behaviour but to describe its functional consequences.

Causal and functional
explanations complement
each other

Although it is important to be clear about the distinction between causal and functional explanations, it is equally valuable to realize that the two kinds of question are complementary and that asking 'why?' questions can often help to understand the answers to 'how?' questions, or vice versa. An example of how causal and functional explanations go hand in hand is illustrated in Fig. 15.3. Prairie dogs (*Cynomys ludovicianus*) are colonial and live in underground tunnels which may be up to 15 m long. The tunnels are usually simple U-shaped passages with an opening to the surface at either end. It has been known for a long time that prairie dogs build little mounds of soil around the two entrances of the tunnel. These mounds were considered to function either as lookout points or to protect the tunnel against floods.

Fig. 15.3 A diagrammatic section of a prairie dog burrow. A typical burrow has two entrances, one with a low, round 'dome' at its entrance and the other with a taller, steeper-sided 'crater'. The different heights and shapes of the burrow entrances cause air to be sucked out of the crater end and therefore in through the dome. From Vogel *et al.* (1973).

However, a closer inspection revealed that the two ends of the burrow have different kinds of mounds. At one end there is a high steep sided 'crater' mound while at the other end there is a low rounded 'dome' (Fig. 15.3). If the mounds are simple lookouts or flood barriers, why should they be different shapes? The answer to this 'why?' question comes from an understanding of how air is exchanged in the tunnel (Vogel *et al.* 1973). A prairie dog living in the long underground tunnel cannot survive without a regular supply of fresh air, and it appears that the mounds around the two entrances are designed to ensure a continuous flow of air through the tunnel. The crater mound is higher and has steeper sides than the dome; as a consequence, air is sucked out of the crater end of the tunnel and into the dome end.

The forces causing the air flow are viscous sucking and the Bernoulli effect. Viscous sucking refers to the fact that when moving air passes a region of stationary air, the still air is dragged along with the current. The effect is larger at the crater end because the crater is higher than the dome and so it is exposed to faster winds. The Bernoulli effect states that the pressure of a steadily moving fluid decreases when its velocity increases. The velocity of air above the crater is greater than above the mound and the crater has very still air inside because of its steep edges. The pressure drop between the inside and outside of the crater is

Prairie dog burrows are designed to generate air flow

therefore higher than in the case of the dome so the Bernoulli effect causes air to be sucked out of the crater end of the tunnel. Vogel *et al.* (1973) demonstrated by means of laboratory experiments with miniature model tunnels and by dropping smoke bombs down real tunnels in the field that the mound system is so effective that it causes the air in the tunnel to change once every 10 min even in a very light breeze. The rate of air exchange is related to wind speed, but is unaffected by wind direction since the mounds are symmetrical. This second feature is important because wind direction is unpredictable in the prairie dog's natural habitat. This example illustrates that a functional question 'what are mounds for?' led to a detailed understanding of a mechanism question 'how do prairie dogs get enough fresh air?'.

Optimality models combine mechanism and function

The optimality models we have encountered in this book bring together mechanisms (in the form of constraints) and functions (the currency) in building up an explanation of behaviour.

A final comment

From natural history to quantitative models

What we have described as behavioural ecology in this book is the present-day equivalent of natural history. It stands in a lineage which gradually developed from detailed descriptions of animal behaviour by naturalists such as Gilbert White and Henri Fabre to experimental studies of natural history by Tinbergen and others. At the moment there is something of a bandwagon in natural history for inventing functional explanations of behaviour, which gives behavioural ecology and sociobiology a bad name. We have tried to avoid this as much as possible (without complete success!) and instead we have emphasized the idea of making testable predictions about adaptation. To illustrate how this approach has developed from studies of natural history let us construct a hypothetical lineage of studies of mating behaviour in dungflies.

A few hundred years ago naturalists would have been satisfied to discover that when two dungflies were seen together, one riding on the other's back, the top one was a male and the one underneath was a female and that the two were mating. A hundred years ago Darwin realized that males in general compete for females. A description of the natural history of dungfly mating at this stage would have included reference to the fact that males are larger than females and that this may be a result of sexual selection. Twenty years ago evolutionary biologists would have stressed the idea that males ride on the backs of females not only to inject sperm, but also for some time after copulation while the female lays her eggs. By guarding the female in this way the male guarantees that his sperm are not displaced by those of another male. In the last ten years behavioural ecologists have started to try and explain why it is that the male rides on the female's back for 40 min and not 10, 20 or 60. In developing a theory to explain why, it has become apparent that the same kind of analysis can be used for bumblebees sucking nectar out of flowers, parents investing in their offspring and many other problems. This gradual reductionist progression from broad description to detailed quantitative analysis and simple generalizations is one of the major themes of development of the natural history lineage.

Increasingly, behavioural ecologists are turning from analyses based purely on costs and benefits to an approach which includes both function and causation. This trend might be justified simply by saying that combining different kinds of explanation gives a more complete picture, but there is more to it than this. Even if one is interested primarily or solely in 'why?' questions, there are two good reasons to include 'how?' in models and data.

First, all cost−benefit models include assumptions about the strategies, or range of behavioural options, open to an individual. The foragers in Chapter 3 had the option of staying in a patch for a certain time, the competitors in Chapter 7 could escalate or retreat, the helpers in Chapter 12 could help relatives or breed, and so on. In each case, characterizing the options available to an individual requires knowledge about behavioural mechanisms. How accurately can foragers tell the time they have spent in a patch? How do competitors assess the size of their opponent? Are helpers able to recognize kin, and if so how? If animals turned out to have no mechanism of timing, no ability to assess opponents, no way of responding selectively to kin, the cost−benefit models based on these strategies would be inappropriate.

The second reason for thinking about mechanisms is that they may directly affect the costs and benefits. If we wanted to understand, for example, why it is that male birds vary in their nuptial plumage we might find that the proximate mechanism controlling plumage colour involves circulating androgens in their blood. We might then ask why it is that all males do not increase their level of androgens to make their plumage bright and hence attract more mates or repel more rivals. To fully answer this, we would need to analyse the mechanisms of androgen action and their multiple effects on the body. We may discover that there are costs of circulating androgens, such as reduced resistance to disease (Folstad & Karter 1992) or increased energy expenditure. In short, the costs and benefits may involve not only things that are measurable by observing behaviour, but also effects that can only be unravelled by understanding mechanisms.

A knowledge of behavioural mechanisms is necessary for defining the strategies available . . .

. . . and for calculating their costs and benefits

Summary

This chapter is in two parts. The first section tries to assess the value and the limitations of the selfish gene and optimality views of evolution. Group selection as an alternative to individual selection is reassessed. The value of optimality arguments can be illustrated by studies of adaptations at the behavioural, physiological and biochemical levels.

In the second part of the chapter, we try to show how different kinds of questions (function and causation) should go hand in hand in studies of behaviour.

Further reading

Maynard Smith (1977b) presents a very clear and thought-provoking summary of unsolved problems in evolutionary theory of relevance to behavioural ecologists using optimality arguments. Particularly valuable is his discussion of what limits the rate of evolution.

Gould (1980) is also a critical and stimulating review of neo-Darwinian evolution. He argues that microevolutionary changes of the type we have discussed in this book may be caused by different mechanisms from those leading to macroevolutionary changes (evolution and extinction of species).

Dawkins (1986) gives an incisive discussion of many current issues in evolutionary theory.

References

Abele, L.G. & Gilchrist, S. 1977. Homosexual rape and sexual selection in acanthocephalan worms. *Science* **197**, 81–3.

Abrahams, M. & Dill, L.M. 1989. A determination of the energetic equivalence of the risk of predation. *Ecology* **70**, 999–1007.

Adams, E.S. Caldwall, R.L. 1990. Deceptive communication in asymmetric fights of the stomatopod crustacean, *Gonodactylus bredini. Anim. Behav.* **39**, 706–16.

Alatalo, R.V., Carlson, A., Lundberg, A. & Ulfstrand, S. 1981. The conflict between male polygamy and female monogamy: the case of the pied flycatcher, *Ficedula hypoleuca. Amer. Natur.* **117**, 738–53.

Alatalo, R.V., Lundberg, A. & Ratti, O. 1990. Male polyterritoriality, and imperfect female choice in the pied flycatcher *Ficedula hypoleuca. Behav. Ecol.* **1**, 171–77.

Alatalo, R.V., Höglund, J. & Lundberg, A. 1991. Lekking in the black grouse — a test of male viability. *Nature* **352**, 155–6.

Alcock, J. 1984. *Animal Behaviour: An Evolutionary Approach*, 3rd edn. Sinauer, Sunderland, Massachusetts.

Alcock, J., Jones, C.E. & Buchmann, S.L. 1977. Male mating strategies in the bee *Centris pallida*, Fox (Anthophoridae: Hymenoptera). *Amer. Natur.* **111**, 145–55.

Alexander, R.D. 1974. The evolution of social behaviour. *Ann. Rev. Ecol. Syst.* **5**, 325–83.

Alexander, R.D. 1975. Natural selection and specialized chorusing behaviour in acoustical insects. In D. Pimentel (ed.), Insects, Science and Society, pp. 35–77. Academic Press, New York.

Alexander, R.D. & Borgia, G. 1979. On the origin and basis of the male–female phenomenon. In M.S. Blum & N.A. Blum (eds), *Sexual Selection and Reproductive Competition in Insects*. Academic Press, New York.

Alexander, R.D. & Sherman, P.W. 1977. Local mate competition and parental investment in social insects. *Science* **196**, 494–500.

Alexander, R. McN. 1975. *The Chordates*. Cambridge University Press, Cambridge.

Altmann, S.A. 1974. Baboons, space, time and energy. *Amer. Zool.* **14**, 221–40.

Andersson, M. 1980. Why are there so many threat displays? *J. theor. Biol.* **86**, 773–81.

Andersson, M. 1982. Female choice selects for extreme tail length in a widowbird. *Nature* **299**, 818–20.

Andersson, M. 1984. The evolution of eusociality. *Ann. Rev. Ecol. Syst.* **15**, 165–89.

Andersson, M. & Wicklund, C.G. 1978. Clumping versus spacing out: experiments on nest predation in fieldfares (*Turdus pilaris*). *Anim. Behav.* **26**, 1207–12.

Aoki, S. 1977. *Colophina clematis* (Homoptera: Pemphigidae), an aphid species with 'soldiers'. *Kontyu, Tokyo* **45**, 276–82.

Arak, A. 1983. Sexual selection by male–male competition in natterjack toad choruses. *Nature* **306**, 261–2.

Arak, A. 1988. Callers and satellites in the natterjack toad: evolutionarily stable decision rules, *Anim. Behav.* **36**, 416–32.

Argyle, M. 1972. Non-verbal communication in human social interaction. In R.A. Hinde (ed.), *Non-verbal Communication*, pp. 243–69. Cambridge University Press, Cambridge.

Arnold, S.J. 1976. Sexual behaviour, sexual interference and sexual defence in the salamanders *Ambystoma maculatum, A. tigrinum* and *Plethodon jordani. Z. Tierpsychol.* **42**, 247–300.

Arnold, S.J. 1981. Behavioural variation in natural populations. II. The inheritance of a feeding response in crosses between geographical races of the garter snake, *Thamnophis elegans. Evolution* **35**, 510–15.

Austad, S.N. 1982. First male sperm priority in the bowl and doily spider, *Frontinella pyramitela. Evolution* **36**, 777–85.

Austad, S.N. 1983. A game theoretical interpretation of male combat in the bowl and doily spider, *Frontinella pyramitela. Anim. Behav.* **31**, 59–73.

Axelrod, R. 1984. *The Evolution of Cooperation*. Basic Books, New York.

Axelrod, R. & Hamilton, W.D. 1981. The evolution of cooperation. *Science* **211**, 1390–6.

Baker, R.R. & Parker, G.A. 1979. The evolution of bird coloration. *Phil. Trans. Roy. Soc. Lond. B* **287**, 63–130.

Baldwin, J.E. & Krebs, H.A. 1981. The evolution of metabolic cycles. *Nature* **291**, 381–2.

Balmford, A. 1991. Mate choice on leks. *Trends Ecol. Evol.* **6**, 87–92.

Barnard, C.J. 1984. *Producers and Scroungers*. Croom Helm, Beckenham.

Barnard, C.J. & Sibly, R.M. 1981. Producers and scroungers: a general model and its application to captive flocks of house sparrows. *Anim. Behav.*

29, 543−50.

Barnard, C.J. & Thompson, D.B.A. 1985. *Gulls and Plovers: The Ecology and Behaviour of Mixed-species Feeding Groups.* Croom Helm, London & Sydney.

Bart, J. & Tornes, A. 1989. Importance of monogamous male birds in determining reproductive success: evidence for house wrens and a review of male removal studies. *Behav. Ecol. Sociobiol.* **24**, 109−16.

Bartz, S.H. 1979. Evolution of eusociality in termites. *Proc. Natn. Acad. Sci. USA* **76**, 5764−8.

Basolo, A. 1990. Female preference predates the evolution of the sword in swordtail fish. *Science* **250**, 808−10.

Bateman, A.J. 1948. Intra-sexual selection in *Drosophila. Heredity* **2**, 349−68.

Bateson, P.P.G. 1983. Rules for changing the rules. In D.S. Bendall (ed.), *Evolution from Molecules to Men*, pp. 483−507. Cambridge University Press, Cambridge.

Bednekoff, P. 1992. *Foraging and Fat reserves in great tits in a variable environment.* D. Phil. thesis, Oxford.

Beehler, B.M. & Foster, M.S. 1988. Hotshots, hotspots and female preferences in the organization of mating systems. *Amer. Natur.* **131**, 203−19.

Beissinger, S.R. & Snyder, N.F.R. 1987. Mate desertion in the snail kite. *Anim. Behav.* **35**, 477−87.

Bell, G. 1980. The costs of reproduction and their consequences. *Amer. Natur.* **109**, 453−64.

Belovsky, G.E. 1978. Diet optimization in a generalist herbivore: the moose. *Theoret. Pop. Biol.* **14**, 105−34.

Belovsky, G.E. 1986a. Generalist herbivore foraging and its role in competitive interactions. *Amer. Zool.* **26**, 51−69.

Belovsky, G.E. 1986b. Optimal foraging and community structure: implications for a guild of generalist grassland herbivores. *Oecologia* **70**, 35−52.

Bensch, S. & Hasselquist, D. 1992. Evidence for active female choice in a polygynous warbler. *Anim. Behav.* **44**, 301−11.

Benton, T.G. & Foster, W.A. 1992. Altruistic housekeeping in a social aphid. *Proc. R. Soc. Lond. B.* **247**, 199−202.

Benzer, S. 1973. Genetic dissection of behavior. *Sci. Am.* **229**, 24−37.

Bercovitch, F. 1988. Coalitions, cooperation and reproductive tactics among adult male baboons. *Anim. Behav.* **36**, 1198−209.

Berthold, P., Wiltschko, W., Miltenberger, H. & Querner, W. 1990a. Genetic transmission of migratory behaviour into a non-migrating population. *Experientia* **46**, 107−8.

Berthold, P., Mohr, G. & Querner, W. 1990b. Steuerung und potentielle Evolutions-geschwindigkeit des obligaten Teilzieher-verhaltens: Ergebnisse eines Zweiweg-selektions experiments mit der Mönschgrassmücke (*Sylvia atricapilla*). *J. Orn.* **131**, 33−45.

Bertram, B.C.R. 1975. Social factors influencing reproduction in wild lions. *J. Zool. Lond.* **177**, 463−82.

Bertram, B.C.R. 1980. Vigilance and group size in ostriches. *Anim. Behav.* **28**, 278−86.

Birkhead, T.R. 1977. The effect of habitat and density on breeding success in common guillemots, *Uria aalge. J. Anim. Ecol.* **46**, 751−64.

Birkhead, T.R. 1979. Mate guarding in the magpie *Pica pica. Anim. Behav.* **27**, 866−74.

Birkhead, T.R. & Clarkson, K. 1980. Mate selection and precopulatory guarding in *Gammarus pulex. Z. Tierpsychol.* **52**, 365−80.

Birkhead, T.R. & Møller, A.P. 1992. *Sperm Competition in Birds: Evolutionary Causes and Consequences.* Academic Press, London.

Birkhead, T.R., Pellatt, J.E. & Hunter, F.M. 1988. Extra-pair copulation and sperm competition in the zebra finch. *Nature.* **334**, 60−2.

Birkhead, T.R., Burke, T., Zann, R., Hunter, F.M. & Krupa, A.P. 1990. Extra-pair paternity and intraspecific brood parasitism in wild zebra finches, *Taeniopygia guttata*, revealed by DNA fingerprinting. *Behav. Ecol. Sociobiol.* **27**, 315−24.

Blurton Jones, N.G. 1968. Observations and experiments on causation of threat displays of the great tit, *Parus major. Anim. Behav. Monogr.* **1**. 75−158.

Borgia, G. 1979. Sexual selection and the evolution of mating systems. In M.S. Blum & N.A. Blum (eds), *Sexual Selection and Reproductive Competition in Insects*, pp. 19−80. Academic Press, New York.

Borgia, G. 1985. Bower quality, number of decorations and mating success of male satin bowerbirds (*Ptilinorynchus violaceus*): an experimental analysis. *Anim. Behav.* **33**, 266−71.

Bowman, R.I. 1979. Adaptive morphology of song dialects in Darwin's finches. *J. Orn. Lpz.* **120**, 353−89.

Bradbury, J.W. 1977. Lek mating behaviour in the hammer-headed bat. *Z. Tierpsychol.* **45**, 225−55.

Bradbury, J.W. & Gibson, R.M. 1983. Leks and mate choice. In P. Bateson (ed.), *Mate Choice*, pp. 109−38. Cambridge University Press. Cambridge.

Bradbury, J.W. & Vehrencamp, S.L. 1977. Social organization and foraging in emballonurid bats. III. Mating systems. *Behav. Ecol. Sociobiol.* **2**, 1−17.

Bradbury, J.W., Gibson, R.M. & Tsai, I.M. 1986.

Hotspots and the evolution of leks. *Anim. Behav.* **34**, 1694–1709.

Bray, O.E., Kennelly, J.J. & Guarino, J.L. 1975. Fertility of eggs produced on territories of vasectomized red-winged blackbirds. *Wilson Bull.* **87**, 187–95.

Brockmann, H.J., & Dawkins, R. 1979. Joint nesting in a digger wasp as an evolutionarily stable preadaptation to social life. *Behaviour* **71**, 203–45.

Brockmann, H.J., Grafen, A. & Dawkins, R. 1979. Evolutionarily stable nesting strategy in a digger wasp. *J. theor. Biol.* **77**, 473–96.

Brodin, A. 1992. Cache dispersion affects retrieval time in hoarding willow tits. *Ornis Scand.* **23**, 7–12.

Brooke, M. de L. 1978. Some factors affecting the laying date, incubation and breeding success of the manx shearwater. *Puffinus puffinus.* *J. Anim. Ecol.* **47**, 477–95.

Brown, C.H. 1982. Ventriloquial and locatable vocalizations in birds. *Z. Tierpsychol.* **59**, 338–50.

Brown, C.R. 1986a. Cliff swallow colonies as information centers. *Science* **234**, 83–5.

Brown, C.R. 1986b. Parasites and their effects on colonies of cliff swallows. *Discovery* **19**, 8–13.

Brown, C.R. 1988. Enhanced foraging efficiency through information centers: a benefit of coloniality in cliff swallows. *Ecology* **69**, 602–13.

Brown, C.R. & Brown, M.B. 1986. Ectoparasitism as a cost of coloniality in cliff swallows (*Hirundo pyrrhonota*). *Ecology* **67**, 1206–18.

Brown, C.R. & Brown, M.B. 1988. A new form of reproductive parasitism in cliff swallows. *Nature* **331**, 66–8.

Brown, J.L. 1964. The evolution of diversity in avian territorial systems. *Wilson Bull.* **76**, 160–9.

Brown, J.L. 1969. The buffer effect and productivity in tit populations. *Amer. Natur.* **103**, 347–54.

Brown, J.L. 1982. Optimal group size in territorial animals. *J. theor. Biol.* **95**, 793–810.

Brown, J.L. 1980. Fitness in complex avian social systems. In H. Markl (ed.), *Evolution of Social Behaviour*, pp. 115–28. Verlag-chemie, Weinheim.

Brown, J.L. 1987. *Helping and Communal Breeding in Birds*. Princeton University Press. Princeton. NJ.

Brown, J.L., Dow, D.D., Brown, E.R. & Brown, S.D. 1978. Effects of helpers on feeding of nestlings in the grey-crowned babbler, *Pomatostomus temporalis*. *Behav. Ecol. Sociobiol.* **4**, 43–60.

Brown, J.L., Brown, E.R., Brown, S.D. & Dow, D.D. 1982. Helpers: effects of experimental removal on reproductive success. *Science* **215**, 421–2.

Bull, J.J. 1980. Sex determination in reptiles. *Q. Rev. Biol.* **55**, 3–21.

Bulmer, M.G. & Taylor, P.D. 1981. Worker–queen conflict and sex ratio theory in social Hymenoptera. *Heredity* **47**, 197–207.

Burke, T. 1989. DNA fingerprinting and other methods for the study of mating success. *Trends Ecol. Evol.* **4**, 139–44.

Burke, T., Davies, N.B., Bruford, M.W. & Hatchwell, B.J. 1989. Parental care and mating behaviour of polyandrous dunnocks *Prunella modularis* related to paternity by DNA fingerprinting. *Nature* **338**, 249–51.

Burley, N. 1986a. Sex ratio manipulation in color-banded populations of zebra finches. *Evolution* **40**, 1191–206.

Burley, N. 1986b. Sexual selection for aesthetic traits in species with biparental care. *Amer. Nat.* **127**, 415–45.

Burley, N., Krantzberg, G. & Radman, P. 1982. Influence of colour-banding on the conspecific preferences of zebra finches. *Anim. Behav.* **27**, 686–98.

Busnel, R.G. & Klasse, A. 1976. *Whistled Languages*. Springer-Verlag, Berlin.

Bygott, J.D., Bertram, B.C.R. & Hanby, J.P. 1979. Male lions in large coalitions gain reproductive advantage. *Nature* **282**, 839–41.

Byrne, R. & Whiten, A. 1988. Tactical deception in primates. *Behav. Brain Sci.* **11**, 233–73.

Cade, W.H. 1979. The evolution of alternative male reproductive strategies in field crickets. In M. Blum & N.A. Blum (eds), *Sexual Selection and Reproductive Competition in Insects*, pp. 343–79. Academic Press, London.

Cade, W.H. 1981. Alternative mating strategies: genetic differences in crickets. *Science* **212**, 563–4.

Cade, W.H. & Wyatt, D.R. 1984. Factors affecting calling behaviour in field crickets, *Teleogryllus* and *Gryllus* (age, weight, density and parasites). *Behaviour* **88**, 61–75.

Caldwell, R.L. 1985. A test of individual recognition in the stomatopod, *Gonodactylus festae*. *Anim. Behav.* **33**, 101–6.

Calvert, W.H., Hedrick, L.E. & Brower, L.P. 1979. Mortality of the monarch butterfly, *Danaus plexippus*: avian predation at five over-wintering sites in Mexico. *Science* **204**, 847–51.

Capranica, R.R., Frishkopf, L.S. & Nevo, E. 1973. Encoding of geographic dialects in the auditory system of the cricket frog. *Science* **182**, 1272–5.

Caraco, T. 1979a. Time budgeting and group size: a theory. *Ecology* **60**, 611–17.

Caraco, T. 1979b. Time budgeting and group size: a test of theory. *Ecology* **60**, 618–27.

Caraco, T. & Wolf, L.L. 1975. Ecological determinants of group sizes of foraging lions. *Amer. Natur.* **109**, 343–52.

Caraco, T., Martindale, S. & Pulliam, H.R. 1980.

Flocking: advantages and disadvantages. *Nature* **285**, 400–1.

Caraco, T., Blanckenhorn, W.U., Gregory, G.M. Newman, J.A., Recer, G.M. & Zwicker, S.M. (1990). Risk-sensitivity: ambient temperature affects foraging choice. *Anim. Behav.* **39**, 338–45.

Carayon, J. 1974. Insemination traumatique hétérosexuelle et homosexuelle chez *Xylocoris maculipennis* (Hem. Anthocoridae). *C.R. Acad. Sci. Paris, D.* **278**, 2803–6.

Caro, T.M. & Bateson, P. 1986. Ontogenetic analysis of alternative tactics. *Anim. Behav.* **34**, 1483–99.

Carpenter, C.R. 1954. Tentative generalisations on the grouping behaviour of non-human primates. *Human Biol.* **26**, 269–76.

Carpenter, F.L. & MacMillen, F.E. 1976. Threshold model of feeding territoriality and test with a Hawaiian honey creeper. *Science.* **194**, 639–42.

Carpenter, F.L., Paton, D.C. & Hixon, M.A. 1983. Weight gain and adjustment of feeding territory size in migrant hummingbirds. *Proc. Natn. Acad. Sci. USA.* **80**, 7259–63.

Cartar, R.V. & Dill, L.M. 1990. Why are bumblebees risk-sensitive foragers? *Behav. Ecol. Sociobiol.* **26**, 121–7.

Caryl, P.G. 1979. Communication by agonistic displays: what can games theory contribute to ethology? *Behaviour* **68**, 136–69.

Caryl, P.G. 1980. Escalated fighting and the war of nerves: games theory and animal combat. In P.P.G. Bateson & P.H. Klopfer (eds), *Perspectives in Ethology*. Plenum Press, New York.

Caryl, P.G. 1982. Animal signals: a reply to Hinde. *Anim. Behav.* **30**, 240–4.

Catchpole, C.K. 1979. *Vocal Communication in Birds*. Edward Arnold, London.

Catchpole, C.K. 1980. Sexual selection and the evolution of complex songs among European warblers of the genus *Acrocephalus*. *Behaviour.* **74**, 149–66.

Catchpole, C.K., Dittami, J. & Leisler, B. 1984. Differential responses to male song repertoires in female songbirds implanted with oestradiol. *Nature* **312**, 563–4.

Chappuis, C. 1971. Un example de l'influence du milieu sur les émissions vocales des oiseaux: l'evolution des chants en forêt équitoriale. *Terre Vie* **25**, 183–202.

Charnov, E.L. 1976. Optimal foraging: the marginal value theorem. *Theoret. Pop Biol.* **9**, 129–36.

Charnov, E.L. 1978. Evolution of eusocial behavior: offspring choice or parental parasitism? *J. theor. Biol.* **75**, 451–66.

Charnov, E.L. 1981. Kin selection and helpers at the nest: effects of paternity and biparental care. *Anim. Behav.* **29**, 631–2.

Charnov, E.L. 1982. *The Theory of Sex Allocation*. Princeton University Press, Princeton, NJ.

Charnov, E.L. & Krebs, J.R. 1974. On clutch size and fitness. *Ibis* **116**, 217–19.

Charnov, E.L. & Skinner, S. 1984. Evolution of host selection and clutch size in parasitoids. *Florida ent.* **67**, 5–21.

Charnov, E.L., Orians, G.H. & Hyatt, K. 1976. The ecological implications of resource depression. *Amer. Natur.* **110**, 247–59.

Cheney, D.L. & Seyfarth R.M. 1990. *How Monkeys See the World*. University of Chicago Press, Chicago.

Clark, A.B. 1978. Sex ratio and local resource competition in a prosimian primate. *Science* **201**, 163–5.

Clayton, D.H. 1991. The influence of parasites on host sexual selection. *Parasitology Today* **7**, 329–34.

Clutton-Brock, T.H. 1975. Feeding behaviour of red colobus and black and white colobus in East Africa. *Folia primatol.* **23**, 165–207.

Clutton-Brock, T.H. 1983. Selection in relation to sex. In D.S. Bendall (ed.), *From Molecules to Men*, pp. 457–81. Cambridge University Press, Cambridge.

Clutton-Brock, T.H. 1989. Mammalian mating systems. *Proc. Roy. Soc. Lond. B.* **236**, 339–72.

Clutton-Brock, T.H. 1991. *The Evolution of Parental Care*. Princeton University Press, Princeton.

Clutton-Brock, T.H. & Albon, S.D. 1979. The roaring of red deer and the evolution of honest advertisement. *Behaviour* **69**, 145–70.

Clutton-Brock, T.H. & Harvey, P.H. 1977. Primate ecology and social organisation. *J. Zool. Lond.* **183**, 1–39.

Clutton-Brock, T.H. & Harvey, P.H. 1979. Comparison and adaptation. *Proc. R. Soc. Lond. B* **205**, 547–65.

Clutton-Brock, T.H., Guinness, F.E. & Albon, S.D. 1982. *Red Deer: The Behaviour and Ecology of Two Sexes*. Chicago University Press, Chicago.

Clutton-Brock, T.H., Albon, S.D. & Guinness, F.E. 1984. Maternal dominance, breeding success and birth sex ratios in red deer. *Nature* **308**, 358–60.

Clutton-Brock, T.H., Hiraiwa-Hasegawa, M. & Robertson, A. 1989. Mate choice on fallow deer leks. *Nature* **340**, 463–5.

Clutton-Brock, T.H. & Vincent, A.C.J. 1991. Sexual selection and the potential reproductive rates of males and females. *Nature* **351**, 58–60.

Conover, D.A. 1984. Adaptive significance of temperature-dependent sex determination in a fish. *Amer. Natur.* **123**, 297–313.

Corbet, P.S. 1962. *A Biology of Dragonflies*. Witherby, London.

Cott, H.B. 1940. *Adaptive Coloration in Animals.* Oxford University Press, Oxford.

Coulson, J.C. 1966. The influence of the pairbond and age on the breeding biology of the kittiwake gull, *Rissa tridactyla. J. Anim. Ecol.* **35**, 269–79.

Cowie, R.J. 1977. Optimal foraging in great tits. *Parus major. Nature* **268**, 137–9.

Cox, C.R. & Le Boeuf, B.J. 1977. Female incitation of mate competition: a mechanism of mate selection. *Amer. Natur.* **111**, 317–35.

Craig, R. 1979. Parental manipulation, kin selection and the evolution of altruism. *Evolution* **33**, 319–34.

Creel, S.R. & Waser, P.M. 1991. Failures of reproductive suppression in dwarf mongooses (*Helogale parvula*): accident or adaptation? *Behav. Ecol.* **2**, 7–15.

Creel, S.R., Monfort, S.L., Wildt, D.E. & Waser, P.M. 1991. Spontaneous lactation is an adaptive result of pseudopregnancy. *Nature* **351**, 660–2.

Crook, J.H. 1964. The evolution of social organisation and visual communication in the weaver birds (Ploceinae). *Behaviour Suppl.* **10**, 1–178.

Crook, J.H. & Gartlan, J.S. 1966. Evolution of primate societies. *Nature* **210**, 1200–3.

Crozier, R.H. & Luykx, P. 1985. The evolution of termites eusociality is unlikely to have been based on a male-haploid analogy. *Amer. Natur.* **126**, 867–9.

Cullen, J.M. 1966. Reduction of ambiguity through ritualization. *Phil. Trans. R. Soc. B* **251**, 363–74.

Cullen, J.M. 1972. Some principles of animal communication. In R.A. Hinde (ed.), *Non-verbal Communication*, pp. 101–22. Cambridge University Press, Cambridge.

Dale, S., Amundsen, T., Lifjeld, J.T. & Slagsvold, T. 1990. Mate sampling behaviour of female pied flycatchers: evidence for active mate choice. *Behav. Ecol. Sociobiol.* **27**, 87–91.

Daly, M. 1979. Why don't male mammals lactate? *J. theor. Biol.* **78**, 325–45.

Darwin, C. 1859. *On the Origin of Species.* Murray, London.

Darwin, C. 1871. *The Descent of Man and Selection in Relation to Sex.* Murray, London.

Davies, N.B. 1989. Sexual conflict and the polygamy threshold. *Anim. Behav.* **38**, 226–34.

Davies, N.B. 1992. *Dunnock Behaviour and Social Evolution.* Oxford University Press, Oxford.

Davies, N.B. & Brooke, M. de L. 1988. Cuckoos versus reed warblers: adaptations and counteradaptations. *Anim. Behav.* **36**, 262–84.

Davies, N.B. & Brooke, M. de L. 1989a. An experimental study of co-evolution between the cuckoo *Cuculus canorus* and its hosts. I. Host egg discrimination. *J. Anim. Ecol.* **58**, 207–24.

Davies, N.B. & Brooke, M. de L. 1989b. An experimental study of co-evolution between the cuckoo *Cuculus canorus* and its hosts. II. Host egg markings, chick discrimination and general discussion. *J. Anim. Ecol.* **58**, 225–36.

Davies, N.B. & Halliday, T.R. 1978. Deep croaks and fighting assessment in toads *Bufo bufo. Nature* **274**, 683–5.

Davies, N.B. & Houston, A.I. 1981. Owners and satellites: the economics of territory defence in the pied wagtail, *Motacilla alba. J. Anim. Ecol.* **50**, 157–80.

Davies, N.B., Hatchwell, B.J., Robson, T. & Burke, T. 1992. Paternity and parental effort in dunnocks *Prunella modularis*: how good are male chick-feeding rules? *Anim. Behav.* **43**, 729–45.

Dawkins, M. 1971. Perceptual changes in chicks: another look at the 'search image' concept. *Anim. Behav.* **19**, 566–74.

Dawkins, R. 1976. *The Selfish Gene.* Oxford University Press, Oxford.

Dawkins, R. 1978. Replicator selection and the extended phenotype. *Z. Tierpsychol.* **47**, 61–76.

Dawkins, R. 1979. Twelve misunderstandings of kin selection. *Z. Tierpsychol.* **51**, 184–200.

Dawkins, R. 1980. Good strategy or evolutionarily stable strategy? In G.W. Barlow & J. Silverberg (eds), *Sociobiology: Beyond Nature/Nurture*, pp. 331–67. Westview Press, Boulder, Colorado.

Dawkins, R. 1982. *The Extended Phenotype.* W.H. Freeman, Oxford.

Dawkins, R. 1986. *The Blind Watchmaker.* Longman, London.

Dawkins, R. & Brockmann, H.J. 1980. Do digger wasps commit the Concorde fallacy? *Anim. Behav.* **28**, 892–6.

Dawkins, R. & Carlisle, T.R. 1976. Parental investment, mate desertion and a fallacy. *Nature* **262**, 131–3.

Dawkins, R. & Krebs, J.R. 1978. Animal signals: information or manipulation? In J.R. Krebs & N.B. Davies (eds), *Behavioural Ecology: An Evolutionary Approach*, 1st edn, pp. 282–309. Blackwell Scientific Publications, Oxford.

Dawkins, R. & Krebs, J.R. 1979. Arms races between and within species. *Proc. R. Soc. Lond. B* **205**, 489–511.

De Groot, P. 1980. Information transfer in a socially roosting weaver bird (*Quelea quelea*: Ploceinae): an experimental study. *Anim. Behav.* **28**, 1249–54.

De Voogt, T.J., Krebs, J.R., Healy, S.D. & Purvis, A. 1992. Evolutionary correlation between song repertoire and a brain nucleus in birds (submitted).

De Vore, I. (ed.). 1965. *Primate Behaviour: Field Studies of Monkeys and Apes.* Holt, Rinehart &

Winston, New York.

Diamond, J.M., Karasov, W.H., Phan, D. & Carpenter, F.L. 1986. Digestive physiology is a determinant of foraging bout frequency in hummingbirds. *Nature* **320**, 62–3.

Dominey, W.J. 1980. Female mimicry in male blue gill sunfish – a genetic polymorphism? *Nature* **284**, 546–8.

Dudai, Y. 1989. *The Neurobiology of Memory*. Oxford University Press, Oxford.

Dudai, Y. & Quinn, W.G. 1980. Genes and learning in *Drosophila*. *Trends in Neurosciences* **3**, 28–30.

Dunbar, R.I.M. 1983. Intraspecific variations in mating strategy. In P.P.G. Bateson & P. Klopfer (eds), *Perspectives in Ethology. Vol. 5*, pp. 385–431. Plenum Press, New York.

Dunbar, R.I.M. 1984. *Reproductive Decisions: An Economic Analysis of Gelada Baboon Social Strategies*. Princeton University Press, Princeton, NJ.

Dunbar, R.I.M. 1988. *Primate Social Systems*. Croom Helm, London.

Duncan, P. & Vigne, N. 1979. The effect of group size in horses on the rate of attacks by blood-sucking flies. *Anim. Behav.* **27**, 623–5.

Dunford, C. 1977. Kin selection for ground squirrel alarm calls. *Amer. Natur.* **111**, 782–5.

Dussourd, D.E., Harvis, C.A., Meinwald, J. & Eisner, T. 1991. Pheromonal advertisement of a nuptial gift by a male moth. *Proc. Natl. Acad. Sci. USA* **88**, 9224–7.

Dybas, H.S. & Lloyd, M. 1974. The habitats of 17 year periodical cicadas (Homoptera: Cicadidae: *Magicicada* spp. *Ecol. Mongr.* **44**, 279–324.

Eberhard, W.G. 1972. Altruistic behaviour in a sphecid wasp: support for kin-selection theory. *Science* **175**, 1390–1.

Eberhard, W.G. 1979. The functions of horns in *Podischnus agenor* (Dynastinae) and other beetles. In M.S. Blum & N.A. Blum (eds), *Sexual Selection and Reproductive competition in Insects*. pp. 231–58. Academic Press, London.

Elgar, M.A. 1986. House sparrows establish foraging flocks by giving chirrup calls if the resources are divisible. *Anim. Behav.* **34**, 169–74.

Elgar, M.A. & Catterall, C.P. 1981. Flocking and predator surveillance in house sparrows: test of an hypothesis. *Anim. Behav.* **29**, 868–72.

Elner, R.W. & Hughes, R.N. 1978. Energy maximization in the diet of the shore crab, *Carcinus maenas. J. Anim. Ecol.* **47**, 103–16.

Emlen, S.T. 1984. Cooperative breeding in birds and mammals. In J.R. Krebs & N.B. Davies (eds), *Behavioural Ecology: An Evolutionary Approach*, 2nd edn, pp. 305–39. Blackwell Scientific Publications, Oxford.

Emlen, S.T. 1991. Evolution of cooperative breeding in birds and mammals. In J.R. Krebs & N.B. Davies (eds), *Behavioural Ecology: an Evolutionary Approach*, 3rd edn, pp. 301–57. Blackwell Scientific Publications, Oxford.

Emlen, S.T. & Oring, L.W. 1977. Ecology, sexual selection and the evolution of mating systems. *Science* **197**, 215–23.

Emlen, S.T. & Wrege, P.H. 1988. The role of kinship in helping decisions among white-fronted bee-eaters. *Behav. Ecol. Sociobiol.* **23**, 305–15.

Emlen, S.T. & Wrege, P.H. 1989. A test of alternate hypotheses for helping behaviour in white-fronted bee-eaters of Kenya. *Behav. Ecol. Sociobiol.* **25**, 303–20.

Emlen, S.T. & Wrege, P.H. 1992. Parent–offspring conflict and the recruitment of helpers among bee eaters. *Nature* **356**, 331–3.

Emlen, S.T., Emlen, J.M. & Levin, S.A. 1986. Sex-ratio selection in species with helpers-at-the-nest. *Amer. Natur.* **127**, 1–8.

Emlen, S.T., Demong, N.J. & Emlen, D.J. 1989. Experimental induction of infanticide in female wattled jacanas. *Auk* **106**, 1–7.

Endler, J.A. 1980. Natural selection on colour patterns in *Poecilia reticulata*. *Evolution* **34**, 76–91.

Endler, J.A. 1983. Natural and sexual selection on color patterns in poeciliid fishes. *Env. Biol. Fish.* **9**, 173–90.

Endler, J.A. 1991. Interactions between predators and prey. In J.R. Krebs & N.B. Davies (eds), *Behavioural Ecology: An Evolutionary Approach*, 3rd. edn pp. 169–96. Blackwell Scientific Publications, Oxford.

Endler, J.A. 1992. Signals, signal conditions and the direction of evolution. *Amer. Natur.* **139**, Suppl. 125–53.

Enquist, M. 1985. Communication during aggressive interactions with particular reference to variation in choice of behaviour. *Anim. Behav.* **33**, 1152–61.

Enquist, M. & Leimar, O. 1983. Evolution of fighting behaviour; decision rules and assessment of relative strength. *J. theor. Biol.* **102**, 387–410.

Enquist, M. & Leimar, O. 1987. Evolution of fighting behaviour; the effect of variation in resource value. *J. theor. Biol.* **127**, 187–205.

Enquist, M. & Leimar, O. 1990. The evolution of fatal fighting. *Anim. Behav.* **39**, 1–9.

Enquist, M., Plane, E. & Roed, J. 1985. Aggressive communication in fulmars (*Fulmaris glacialis*) competing for food. *Anim. Behav.* **33**, 1107–20.

Enquist, M., Leimar, O., Ljungberg, T., Mallner, Y. & Segerdahl, N. 1990. A test of the sequential assessment game: fighting in the cichlid fish

Nannacara anomala. Anim. Behav. **40**, 1–14.

Erichsen, J.T., Krebs, J.R. & Houston, A.I. 1980. Optimal foraging and cryptic prey. *J. Anim. Ecol.* **49**, 271–6.

Erickson, C.J. & Zenone, P.G. 1976. Courtship differences in male ring doves: avoidance of cuckoldry? *Science* **192**, 1353–4.

Evans, H.E. 1977. Extrinsic and intrinsic factors in the evolution of insect sociality. *BioScience* **27**, 613–17.

Evans, M.R. & Hatchwell, B.J. 1992. An experimental study of male adornment in the scarlet-tufted malachite sunbird: I. The role of pectoral tufts in territorial defence. *Behav. Ecol. Sociobiol.* **29**, 413–19.

Ewald, P.W. & Rohwer, S. 1982. Effects of supplemental feeding on timing of breeding, clutch size and polygamy in red winged blackbirds. *Agelaius phoeniceus. J. Anim. Ecol.* **51**, 429–50.

Ezaki, Y. 1990. Female choice and the causes and adaptiveness of polygyny in great reed warblers. *J. Anim. Ecol.* **59**, 103–19.

Feare, C. 1984. *The Starling.* Oxford University Press, Oxford.

Ferguson, M.W.J. & Joanen, T. 1982. Temperature of egg incubation determines sex in *Alligator mississippiensis. Nature* **296**, 850–3.

Fischer, E.A. 1980. The relationship between mating system and simultaneous hermaphroditism in the coral reef fish, *Hypoplectrus nigricans. Anim. Behav.* **28**, 620–33.

Fisher, R.A. 1930. *The Genetical Theory of Natural Selection.* Clarendon Press, Oxford.

Fitzgibbon, C.D. 1989. A cost to individuals with reduced vigilance in groups of Thompson's gazelles hunted by cheetahs. *Anim. Behav.* **37**, 508–10.

Fitzgibbon, C.D. & Fanshaw, J.H. 1988. Stotting in Thompson's gazelles: an honest signal of condition. *Behav. Ecol. Sociobiol.* **23**, 69–74.

Folstad, I. & Karter, A.J. 1992. Parasites, bright males and the immunocompetence handicap. *Amer. Natur.* **139**, 603–22.

Foster, S.A. 1985. Group foraging by a coral reef fish: a mechanism for gaining access to defended resources. *Anim. Behav.* **33**, 782–79.

Foster, W.A. 1990. Experimental evidence for effective and altruistic colony defence against natural predators by soldiers of the gall-forming aphid *Pemphigus spyrothecae* (Hemiptera: Pemphigidae). *Behav. Ecol. Sociobiol.* **27**, 421–30.

Foster, W.A. & Treherne, J.E. 1981. Evidence for the dilution effect in the selfish herd from fish predation on a marine insect. *Nature* **295**, 466–7.

Fretwell, S.D. 1972. *Populations in a Seasonal Environment.* Princeton University Press, Princeton, NJ.

Fricke, H.W. 1979. Mating system, resource defence and sex change in the anemonefish, *Amphiprion akallopisos. Z. Tierpsychol.* **50**, 313–26.

Fricke, H.W. & Fricke, S. 1977. Monogamy and sex change by aggressive dominance in coral reef fish. *Nature* **266**, 830–2.

Frisch, K. von. 1967. *The Dance Language and Orientation of Bees.* Belknap Press, Cambridge, MA.

Gadagkar, R. 1990. Evolution of eusociality: the advantage of assured fitness returns. *Phil. Trans. R. Soc. Lond. B.* **329**, 17–25.

Galef, B.G. & Wigmore, S.W. 1983. Transfer of information concerning distant foods: a laboratory investigation of the 'information centre hypothesis'. *Anim. Behav.* **31**, 748–58.

Geist, V. 1966. The evolution of horn-like organs. *Behaviour* **27**, 175–213.

Geist, V. 1974. On fighting strategies in animal conflict. *Nature* **230**, 354.

Ghiselin, M.T. 1969. The evolution of hermaphroditism among animals. *Q. Rev. Biol.* **44**, 189–208.

Gibbs, H.L., Weatherhead, P.J., Boag, P.T., White, B.N., Tabak, L.M. & Hoysak, D.J. 1990. Realized reproductive success of polygynous red-winged blackbirds revealed by DNA markers. *Science* **250**, 1394–7.

Gibson, R.M. & Höglund, J. 1992. Copying and sexual selection. *Trends Ecol. Evol.* (in press).

Gibson, R.M., Bradbury, J.W. & Vehrencamp, S.L. 1991. Mate choice in lekking sage grouse revisited: the roles of vocal display, female site fidelity, and copying. *Behav. Ecol.* **2**, 165–80.

Gilbert, L.E. 1976. Postmating female odor in *Heliconius* butterflies: a male contributed antiaphrodisiac? *Science* **193**, 419–20.

Gill, F.B. & Wolf, L.L. 1975. Economics of feeding territoriality in the golden-winged sunbird. *Ecology* **56**, 333–45.

Gilliam, J.F. 1982. *Foraging under mortality risk in size-structured populations.* Ph.D. thesis, Michigan State University.

Giraldeau, L.A. & Gillis, D. 1985. Optimal group size can be stable: a reply to Sibly. *Anim. Behav.* **33**, 666–7.

Gittleman, J.L. & Harvey, P.H. 1980. Why are distasteful prey not cryptic? *Nature* **286**, 149–50.

Goldthwaite, R.O., Coss, R.G. & Owings, D.H. 1990. Evolutionary dissipation of an antisnake system: differential behavior by California and arctic ground squirrels in above- and below-ground contexts. *Behaviour* **112**, 246–69.

Goss-Custard, J.D. 1970. Feeding dispersion in some overwintering wading birds. In J.H. Crook (ed.), *Social Behaviour in Birds and Mammals.*

pp. 3–34. Academic Press, London.

Goss-Custard, J.D. 1976. Variation in the dispersion of redshank (*Tringa totanus*) on their winter feeding grounds. *Ibis* **118**, 257–63.

Gould, S.J. 1966. Allometry and size in ontogeny and phylogeny. *Biol. Rev.* **41**, 587–640.

Gould, S.J. 1978. *Ever Since Darwin: Reflections in Natural History*. Andre Deutsch, London.

Gould, S.J. 1980. Is a new and general theory of evolution emerging? *Paleobiology* **6**, 119–30.

Gould, S.J. & Lewontin, R.C. 1979. The spandrels of San Marco and the Panglossian paradigm: a critique of the adaptationist programme. *Proc. R. Soc. Lond. B.* **205**, 581–98.

Grafen, A. 1986a. A geometric view of relatedness. In R. Dawkins & M. Ridley (eds), *Oxford Surveys in Evolutionary Biology*, Vol. 2, pp. 29–89. Oxford University Press, Oxford.

Grafen, A. 1986b. Split sex ratios and the evolutionary origins of eusociality. *J. theor. Biol.* **122**, 95–121.

Grafen, A. 1990a. Biological signals as handicaps. *J. Theor. Biol.* **144**, 517–46.

Grafen, A. 1990b. Sexual selection unhandicapped by the Fisher process. *J. theor. Biol.* **144**, 473–516.

Grafen, A. 1990c. Do animals really recognize kin? *Anim. Behav.* **39**, 42–54.

Grafen, A. 1991. Modelling in behavioural ecology. In J.R. Krebs & N.B. Davies (eds), *Behavioural Ecology: An Evolutionary Approach*, 3rd edn, pp. 5–31. Blackwell Scientific Publications, Oxford.

Greenberg, I. 1979. Genetic component of kin recognition in primitively social bees. *Science* **206**, 1095–7.

Greenlaw, J.S. & Post, W. 1985. Evolution of monogamy in seaside sparrows *Ammodramus maritimus*: tests of hypotheses. *Anim. Behav.* **33**, 373–83.

Greenwood, P.J. 1980. Mating systems, philopatry and dispersal in birds and mammals. *Anim. Behav.* **28**, 1140–62.

Greenwood, P.J., Harvey, P.H. & Perrins, C.M. 1978. Inbreeding and dispersal in the great tit. *Nature* **271**, 52–4.

Grey, R. 1987. Faith and foraging: a critique of optimal foraging theory. In A.C. Kamil, J.R. Krebs & H.R. Pulliam (eds), *Foraging Behavior*. Plenum Press, New York.

Grosberg, R.K. & Quinn, J.F. 1986. The genetic control and consequences of kin recognition by the larvae of a colonial marine invertebrate. *Nature* **322**, 456–9.

Gross, M.R. 1982. Sneakers, satellites and parentals: polymorphic mating strategies in North American

sunfishes. *Z. Tierpsychol.* **60**, 1–26.

Gross, M.R. 1985. Disruptive selection for alternative life histories in salmon. *Nature* **313**, 47–8.

Gross, M.R. 1992. Alternative reproductive tactics and strategies. *Behav. Ecol.* (in press).

Gross, M.R. & Charnov, E.L. 1980. Alternative male life histories in bluegill sunfish. *Proc. Natn. Acad. Sci. USA* **77**, 6937–40.

Gross, M.R. & Sargent, R.C. 1985. The evolution of male and female parental care in fishes. *Amer. Zool.* **25**, 807–22.

Gross, M.R. & Shine, R. 1981. Parental care and mode of fertilization in ectothermic vertebrates. *Evolution* **35**, 775–93.

Guilford, T. 1986. How do 'warning colours' work? Conspicuousness may reduce recognition errors in experienced predators. *Anim. Behav.* **34**, 286–8.

Guilford, T. & Dawkins, M.S. 1991. Receiver psychology and the evolution of animal signals. *Anim. Behav.* **39**, 706–16.

Gustafsson, L. & Sutherland, W.J. 1988. The costs of reproduction in the collared flycatcher *Ficedula albicollis*. *Nature* **335**, 813–15.

Gwynne, D.T. 1982. Mate selection by female katydids (Orthoptera Tettigoniidae. *Conocephalus nigropleurum*). *Anim. Behav.* **30**, 734–8. .

Haas, V. 1985. Colonial and single breeding in fieldfares. *Turdus pilaris* L.: a comparison of nesting success in early and late broods. *Behav. Ecol. Sociobiol.* **16**, 119–24.

Haldane, J.B.S. 1953. Animal populations and their regulation. *Penguin Modern Biology* **15**, 9–24.

Hamilton, W.D. 1964. The genetical evolution of social behaviour. I, II. *J. theor. Biol.* **7**, 1–52.

Hamilton, W.D. 1967. Extraordinary sex ratios. *Science* **156**, 477–88.

Hamilton, W.D. 1971. Geometry for the selfish herd. *J. theor. Biol.* **31**, 295–311.

Hamilton, W.D. 1972. Altruism and related phenomena, mainly in social insects. *Ann. Rev. Ecol. Syst.* **3**, 193–232.

Hamilton, W.D. 1979. Wingless and fighting males in fig wasps and other insects. In M.S. Blum & N.A. Blum (eds), *Sexual Selection and Reproductive Competition in Insects*, pp. 167–220. Academic Press, London.

Hamilton, W.D. & Zuk, M. 1982. Heritable true fitness and bright birds: a role for parasites? *Science* **218**, 384–7.

Hammerstein, P. & Parker, G.A. 1982. The asymmetric war of attrition. *J. theor. Biol.* **96**, 647–82.

Hanken, J. & Sherman, P.W. 1981. Multiple paternity in Belding's ground squirrel litters. *Science* **212**, 351–3.

Harcourt, A.H., Harvey, P.H., Larson, S.G. & Short, R.V. 1981. Testis weight, body weight and breeding system in primates. *Nature* **293**, 55−7.

Harper, D.G.C. 1982. Competitive foraging in mallards: 'ideal free' ducks. *Anim. Behav.* **30**, 575−84.

Harvey, P.H. 1985. Intrademic group selection and the sex ratio. In R.M. Sibly & R.H. Smith (eds), *Behavioural Ecology: Ecological Consequences of Adaptive Behaviour*, pp. 59−73. 25th Symposium of the British Ecological Society. Blackwell Scientific Publications, Oxford.

Harvey, P.H. & Pagel, M.D. 1991. *The Comparative Method in Evolutionary Biology*. Oxford University Press, Oxford.

Harvey, P. & Purvis, A. 1991. Comparative methods for explaining adaptations. *Nature* **351**, 619−24.

Harvey, P.H., Kavanagh, M. & Clutton-Brock, T.H. 1978. Sexual dimorphism in primate teeth. *J. Zool. Lond.* **186**, 475−86.

Harvey, P.H., Bull, J.J. & Paxton, R.J. 1983. Looks pretty nasty. *New Sci.* **97**, 26−7.

Harvey, P.H., Bull, J.J., Pemberton, M. & Paxton, R.J. 1982. The evolution of aposematic coloration in distasteful prey: a family model. *Amer. Natur.* **119**, 710−19.

Hassell, M.P. 1971. Mutual interference between searching insect parasites. *J. Anim. Ecol.* **40**, 473−86.

Heller, R. & Milinski, M. 1979. Optimal foraging of sticklebacks on swarming prey. *Anim. Behav.* **27**, 1127−41.

Hinde, R.A. 1956. The biological significance of the territories of birds. *Ibis* **98**, 340−69.

Hinde, R.A. 1970. *Animal Behaviour: A Synthesis of Ethology and Comparative Psychology*, 2nd edn. McGraw-Hill, New York.

Hinde, R.A. 1974. *Biological Bases of Human Social Behaviour*. McGraw-Hill, New York.

Hinde, R.A. 1981. Animal signals: ethological and games-theory approaches are not incompatible. *Anim. Behav.* **29**, 535−42.

Hixon, M.A. 1982. Energy maximizers and time minimizers: theory and reality. *Amer. Natur.* **119**, 596−9.

Hixon, M.A., Carpenter, F.L. & Paton, D.C. 1983. Territory area, flower density and time budgeting in hummingbirds: an experimental and theoretical analysis. *Amer. Natur.* **122**, 366−91.

Hobbs, N.T. 1990. Diet selection by generalist herbivores: a test of the linear programming model. In R.N. Hughes (ed.), *Behavioural Mechanisms of Food Selection*, NATO ASI series Vol. 20, pp. 395−414. Springer-Verlag, Berlin.

Hodges, C.M. & Wolf, L.L. 1981. Optimal foraging bumblebees? why is nectar left behind in flowers?

Behav. Ecol. Sociobiol. **9**, 41−4.

Höglund, J. 1989. Size and plumage dimorphism in lek-breeding birds: a comparative analysis. *Amer. Natur.* **134**, 72−87.

Högstedt, G. 1980. Evolution of clutch size in birds: adaptive variation in relation to territory quality. *Science* **210**, 1148−50.

Hölldobler, B. 1971. Communication between ants and their guests. *Sci. Am.* **224**, 86−93.

Hölldobler, B. 1977. Communication in social Hymenoptera. In T.A. Seboek (ed.), *How Animals Communicate*, pp. 418−71. Indiana University Press, Bloomington & London.

Holmes, W.G. & Sherman, P.W. 1982. The ontogeny of kin recognition in two species of ground squirrels. *Amer. Zool.* **22**, 491−517.

Hoogland, J.L. 1979a. The effect of colony size on individual alertness of prairie dogs (Sciuridae: Cynomys spp.) *Anim. Behav.* **27**, 394−407.

Hoogland, J.L. 1979b. Aggression, ectoparasitism and other possible costs of prairie dog (Sciuridae: Cynomys spp.) coloniality. *Behaviour* **69**, 1−35.

Hoogland, J.L. 1983. Nepotism and alarm calling in the black-tailed prairie dog, Cynomys ludovicianus. *Anim. Behav.* **31**, 472−9.

Houde, A.E. 1988. Genetic differentiation in female choice between two guppy populations. *Anim. Behav.* **36**, 511−16.

Houde, A.E. & Endler, J.A. 1990. Correlated evolution of female mating preferences and male colour patterns in the guppy *Poecilia reticulata*. *Science* **248**, 1405−8.

Houde, A.E. & Torio, A.J. 1992. Effect of parasite infection on male colour pattern and female choice in guppies. *Behav. Ecol.* **3**, 346−51.

Houston, A.I. & McNamara, J.M. 1982. A sequential approach to risk-taking. *Anim. Behav.* **30**, 1260−1.

Houston, A.I. & McNamara, J.M. 1985. The choice of two prey types that minimises the probability of starvation. *Behav. Ecol. Sociobiol.* **17**, 135−41.

Houston, A.I., Clark, C.W., McNamara, J.M. & Mangel, M. 1988. Dynamic models in behavioural and evolutionary ecology. *Nature* **332**, 29−34.

Houtman, A.M. 1992. Female zebra finches choose attractive partners for extra-pair copulations. *Proc. R. Soc. B.* **249**, 3−6.

Howard, R.D. 1978a. Factors influencing early embryo mortality in bullfrogs. *Ecology* **59**, 789−98.

Howard, R.D. 1978b. The evolution of mating strategies in bullfrogs. Rana catesbeiana. *Evolution* **32**, 850−71.

Hrdy, S.B. 1977. *The Langurs of Abu: Female and Male Strategies of Reproduction*. Harvard University Press, Cambridge, Mass.

Hughes, R.N. (ed.). 1990. *Behavioural Mechanisms of Food Selection*. NATO ASI series series G: Ecological sciences, Vol. 20. Springer-Verlag, Berlin.

Hunter, M.L. & Krebs, J.R. 1979. Geograpical variation in the song of the great tit (*Parus major*) in relation to ecological factors. *J. Anim. Ecol.* **48**, 759–85.

Hurly, T.A. 1992. Energetic reserves of marsh tits (*Parus palustris*): food and fat storage in response to variable food supply. *Behav. Ecol.* **3**, 181–8.

Huxley, J.S. 1966. A discussion of ritualisation of behaviour in animals and man: introduction. *Phil. Trans. R. Soc. B.* **251**, 247–71.

Ims, R.A. 1987. Responses in spatial organization and behaviour to manipulations of the food resource in the vole *Clethrionomys rufocanus*. *J. Anim. Ecol.* **56**, 585–96.

Ims, R.A. 1988. Spatial clumping of sexually receptive females induces space sharing among male voles. *Nature* **335**, 541–3.

Inman, A.J. 1990. Group foraging in starlings: distributions of unequal competitors. *Anim. Behav.* **40**, 801–10.

Iwasa, Y., Pomiankowski, A. & Nee, S. 1991. The evolution of costly mate preferences II. The 'handicap' hypothesis. *Evolution* **45**, 1431–1442.

Jakobsson, S., Radesäter, T. & Järvi, T. 1979. On the fighting behaviour of *Nannacara anomala* (Pisces, Cichlidae) males. *Z. Tierpsychol.* **49**, 210–20.

Jamieson, I.G. 1989. Behavioural heterochrony and the evolution of birds' helping at the nest: an unselected consequence of communal breeding? *Amer. Natur.* **133**, 394–406.

Jamieson, I.G. & Craig, J.L. 1987. Critique of helping behaviour in birds: a departure from functional explanations. In P. Bateson & P. Klopfer (eds), *Perspectives in Ethology*, Vol. 7, pp. 79–98. Plenum Press, New York.

Jarman, P.J. 1974. The social organization of antelope in relation to their ecology. *Behaviour* **48**, 215–67.

Jarvis, J.U.M. 1981. Eusociality in a mammal: cooperative breeding in naked mole rat colonies. *Science* **212**, 571–3.

Jarvis, M.J.F. 1974. The ecological significance of clutch size in the South African gannet (*Sula capensis*, Lichenstein). *J. Anim. Ecol.* **43**, 1–17.

Jeffreys, A.J., Wilson, V. & Thein, S.L. 1985. Hypervariable 'minisatellite' regions in human DNA. *Nature* **314**, 67–73.

Jenni, D.A. & Collier, G. 1972. Polyandry in the American jacana, *Jacana spinosa*. *Auk* **89**, 743–65.

Jennings, T. & Evans, S.M. 1980. Influence of position in the flock and flock size on vigilance in the starling. *Sturnus vulgaris*. *Anim. Behav.* **30**, 634–5.

Kacelnik, A. 1984. Central place foraging in Starlings (*Sturnus vulgaris*). I. Patch residence time. *J. Anim. Ecol.* **53**, 283–99.

Kacelnik, A. & Cuthill, I.C. 1987. Starlings and optimal foraging: modelling in fractal world. In A.C. Kamil, J.R. Krebs & H.R. Pulliam (eds), *Foraging Behavior*. Plenum Press, New York.

Kacelnik, A., Krebs, J.R. & Bernstein, C. 1992. The ideal free distribution and predator–prey populations. *Trends Ecol. Evol.* **7**, 50–5.

Kamil, A.C., Krebs, J.R. & Pulliam, H.R. (eds). 1987. *Foraging Behavior*, Plenum Press, New York.

Kempenaers, B. Verheyen, G.R., Van den Broeck, M., Burke, T., Van Broeckhoven, C. & Dhondt, A.A. 1992. Extra-pair paternity results from female preference for high-quality males in the blue tit. *Nature* **357**, 494–6.

Kenward, R.E. 1978. Hawks and doves: factors affecting success and selection in goshawk attacks on wood-pigeons. *J. Anim. Ecol.* **47**, 449–60.

Kirkpatrick, M. & Ryan, M.J. 1991. The evolution of mating preference and the paradox of the lek. *Nature* **350**, 33–8.

Kleiman, D.G. 1977. Monogamy in mammals. *Q. Rev. Biol.* **52**, 39–69.

Kodric-Brown, A. 1989. Dietary carotenoids and male mating success in the guppy: an environmental component to female choice. *Behav. Ecol. Sociobiol.* **25**, 393–401.

Kodric-Brown, A. & Brown, J.H. 1984. Truth in advertising: the kinds of traits favoured by sexual selection. *Amer. Natur.* **124**, 309–23.

Koenig, W.D. 1990. Opportunity of parentage and nest destruction in polygynandrous acorn woodpeckers: an experimental study. *Behav. Ecol.* **1**, 55–61.

Koenig, W.D., Mumme, R.L. & Pitelka, F.A. 1984. The breeding system of the acorn woodpecker in central coastal California. *Z. Tierpsychol.* **65**, 289–308.

Koenig, W.D., Pitelka, F.A., Carmen, W.J., Mumme, R.L. & Stanback, M.T. 1992. The evolution of delayed dispersal in cooperative breeders. *Q. Rev. Biol.* **67**, 111–50.

Komdeur, J. 1992. Importance of habitat saturation and territory quality for the evolution of cooperative breeding in the Seychelles Warbler, *Nature* **358**, 493–5.

Kramer, D.L. 1985. Are colonies supraoptimal groups? *Anim. Behav.* **33**, 1031.

Krebs, J.R. 1971. Territory and breeding density in the great tit. *Parus major* L. *Ecology* **52**, 2–22.

Krebs, J.R. 1982. Territorial defence in the great tit. *Parus major*: do residents always win? *Behav.*

Ecol. Sociobiol. **11**, 185–94.

Krebs, J.R. & Dawkins, R. 1984. Animal signals: mind reading and manipulation. In J.R. Krebs & N.B. Davies (eds), *Behavioural Ecology: An Evolutionary Approach*, 2nd edn, pp. 380–402. Blackwell Scientific Publications, Oxford.

Krebs, J.R. & Kacelnik, A. 1991. Decision-making. In J.R. Krebs & N.B. Davies (eds), *Behavioural Ecology: an Evolutionary Approach*, 3rd edn, pp. 105–36. Blackwell Scientific Publications, Oxford.

Krebs, J.R., MacRoberts, M.H. & Cullen, J.M. 1972. Flocking and feeding in the great tit *Parus major*: an experimental study. *Ibis* **114**, 507–30.

Krebs, J.R., Ashcroft, R. & Webber, M. 1978. Song repertoires and territory defence in the great tit. *Parus major. Nature* **271**, 539–42.

Krebs, J.R., Erichsen, J.T., Webber, M.I. & Charnov, E.L. 1977. Optimal prey selection in the great tit. *Parus major. Anim. Behav.* **25**, 30–8.

Kroodsma, D.E. 1976. Reproductive development in a female song-bird: differential stimulation by quality of male song. *Science* **192**, 574–5.

Kruuk, H. 1964. Predators and anti-predator behaviour of the black headed gull, *Larus ridibundus. Behaviour Suppl.* **11**, 1–129.

Kruuk, H. 1972. *The Spotted Hyena*. University of Chicago Press, Chicago.

Lack, D. 1966. *Population Studies of Birds*. Clarendon Press, Oxford.

Lack, D. 1968. *Ecological Adaptations for Breeding in Birds*. Methuen, London.

Lacy, R.C. 1980. The evolution of eusociality in termites: a haplodiploid analogy. *Amer. Natur.* **116**, 449–51.

Lande, R. 1981. Models of speciation by sexual selection of polygenic traits. *Proc. Natn. Acad. Sci. USA* **78**, 3721–5.

Lank, D.B., Oring, L.W. & Maxson, S.J. 1985. Mate and nutrient limitation of egg laying in a polyandrous shorebird. *Ecology* **66**, 1513–24.

Lazarus, J. & Metcalfe, N.B. 1990. Tit for tat cooperation in sticklebacks: a critique of Milinski. *Anim. Behav.* **39**, 987–8.

Le Boeuf, B.J. 1972. Sexual behaviour in the northern elephant seal. *Mirounga angustirostris. Behaviour* **41**, 1–26.

Le Boeuf, B.J. 1974. Male–male competition and reproductive success in elephant seals. *Amer. Zool.* **14**, 163–76.

Le Boeuf, B.J. & Reiter, J. 1988. Lifetime reproductive success in Northern elephant seals. In T.H. Clutton-Brock (ed.), *Reproductive Success*. 344–62. Chicago University Press, Chicago.

Leonard, M.L. & Picman, J. 1987. Female settlement in marsh wrens: is it affected by other females?

Behav. Ecol. Sociobiol. **21**, 135–40.

Lessells, C.M. 1991. The evolution of life histories. In J.R. Krebs & N.B. Davies (eds), *Behavioural Ecology: An Evolutionary Approach*, 3rd edn, pp. 32–68. Blackwell Scientific Publications, Oxford.

Leuthold, W. 1966. Variations in territorial behaviour of Uganda Kob. *Adenota kob thomasi. Behaviour* **27**, 215–58.

Lightbody, J.P. & Weatherhead, P.J. 1988. Female settling patterns and polygyny: tests of a neutral-mate-choice hypothesis. *Amer. Natur.* **132**, 20–33.

Lill, A. 1974. Sexual behaviour of the lek-forming white-bearded manakin (*Manacus manacus trinitatis*). *Z. Tierpsychol.* **36**, 1–36.

Lima, S. 1984. Downy woodpecker foraging behavior: efficient sampling in simple stochastic environments. *Ecology* **65**, 166–74.

Lima, S.L. 1986. Predation risk and unpredictable feeding conditions: determinants of body mass in birds. *Ecology* **67**, 377–85.

Lima, S., Valone, T.J. & Caraco, T. 1985. Foraging-efficiency–predation-risk tradeoff in the grey squirrel. *Anim. Behav.* **33**, 155–65.

Lloyd, M. & Dybas, H.S. 1966. The periodical cicada problem: II Evolution. *Evolution* **20**, 466–505.

Lorenz, K. 1966. *On Aggression*. Methuen, London.

Lotem, A., Nakamura, H. & Zahavi, A. 1992. Rejection of cuckoo eggs in relation to host age: a possible evolutionary equilibrium. *Behav. Ecol.* **3**, 128–32.

Lyon, B.E. & Montgomerie, R.D. 1985. Conspicuous plumage of birds: sexual selection or unprofitable prey? *Anim. Behav.* **33**, 1038–40.

Lyon, B.E., Montgomerie, R.D. & Hamilton, L.D. 1987. Male parental care and monogamy in snow buntings. *Behav. Ecol. Sociobiol.* **20**, 377–82.

McClintock, M.K. 1971. Menstrual synchrony and suppression. *Nature* **229**, 244–5.

Macdonald, D.W. & Moehlman, P.D. 1986. Cooperation, altruism and restraint in the reproduction of carnivores. In P. Bateson & P. Klopfer (eds), *Perspectives in Ethology*, Vol. 5, pp. 433–67. Plenum Press, New York.

McKinney, F., Derrickson, S.R. & Mineau, P. 1983. Forced copulation in waterfowl, *Behaviour* **86**, 250–94.

Mackinnon, J. 1974. The ecology and behaviour of wild orangutans. *Pongo pygmaeus. Anim. Behav.* **22**, 3–74.

MacNally, R. & Young, D. 1981. Song energetics of the bladder cicada *Cystosoma saundersii. J. Exp. Biol* **90**, 185–96.

McNamara, J.M. & Houston, A.I. 1990. The value of fat reserves and the tradeoff between starvation and predation. *Acta Biotheor.* **38**, 37–61.

McNamara, J.M., Houston, A.I. & Krebs, J.R. 1990. Why hoard? The economics of food storing in tits. *Behav. Ecol.* **1**, 12–23.

McPhail, J.D. 1969. Predation and the evolution of a stickleback. *Gasterosteus. J. Fish. Res. Bd. Canada.* **26**, 3183–208.

Magurran, A.E. 1990. The adaptive significance of schooling as an antipredator defence in fish. *Annals Zool. Fenn.* **27**, 51–66.

Magurran, A.E. & Seghers, B.H. 1991. Variation in schooling and aggression amongst guppy (*Poecilia reticulata*) populations in Trinidad. *Behaviour* **118**, 214–34.

Magurran, A.E., Seghers, B.H., Carvalho, G.R. & Shaw, P.W. 1992. Behavioural consequences of an artificial introduction of guppies (*Poecilia reticulata*) in N. Trinidad: evidence for the evolution of antipredator behaviour in the wild. *Proc. Roy. Soc. B.* **248**, 117–22.

Major, P.F. 1978. Predator–prey interactions in two schooling fishes. *Caranx ignobilis* and *Stolephorus purpureus. Anim. Behav.* **26**, 760–77.

Manning, A. 1961. Effects of artificial selection for mating speed in *Drosophila melanogaster. Anim. Behav.* **9**, 82–92.

Marden, J.H. & Waage, J.K. 1990. Escalated damselfly territorial contests are energetic wars of attrition. *Anim. Behav.* **39**, 954–9.

Marler, P. 1955. Characteristics of some animal calls. *Nature* **176**, 6–8.

Marten, K. & Marler, P. 1977. Sound transmission and its significance for animal vocalization. I. Temperate habitats. *Behav. Ecol. Sociobiol.* **2**, 271–90.

Marten, K., Quine, D. & Marler, P. 1977. Sound transmission and its significance for animal vocalization. II. Tropical forest habitats, *Behav. Ecol. Sociobiol.* **2**, 291–302.

May, R.M. & Robinson, S.K. 1985. Population dynamics of avian brood parasitism. *Amer. Natur.* **126**, 475–84.

Maynard Smith, J. 1964. Group selection and kin selection. *Nature* **201**, 1145–7.

Maynard Smith, J. 1974. The theory of games and the evolution of animal conflicts. *J. theor. Biol.* **47**, 209–21.

Maynard Smith, J. 1976a. Group selection. *Q. Rev. Biol.* **51**, 277–83.

Maynard Smith, J. 1976b. Evolution and the theory of games. *Amer. Sci.* **64**, 41–5.

Maynard Smith, J. 1977a. Parental investment – a prospective analysis. *Anim. Behav.* **25**, 1–9.

Maynard Smith, J. 1977b. The limitations of evolutionary theory. In R. Duncan & M. Weston-Smith (eds), *The Encyclopaedia of Ignorance: Life Sciences and Earth Sciences*, pp. 235–42.

Pergamon Press, Oxford.

Maynard Smith, J. 1978. Optimization theory in evolution. *Ann. Rev. Ecol. Syst.* **9**, 31–56.

Maynard Smith, J. 1980. A new theory of sexual investment. *Behav. Ecol. Sociobiol.* **7**, 247–51.

Maynard Smith, J. 1982. *Evolution and the Theory of Games.* Cambridge University Press, Cambridge.

Maynard Smith, J. & Ridpath, M.G. 1972. Wife sharing in the Tasmanian native hen *Tribonyx mortierii*: a case of kin selection? *Amer. Natur.* **106**, 447–52.

Mazur, J. 1984. Tests of an equivalence rule for fixed and variable reinforcer delays. *J. Exp. Psychol. Anim. Behav. Proc.* **10**, 426–36.

Metcalf, R.A. 1980. Sex ratios, parent offspring conflict, and local competition for mates in the social wasps *Polistes metricus* and *Polistes variatus. Amer. Natur.* **116**, 642–54.

Metcalf, R.A. & Whitt, G.S. 1977. Relative inclusive fitness in the social wasp *Polistes metricus. Behav. Ecol. Sociobiol.* **2**, 353–60.

Michener, C.D. 1974. *The Social Behavior of the Bees.* Belknap Press, Harvard.

Milinski, M. 1979. An evolutionarily stable feeding strategy in sticklebacks. *Z. Tierpsychol.* **51**, 36–40.

Milinski, M. 1984a. Competitive resource sharing: an experimental test of a learning rule of ESSs. *Anim. Behav.* **32**, 233–42.

Milinski, M. 1984b. A predator's cost of overcoming the confusion effect of swarming prey. *Anim. Behav.* **32**, 1157–62.

Milinski, M. 1987. Tit for tat in sticklebacks and the evolution of cooperation. *Nature* **327**, 15–17.

Milinski, M. & Bakker, T.C.M. 1990. Female sticklebacks use male coloration in mate choice and hence avoid parasitized males. *Nature* **344**, 330–3.

Milinski, M. & Heller, R. 1978. Influence of a predator on the optimal foraging behaviour of sticklebacks (*Gasterosteus aculeatus*). *Nature* **275**, 642–4.

Milinski, M. & Parker, G.A. 1991. Competition for resources. In J.R. Krebs & N.B. Davies (eds), *Behavioural Ecology: an Evolutionary Approach*, 3rd edn, pp. 137–68. Blackwell Scientific Publications, Oxford.

Milinski, M., Pfluger, D., Külling, D. & Kettler, R. 1990a. Do sticklebacks cooperate repeatedly in reciprocal pairs? *Behav. Ecol. Sociobiol.* **27**, 17–21.

Milinski, M., Külling, D. & Kettler, R. 1990b. Tit for tat: sticklebacks 'trusting' a cooperating partner. *Behav. Ecol.* **1**, 7–10.

Mock, D.W. 1984. Siblicidal aggression and resource monopolization in birds. *Science* **225**, 731–3.

Mock, D.W. 1985. Siblicidal brood reduction: the prey-size hypothesis. *Amer. Natur.* **125**, 327–343.

Modell, W. 1969. Horns and antlers. *Sci. Am.* **220**, 114–22.

Moehlman, P.D. 1979. Jackal helpers and pup survival. *Nature* **277**, 382–3.

Moksnes, A., Røskaft, E., Braa, A.T., Korsnes, L., Lampe, H.M. & Pedersen, H.Ch. 1991. Behavioural responses of potential hosts towards artificial cuckoo eggs and dummies. *Behaviour* **116**, 64–89.

Møller, A.P. 1988. Female choice selects for male sexual tail ornaments in the monogamous swallow. *Nature* **332**, 640–2.

Møller, A.P. 1989. Viability costs of male tail ornaments in a swallow. *Nature* **339**, 132–5.

Møller, A.P. 1990. Effects of a haematophagous mite on the barn swallow *Hirundo rustica*: a test of the Hamilton and Zuk hypothesis. *Evolution* **44**, 771–84.

Møller, A.P. & Birkhead, T.R. 1992. A pairwise comparative method as illustrated by copulation frequency in birds. *Amer. Natur.* **139**, 644–56.

Montgomerie, R.D. & Weatherhead, P.J. 1988. Risks and rewards of nest defence by parent birds. *Q. Rev. Biol.* **63**, 167–87.

Moodie, G.E.E. 1972. Predation, natural selection and adaptation in an unusual three spined stickleback. *Heredity* **28**, 155–67.

Morris, D. 1957. Typical intensity and its relation to the problem of ritualisation. *Behaviour* **11**, 1–12.

Morton, E.S. 1975. Ecological sources of selection on avian sounds. *Amer. Natur.* **109**, 17–34.

Mumme, R.L. 1992. Do helpers increase reproductive success? An experimental analysis in the Florida scrub jay. *Behav. Ecol. Sociobiol.* **31**, 319–28.

Myers, J.P., Connors, P.G. & Pitelka, F.A. 1979. Territory size in wintering sanderlings: the effects of prey abundance and intruder density. *Auk* **96**, 551–61.

Nakamura, H. 1990. Brood parasitism by the cuckoo *Cuculus canorus* in Japan and the start of new parasitism on the azure-winged magpie *Cyanopica cyana*. *Jap. J. Ornithol.* **39**, 1–18.

Nakatsuru, K. & Kramer, D.L. 1982. Is sperm cheap? Limited male fertility and female choice in the lemon tetra (Pisces: Characidae). *Science* **216**, 753–5.

Neill, S.R., St. J. & Cullen, J.M. 1974. Experiments on whether schooling by their prey affects the hunting behaviour of cephalods and fish predators. *J. Zool. Lond.* **172**, 549–69.

Nelson, J.B. 1964. Factors influencing clutch-size and chick growth in the North Atlantic gannet. *Sula bassana. Ibis* **106**, 63–77.

Nisbet, I.C.T. 1977. Courtship feeding and clutch size in common terns *Sterna hirundo*. In B. Stonehouse & C.M. Perrins (eds), *Evolutionary Ecology*, pp. 101–9. London, Macmillan.

Nottebohm, F. 1975. Continental patterns of song variability in *Zontrichia capensis*: some possible ecological correlates. *Amer. Natur.* **109**, 605–24.

Nur, N. 1988. The consequences of brood size for breeding blue tits. III Measuring the cost of reproduction: survival, future fecundity and differential dispersal. *Evolution* **42**, 351–62.

Nur, N. & Hasson, O. 1984. Phenotypic plasticity and the handicap principle. *J. theor. Biol.* **110**, 275–97.

Orians, G.H. 1969. On the evolution of mating systems in birds and mammals. *Amer. Natur.* **103**, 589–603.

Orians, G.H. 1980. *Some Adaptations of Marsh-Nesting Blackbirds*. Princeton University Press, Princeton, New Jersey.

Oring, L.W. 1982. Avian mating systems. In D.S. Farner & J.R. King (eds), *Avian Biology*, Vol. 6. Academic Press, London.

Oster, G.F. & Wilson, E.O. 1978. *Caste and Ecology in the Social Insects*. Princeton University Press, Princeton, NJ.

Owen, D.F. 1954. The winter weights of titmice *Ibis* **96**, 299–309.

Owen, R.E. & Plowright, R.C. 1982. Worker–queen conflict and male parentage in bumblebees. *Behav. Ecol. Sociobiol.* **11**, 91–9.

Owen-Smith, N. 1977. On territoriality in ungulates and an evolutionary model. *Q. Rev. Biol.* **52**, 1–38.

Owings, D.H. & Coss, R.G. 1977. Snake mobbing by California ground squirrels: adaptive variation and ontogeny. *Behaviour* **62**, 50–69.

Packer, C. 1977. Reciprocal altruism in *Papio anubis. Nature* **265**, 441–3.

Packer, C. 1979. Inter-troop transfer and inbreeding avoidance in *Papio anubis. Anim. Behav.* **27**, 1–36.

Packer, C. & Pusey, A.E. 1982. Cooperation and competition within coalitions of male lions: kin selection or game theory? *Nature* **296**, 740–2.

Packer, C. & Pusey, A.E. 1983a. Male takeovers and female reproductive parameters: a simulation of oestrus synchrony in lions (*Panthera leo*). *Anim. Behav.* **31**, 334–40.

Packer, C. & Pusey, A.E. 1983b. Adaptations of female lions to infanticide by incoming males. *Amer. Natur.* **121**, 716–28.

Packer, C., Gilbert, D.A., Pusey, A.E. & O'Brien, S.J. 1991. A molecular genetic analysis of kinship and cooperation in African lions. *Nature* **351**, 562–5.

Page, G. & Whitacre, D.F. 1975. Raptor predation on wintering shorebirds. *Condor* **77**, 73–83.

Page, R.E. Jr., Robinson, G.E. & Fondrk, M.K. 1989. Genetic specialists, kin recognition and nepotism in honey-bee colonies. *Nature* **338**, 576–9.

Parker, G.A. 1970. The reproductive behaviour and the nature of sexual selection in *Scatophaga stercoraria* L. (Diptera: Scatophagidae). II. The fertilization rate and the spatial and temporal relationships of each sex around the site of mating and oviposition. *J. Anim. Ecol.* **39**, 205–28.

Parker, G.A. 1974. Assessment strategy and the evolution of animal conflicts. *J. theor. Biol.* **47**, 223–43.

Parker, G.A. 1978. Searching for mates. In J.R. Krebs & N.B. Davies (eds), *Behavioural Ecology: An Evolutionary Approach*, 1st edn, pp. 214–44. Blackwell Scientific Publications, Oxford.

Parker, G.A. 1979. Sexual selection and sexual conflict. In M.S. Blum & N.A. Blum (eds), *Sexual Selection and Reproductive Competition in Insects*, pp. 123–66. Academic Press, New York.

Parker, G.A. & Knowlton, N. 1980. The evolution of territory size. Some ESS models. *J. theor. Biol.* **84**, 445–76.

Parker, G.A. & Rubenstein, D.I. 1981. Role assessment, reserve strategy and the acquisition of information in asymmetric animal contests. *Anim. Behav.* **29**, 221–40.

Parker, G.A. & Sutherland, W.J. 1986. Ideal free distributions when individuals differ in competitive ability: phenotype-limited ideal free models. *Anim. Behav.* **34**, 1222–42.

Parker, G.A., Baker, R.R. & Smith, V.C.F. 1972. The origin and evolution of gamete dimorphism and the male–female phenomenon. *J. theor. Biol.* **36**, 529–53.

Partridge, B.L. & Pitcher, T.J. 1979. Evidence against a hydrodynamic function for fish schools. *Nature* **279**, 418–19.

Partridge, L. 1980. Mate choice increases a component of offspring fitness in fruit flies. *Nature* **283**, 290–1.

Partridge, L. 1983. Genetics and behaviour. In T.R. Halliday & P.J.B. Slater (eds), *Animal Behaviour: Genes, Development and Learning*, pp. 11–51. Blackwell Scientific Publications, Oxford.

Paton, D. 1986. Communication by agonistic displays: II. Perceived information and the definition of agonistic displays. *Behaviour* **99**, 157–75.

Paton, D. & Caryl, P.G. 1986. Communication by agonistic displays: I. Variation in information content between samples. *Behaviour* **98**, 213–39.

Patterson, I.J. 1965. Timing and spacing of broods in the black-headed gull, *Larus ridibundus*. *Ibis* **107**, 433–59.

Perrill, S.A., Gerhardt, H.C. & Daniel, R. 1982. Mating strategy shifts in male green treefrogs, *Hyla cinerea*: an experimental study. *Anim. Behav.* **30**, 43–8.

Perrins, C.M. 1965. Population fluctuations and clutch size in the great tit. *Parus major*. L. *J. Anim. Ecol.* **34**, 601–47.

Perrins, C.M. 1979. *British Tits*. New Naturalist Series. Collins. London.

Petrie, M. 1983. Female moorhens compete for small fat males. *Science* **220**, 413–15.

Petrie, M. & Møller, A.P. 1991. Laying eggs in others' nests: intraspecific brood parasitism in birds. *Trends Ecol. Evol.* **6**, 315–20.

Petrie, M., Halliday, T. & Sanders, C. 1991. Peahens prefer peacocks with elaborate trains. *Anim. Behav.* **41**, 323–31.

Pettifor, R.A., Perrins, C.M. & McCleery, R.H. 1988. Individual optimization of clutch size in great tits. *Nature* **336**, 160–2.

Pianka, E.R. & Parker, W.S. 1975. Age-specific reproductive tactics. *Amer. Natur.* **109**, 453–64.

Pietrewicz, A.T. & Kamil, A.C. 1981. Search images and the detection of cryptic prey: an operant approach. In A.C. Kamil & T.D. Sargent (eds), *Foraging Behavior: Ecological, Ethological and Psychological Approaches*, pp. 311–32. Garland STPM Press, New York.

Pitcher, T.J., Magurran, A.E. & Winfield, I.J. 1982. Fish in larger shoals find food faster. *Behav. Ecol. Sociobiol.* **10**, 149–51.

Plowright, R.C., Fuller, G.A. & Paloheimo, J.E. 1989. Shell-dropping by Northwestern crows: a re-examination of an optimal foraging study. *Can. J. Zool.* **67**, 770–71.

Pomiankowski, A., Iwasa, Y. & Nee, S. 1991. The evolution of costly mate preferences. I. Fisher and biased mutation. *Evolution* **45**, 1422–30.

Poran, N.S. & Coss, R.G. 1990. Development of antisnake defences in California ground squirrels *Spermophilus beecheyi*: I. Behavioural and immunological relationships. *Behaviour* **112**, 222–45.

Power, M.E. 1984. Habitat quality and the distribution of algae-grazing catfish in a Panamanian stream. *J. Anim. Ecol.* **53**, 357–74.

Prins, H.H.Th., Ydenberg, R.C. & Drent, R.H. 1980. The interaction of brent geese *Branta bernicla* and sea plantain *Plantago maritima* during spring staging: field observations and experiments. *Acta bot. neerl.* **29**, 585–96.

Pruett-Jones, S.G. & Lewis, M.J. 1990. Habitat limitation and sex ratio promote delayed dispersal in superb fairy wrens. *Nature* **348**, 541–2.

Pulliam, H.R. 1976. The principle of optimal behavior and the theory of communities. In P.H.

Klopfer & P.P.G. Bateson (eds), *Perspectives in Ethology*, pp. 311–32. Plenum Press, New York.

Pulliam, H.R. & Caraco, T. 1984. Living in groups: is there an optimal group size? In J.R. Krebs & N.B. Davies (eds), *Behavioural Ecology: An Evolutionary Approach*, 2nd edn, pp. 122–47. Blackwell Scientific Publications, Oxford.

Pulliam, H.R., Pyke, G.H. & Caraco, T. 1982. The scanning behavior of juncos: a game-theoretical approach. *J. theor. Biol.* **95**, 89–103.

Queller, D.C. 1989. The evolution of eusociality: reproductive head starts of workers. *Proc. Natl. Acad. Sci. USA* **86**, 3224–6.

Rabenold, K.N. 1984. Cooperative enhancement of reproductive success in tropical wren societies. *Ecology* **65**, 871–85.

Rabenold, P.P., Rabenold, K.N., Piper, W.H., Haydock, J. & Zack, S.W. 1990. Shared paternity revealed by genetic analysis in cooperatively breeding tropical wrens. *Nature* **348**, 538–40.

Ralls, K., Brugger, K. & Ballou, J. 1979. Inbreeding and juvenile mortality in small populations of ungulates. *Science* **206**, 1101–3.

Ratnieks, F.L.W. 1988. Reproductive harmony via mutual policing by workers in eusocial Hymenoptera. *Amer. Natur.* **132**, 217–36.

Ratnieks, F.L.W. & Visscher, P.W. 1989. Worker policing in the honeybee. *Nature* **342**, 796–7.

Reboreda, J.C. & Kacelnik, A. 1990. On cooperation, tit for tat and mirrors. *Anim. Behav.* **40**, 1188–91.

Reed, T.R. 1982. Interspecific territoriality in the chaffinch and great tit on islands and the mainland of Scotland: playbacks and removal experiments. *Anim. Behav.* **30**, 171–81.

Reeve, H.K. 1992. Queen activation of lazy workers in colonies of the eusocial naked mole rat. *Nature* **358**, 147–49.

Reyer, H.-U. 1980. Flexible helper structure as an ecological adaptation in the pied kingfisher, *Ceryle rudis rudis.* L. *Behav. Ecol. Sociobiol.* **6**, 219–27.

Reyer, H.-U. 1984. Investment and relatedness: a cost/benefit analysis of breeding and helping in the pied kingfisher (*Ceryle rudis*). *Anim. Behav.* **32**, 1163–78.

Reyer, H.-U. & Westerterp, K. 1985. Parental energy expenditure as a proximate cause of helper recruitment in the pied kingfisher (*Ceryle rudis*). *Behav. Ecol. Sociobiol.* **17**, 363–70.

Reynolds, J.D. 1987. Mating system and nesting biology of the red-necked phalarope *Phalaropus lobatus*: what constrains polyandry? *Ibis* **129**, 225–42.

Ridley, M. 1983. *The Explanation of Organic Diversity*. Clarendon Press, Oxford.

Rippin, A.B. & Boag, D.A. 1974. Spatial organization among male sharp-tailed grouse on arenas. *Can. J. Zool.* **52**, 591–7.

Robinson, S.K. 1986. The evolution of social behaviour and mating systems in the blackbirds (Icterinae). In D.I. Rubenstein & R.W. Wrangham (eds), *Ecological Aspects of Social Evolution*, pp. 175–200. Princeton University Press, Princeton.

Rohwer, S. & Rohwer, F.C. 1978. Status signalling in Harris sparrows: experimental deceptions achieved. *Anim. Behav.* **26** 1012–22.

Rohwer, S. & Spaw, C.D. 1988. Evolutionary lag versus bill-size constraints: a comparative study of the acceptance of cowbird eggs by old hosts. *Evol. Ecol.* **2**, 27–36.

Rood, J.P. 1978. Dwarf mongoose helpers at the den. *Z. Tierpsychol.* **48**, 277–87.

Rood, J.P. 1990. Group size, survival, reproduction and routes to breeding in dwarf mongooses. *Anim. Behav.* **39**, 566–72.

Roper, T.J. & Redston, S. 1987. Conspicuousness of distasteful prey affects the strength and durability of one-trial avoidance learning. *Anim. Behav.* **35**, 739–47.

Rosenzweig, M.L. 1986. Hummingbird isolegs in an experimental system. *Behav. Ecol. Sociobiol.* **19**, 313–22.

Røskaft, E., Jarvi, T., Bakken, M., Bech, C. & Reinertsen, R.F. 1986. The relationship between social status and resting metabolic rate in great tits (*Parus major*) and pied flycatchers (*Ficedula hypoleuca*). *Anim. Behav.* **34**, 838–42.

Rothstein, S.I. 1974. Mechanisms of avian egg recognition: possible learned and innate factors. *Auk* **91**, 796–807.

Rothstein, S.I. 1979. Gene frequencies and selection for inhibitory traits, with special emphasis on the adaptiveness of territoriality. *Amer. Natur.* **113**, 317–31.

Rothstein, S.I. 1982. Mechanisms of avian egg recognition: which egg parameters elicit responses by rejector species? *Behav. Ecol. Sociobiol.* **11**, 229–39.

Rothstein, S.I. 1986. A test of optimality: egg recognition in the eastern phoebe. *Anim. Behav.* **34**, 1109–19.

Rothstein, S.I. 1990. A model system for coevolution: avian brood parasitism. *Ann. Rev. Ecol. Syst.* **21**, 481–508.

Royama, T. 1970. Factors governing the hunting behaviour and selection of food by the great tit, *Parus major. J. Anim. Ecol.* **39**, 619–68.

Rubenstein, D.I. 1986. Ecology and sociality in horses and zebras. In D.I. Rubenstein & R.W. Wrangham (eds), *Ecological Aspects of Social Evolution*, pp. 282–302. Princeton University

Press, Princeton.

Rubenstein, D.I. & Hack, M.A. 1992. Horse signals: the sounds and scents of fury. *Evol. Ecol.* **6**, 254–60.

Rutberg, A.T. 1983. The evolution of monogamy in primates. *J. theor. Biol.* **104**, 93–112.

Ryan, M.J., Tuttle, M.D. & Taft, L.K. 1981. The costs and benefits of frog chorusing behavior. *Behav. Ecol. Sociobiol.* **8**, 273–8.

Ryan, M.J., Tuttle, M.D. & Rand, A.S. 1982. Bat predation and sexual advertisement in a neotropical anuran. *Amer. Natur.* **119**, 136–9.

Ryan, M.J., Fox, J.H., Wikzynski, W. & Rand, A.S. 1990. Sexual selection for sensory exploitation in the frog. *Physalaemus pustulosus. Nature* **343**, 66–8.

Sargent, T.D. 1981. Antipredator adaptations of underwing moths. In A.C. Kamil & T.D. Sargent (eds), *Foraging Behavior: Ecological, Ethological and Psychological Approaches.* pp. 259–84. Garland STPM Press, New York.

Schlenoff, D.H. 1985. The startle responses of blue jays to *Catocala* (Lepidoptera: Noctuidae) prey models. *Anim. Behav.* **33**, 1057–67.

Schmid-Hempel, P. 1986. Do honeybees get tired? The effect of load weight on patch departure. *Anim. Behav.* **34**, 1243–50.

Schmid-Hempel, P. 1987. Efficient nectar collection by honeybees. I. Economic models. *J. Anim. Ecol.* **56**, 209–18.

Schmid-Hempel, P. & Wolf, T.J. 1988. Foraging effort and life span in a social insect. *J. Anim. Ecol.* **57**, 509–22.

Schmid-Hempel, P., Kacelnik, A. & Houston, A.I. 1985. Honeybees maximise efficiency by not filling their crop. *Behav. Ecol. Sociobiol.* **17**, 61–6.

Schoener, T.W. 1983. Simple models of optimal feeding-territory size: a reconciliation. *Amer. Natur.* **121**, 608–29.

Searcy, W.A. & Yasukawa, K. 1989. Alternative models of territorial polygyny in birds. *Amer. Natur.* **134**, 323–43.

Seger, J. 1983. Partial bivoltinism may cause alternating sex-ratio biasses that favour eusociality. *Nature* **301**, 59–62.

Seger, J. 1991. Cooperation and conflict in social insects. In J.R. Krebs & N.B. Davies (eds), *Behavioural Ecology: an Evolutionary Approach,* 3rd edn, pp. 338–73. Blackwell Scientific Publications, Oxford.

Seghers, B.H. 1974. Schooling behaviour in the guppy *Poecilia reticulata*: an evolutionary response to predation. *Evolution* **28**, 486–9.

Selander, R.K. 1972. Sexual selection and dimorphism in birds. In B. Campbell (ed.), *Sexual Selection and the Descent of Man,* pp. 180–230. Aldine, Chicago.

Semler, D.E. 1971. Some aspects of adaptation in a polymorphism for breeding colours in the three-spine stickleback (*Gasterosteus aculeatus* L.). *J. Zool. Lond.* **165**, 291–302.

Seyfarth, R.M. & Cheney, D.L. 1984. Grooming, alliances and reciprocal altruism in vervet monkeys. *Nature* **308**, 541–3.

Seyfarth, R.M., Cheney, D.L. & Marler, P. 1980. Vervet monkey alarm calls: evidence of predator classification and semantic communication. *Science* **210**, 801–3.

Shalter, M.D. 1978. Localisation of passerine seet and mobbing calls by goshawks and pygmy owls. *Z. Tierpsychol.* **46**, 260–7.

Shepher, J. 1971. Mate selection among second generation kibbutz adolescents and adults: incest avoidance and negative imprinting. *Arch. Sex. Behav.* **1**, 293–307.

Sherman, P.W. 1977. Nepotism and the evolution of alarm calls. *Science* **197**, 1246–53.

Sherman, P.W. 1981a. Reproductive competition and infanticide in Belding's ground squirrels and other animals. In R.D. Alexander & D.W. Tinkle (eds), *Natural Selection and Social Behaviour: Recent Research and New Theory,* pp. 311–31. Chiron Press, New York.

Sherman, P.W. 1981b. Kinship, demography and Belding's ground squirrel nepotism. *Behav. Ecol. Sociobiol.* **8**, 251–9.

Sherman, P.W., Jarvis, J.U.M. & Alexander, R.D. (eds) 1991. *The Biology of the Naked Mole Rat.* Princeton University Press.

Sherman, P.W., Jarvis, J.U.M. & Braude, S.H. 1992. Naked mole rats. *Sci. Amer.* **267**(2), 72–78.

Shields, W.M. 1980. Ground squirrel alarm calls: nepotism or parental care? *Amer. Natur.* **116**, 599–603.

Shuster, S.M. & Wade, M.J. 1991. Equal mating success among male reproductive strategies in a marine isopod. *Nature* **350**, 608–10.

Sibley, C.G. & Monroe, B.L. 1990. *Distribution and Taxonomy of Birds of the World.* Yale University Press, New Haven Connecticut.

Sibly, R.M. 1983. Optimal group size is unstable. *Anim. Behav.* **31**, 947–8.

Sigurjonsdottir, H. & Parker, G.A. 1981. Dung fly struggles: evidence for assessment strategy. *Behav. Ecol. Sociobiol.* **8**, 219–30.

Sillén-Tullberg, B. 1985. Higher survival of an aposematic than of a cryptic form of a distasteful bug. *Oecologia* **67**, 411–15.

Sillén-Tullberg, B. 1988. Evolution of gregariousness in aposematic butterfly larvae: a phylogenetic analysis. *Evolution* **42**, 293–305.

Silverin, B. 1980. Effects of long-acting testosterone treatment on free-living pied flycatchers *Fidecula hypoleuca*. *Anim. Behav.* **28**, 906–12.

Silverman, H.B. & Dunbar, M.J. 1980. Aggressive tusk use by the narwhal. *Monodon monoceros* L. *Nature* **284**, 57–8.

Simon, C. 1979. Debut of the seventeen-year-old cicada. *Nat. Hist.* **88** 38–45.

Sinclair, A.R.E. 1977. *The African Buffalo*. University of Chicago Press, Chicago.

Slatkin, M. & Maynard Smith, J. 1979. Models of coevolution. *Q. Rev. Biol.* **54**, 233–63.

Smith, J.N.M., Yom-Tov, Y. & Moses, R. 1982. Polygyny, male parental care and sex ratios in song sparrows: an experimental study. *Auk* **99**, 555–64.

Smith, R.L. 1979. Repeated copulation and sperm precedence: paternity assurance for a male brooding water bug. *Science* **205**, 1029–31.

Smith, S.M. 1977. Coral snake pattern rejection and stimulus generalisation by naïve great kiskadees (Aves: Tyrannidae). *Nature* **265**, 535–6.

Soler, M. & Møller, A.P. 1990. Duration of sympatry and co-evolution between the great spotted cuckoo and its magpie host. *Nature* **343**, 748–50.

Southwood, T.R.E. 1981. Bionomic strategies and population parameters. In R.M. May (ed.), *Theoretical Ecology*, 2nd edn, pp. 30–52. Blackwell Scientific Publications, Oxford.

Stacey, P.B. & Koenig, W.D. 1984. Cooperative breeding in the acorn woodpecker. *Sci. Am.* **251**, 100–7.

Stacey, P.B. & Ligon, J.D. 1991. The benefits-of-philopatry hypothesis for the evolution of cooperative breeding: variation in territory quality and group size effects. *Amer. Natur.* **137**, 831–46.

Stander, P.E. 1992. Cooperative hunting in lions: the role of the individual. *Behav. Ecol. Sociobiol.* **29**, 445–54.

Stenmark, G., Slagsvold, T. & Lifjeld, J.T. 1988. Polygyny in the pied flycatcher *Ficedula hypoleuca*: a test of the deception hypothesis. *Anim. Behav.* **36**, 1646–57.

Stephens, D.W. & Krebs, J.R. 1986. *Foraging Theory*. Princeton University Press, Princeton, NJ.

Strassmann, J.E., Queller, D.C., Solis, C.R. & Hughes, C.R. 1991. Relatedness and queen number in the neotropical wasp. *Anim. Behav.* **42**, 461–70.

Shykoff, J.A. & Schmid Hempel, P. 1991. Parasites and the advantage of genetic variability within social insect colonies. *Proc. Roy. Soc. Lond.* B **243**, 59–62.

Stubblefield, J.W. & Charnov, E.L. 1986. Some conceptual issues in the origin of eusociality. *Heredity* **57**, 181–7.

Sutherland, W.J. & Parker, G.A. 1985. Distribution of unequal competitors. In R.M. Sibly & R.H. Smith (eds), *Behavioural Ecology: Ecological Consequences of Adaptive Behaviour*. Blackwell Scientific Publications, Oxford.

Syren, R.M. & Luykx, P. 1977. Permanent segmental interchange complex in the termite *Incisitermes schwarzi*. *Nature* **266**, 167–8.

Taborsky, M. 1984. Broodcare helpers in the cichlid fish *Lamprologus brichardi*: their costs and benefits. *Anim. Behav.* **32**, 1236–52.

Taborsky, M. & Limberger, D. 1981. Helpers in fish. *Behav. Ecol. Sociobiol.* **8**, 143–5.

Thornhill, R. 1976. Sexual selection and nuptial feeding behaviour in *Bittacus apicalis* (Insecta: Mecoptera). *Amer. Natur.* **110**, 529–48.

Thornhill, R. 1980. Rape in *Panorpa* scorpionflies and a general rape hypothesis. *Anim. Behav.* **28**, 52–9.

Thornhill, R. 1981. *Panorpa* (Mecoptera: Panorpidae) scorpionflies: systems for understanding resource-defence polygyny and alternative male reproductive efforts. *Ann. Rev. Ecol. Syst.* **12**, 355–86.

Thornhill, R. & Alcock, J. 1983. *The Evolution of Insect Mating Systems*. Harvard University Press, Cambridge, Massachusetts.

Tinbergen, L. 1960. The natural control of insects in pinewoods. I. Factors influencing the intensity of predation by song birds. *Archs. Neerl. Zool* **13**, 265–343.

Tinbergen, N. 1953. *The Herring Gull's World*. New Naturalist Series. Collins, London.

Tinbergen, N. 1957. The functions of territory. *Bird Study* **4**, 14–27.

Tinbergen, N. 1963. On aims and methods of ethology. *Z. Tierpsychol.* **20**, 410–33.

Tinbergen, N. 1974. *Curious Naturalists*. Penguin Education, Harmondsworth.

Tinbergen, N., Impekoven, M. & Franck, D. 1967. An experiment on spacing out as a defence against predators. *Behaviour* **28**, 307–21.

Trivers, R.L. 1971. The evolution of reciprocal altruism. *Q. Rev. Biol.* **46**, 35–57.

Trivers, R.L. 1972. Parental investment and sexual selection. In B. Campbell (ed.), *Sexual Selection and the Descent of Man*, pp. 139–79. Aldine, Chicago.

Trivers, R.L. 1974. Parent–offspring conflict. *Amer. Zool.* **14**, 249–64.

Trivers, R.L. 1985. *Social Evolution*. Benjamin Cummings, Menlo Park.

Trivers, R.L. & Hare, H. 1976. Haplodiploidy and the evolution of social insects. *Science* **191**, 249–63.

Trune, D.R. & Slobodchikoff, C.N. 1976. Social effects of roosting on the metabolism of the pallid bat. *Antrozous pallidus*. *J. Mammal.* **57**, 656–63.

Valen, L.M. van 1980. Patch selection benefactors and a revitalisation of ecology. *Evolutionary Theory* **4**, 231–3.

VanderWall, S.B. 1990. *Food Hoarding in Animals*. University of Chicago Press, Chicago.

Vehrencamp, S.L. 1977. Relative fecundity and parental effort in communally nesting anis. *Crotophaga sulcirostris*. *Science* **197**, 403–5.

Vehrencamp, S.L. 1978. The adaptive significance of communal nesting in groove-billed anis. *Crotophaga sulcirostris*. *Behav. Ecol. Sociobiol.* **4**, 1–33.

Vehrencamp, S.L. 1983. A model for the evolution of despotic versus egalitarian societies. *Anim. Behav.* **31**, 667–82.

Verner, J. 1977. On the adaptive significance of territoriality. *Amer. Natur.* **111**, 769–75.

Verner, J. & Willson, M.F. 1966. The influence of habitats on mating systems of North American passerine birds. *Ecology* **47**, 143–7.

Vogel, S., Ellington, C.P. & Kilgore, D.L. 1973. Wind-induced ventilation of the burrows of the prairie dog *Cynomys ludovicianus*. *J. Comp. Physiol.* **85**, 1–14.

Waage, J.K. 1979. Dual function of the damselfly penis: sperm removal and transfer. *Science* **203**, 916–18.

Waage, J.K. 1988. Confusion over residency and the escalation of damselfly territorial disputes. *Anim. Behav.* **36**, 586–95.

De Waal, F.B.M. 1986. Deception in the natural communication of Chimpanzees. In R.W. Mitchell & N.S. Thompson (eds), *Deception: Perspectives on human and non-human deceit*. SUNY Press, Albany.

Wade, M.J. 1978. A critical review of the models of group selection. *Q. Rev. Biol.* **53**, 101–4.

Ward, P.S. 1983. Genetic relatedness and colony organisation in a species complex of pomerine ants. II. Patterns of sex ratio investment. *Behav. Ecol. Sociobiol.* **12**, 301–7.

Ward, P. & Zahavi, A. 1973. The importance of certain assemblages of birds as 'information-centres' for food finding. *Ibis* **115**, 517–34.

Warner, R.R. 1975. The adaptive significance of sequential hermaphroditism in animals. *Amer. Natur.* **109**, 61–82.

Warner, R.R. 1978. The evolution of hermaphroditism and unisexuality in aquatic and terrestrial vertebrates. In E.S. Reese & F.J. Lighter (eds), *Contrasts in Behaviour*, pp. 77–101. Wiley, New York.

Warner, R.R. 1987. Female choice of sites versus mates in a coral reef fish *Thalassoma bifasciatum*. *Anim. Behav.* **35**, 1470–8.

Warner, R.R. 1990. Male versus female influences on mating site determination in a coral reef fish. *Anim. Behav.* **39**, 540–8.

Warner, R.R. & Hoffman, S.G. 1980. Local population size as a determinant of mating system and sexual competition in two tropical marine fishes (*Thalassoma* spp.). *Evolution* **34**, 508–18.

Warner, R.R., Robertson, D.R. & Leigh, E.G. 1975. Sex change and sexual selection. *Science* **190**, 633–8.

Waser, P.M. & Waser, M.S. 1977. Experimental studies of primate vocalisation: specializations for long distance propagation. *Z. Tierpsychol.* **43**, 239–63.

Watson, A. 1967. Territory and population regulation in the red grouse. *Nature* **215**, 1274–5.

Watson, A. 1970. Territorial and reproductive behaviour of red grouse. *J. Reprod. Fert. Suppl.* **11**, 3–14.

Weatherhead, P.J. & Robertson, R.J. 1979. Offspring quality and the polygyny threshold: the 'sexy son hypothesis'. *Amer. Natur.* **113**, 201–8.

Weihs, D. 1973. Hydrodynamics of fish schooling. *Nature* **241**, 290–1.

Wells, K.D. 1977. The social behaviour of anuran amphibians. *Anim. Behav.* **25**, 666–93.

Werner, E.E., Gilliam, J.F., Hall, D.J. & Mittelbach, G.E. 1983. An experimental test of the effects of predation risk on habitat use in fish. *Ecology* **64**, 1540–8.

Werren, J.H. 1980. Sex ratio adaptations to local mate competition in a parasitic wasp. *Science* **208**, 1157–9.

West Eberhard, M.J. 1975. The evolution of social behaviour by kin selection. *Q. Rev. Biol.* **50**, 1–33.

West Eberhard, M.J. 1978a. Temporary queens in *Metapolybia* wasps: non-reproductive helpers without altruism? *Science* **200**, 441–3.

West Eberhard, M.J. 1978b. Polygyny and the evolution of social behavior in wasps. *J. Kans. Ent. Soc.* **51**, 832–56.

Whitfield, D.P. 1986. Plumage variability and territoriality in breeding turnstone. *Arenaria interpres*: status signalling or individual recognition? *Anim. Behav.* **34**, 1471–82.

Whitham, T.G. 1978. Habitat selection by *Pemphigus* aphids in response to resource limitation and competition. *Ecology* **59**, 1164–76.

Whitham, T.G. 1979. Territorial behaviour of *Pemphigus* gall aphids. *Nature* **279**, 324–5.

Whitham, T.G. 1980. The theory of habitat selection examined and extended using *Pemphigus* aphids. *Amer. Natur.* **115**, 449–66.

Whitney, C.L. & Krebs, J.R. 1975a. Mate selection in Pacific tree frogs. *Nature* **255**, 325–6.

Whitney, C.L. & Krebs, J.R. 1975b. Spacing and

calling in Pacific tree frogs. *Hyla regilla. Can. J. Zool.* **53**, 1519–27.

Wiley, R.H. 1973. Territoriality and non-random mating in the sage grouse. *Centrocercus urophasianus. Anim. Behav. Monogr.* **6**, 87–169.

Wiley, R.H. 1983. The evolution of communication: information and manipulation. In T.R. Halliday & P.J.B. Slater (eds), *Animal Behaviour. Vol. 2. Communication*, pp. 156–89. Blackwell Scientific Publications. Oxford.

Wiley, R.H. & Richards, D.G. 1978. Physical constraints on acoustic communication in the atmosphere: implications for the evolution of animal vocalizations. *Behav. Ecol. Sociobiol.* **3**, 69–94.

Wilkinson, G.S. 1984. Reciprocal food sharing in the vampire bat. *Nature* **308**, 181–4.

Wilkinson, P.F. & Shank, C.C. 1977. Rutting-fight mortality among musk oxen on Banks Island, Northwest Territories, Canada. *Anim. Behav.* **24**, 756–8.

Williams, G.C. 1966. *Adaptation and Natural Selection.* Princeton University Press, Princeton, NJ.

Williams, G.C. 1975. *Sex and Evolution.* Princeton University Press, Princeton, NJ.

Williams, G.C. 1979. The question of adaptive sex ratio in outcrossed vertebrates. *Proc. R. Soc. Lond.* B **205**, 567–80.

Wilson, D.S. 1980. *The Natural Selection of Populations and Communities.* Benjamin/Cummings. Menlo Park, California.

Wilson, E.O. 1971. *The Insect Societies.* Belknap Press, Harvard.

Wilson, E.O. 1975. *Sociobiology: The New Synthesis.* Belknap Press, Harvard.

Wilson, E.O. 1980. Caste and division of labor in leaf-cutter ants (Hymenoptera, Formicidae: *Atta*). II. The ergonomic optimization of leaf cutting. *Behav. Ecol. Sociobiol.* **7**, 157–65.

Wolf, L., Ketterson, E.D. & Nolan, V. Jr. 1990. Behavioural response of female dark-eyed juncos to experimental removal of their mates: implications for the evolution of male parental care. *Anim. Behav.* **39**, 125–34.

Wolf, T.J. & Schmid-Hempel, P. 1989. Extra loads and foraging lifespan in honeybee workers. *J. Anim. Ecol.* **58**, 943–54.

Woolfenden, G.E. & Fitzpatrick, J.W. 1978. The inheritance of territory in group breeding birds. *BioScience* **28**, 104–8.

Woolfenden, G.E. & Fitzpatrick, J.W. 1984. *The Florida Scrub Jay.* Princeton University Press, Princeton, NJ.

Wynne-Edwards, V.C. 1962. *Animal Dispersion in Relation to Social Behaviour.* Oliver & Boyd, Edinburgh.

Wynne-Edwards, V.C. 1986. *Evolution Through Group Selection.* Blackwell Scientific Publications, Oxford.

Yamagishi, S. & Fujioka, M. 1986. Heavy brood parasitism by the common cuckoo *Cuculus canorus* on the azure-winged magpie *Cyanopica cyana. Tori* **34**, 91–6.

Yom-Tov, Y. 1980. Intraspecific nest parasitism in birds. *Biol. Rev.* **55**, 93–108.

Zach, R. 1979. Shell dropping: decision making and optimal foraging in Northwestern crows. *Behaviour* **68**, 106–17.

Zahavi, A. 1974. Communal nesting by the Arabian babbler: a case of individual selection. *Ibis* **116**, 84–7.

Zahavi, A. 1975. Mate selection – a selection for a handicap. *J. theor. Biol.* **53**, 205–14.

Zahavi, A. 1977. The cost of honesty (further remarks on the handicap principle). *J. theor. Biol.* **67**, 603–5.

Zahavi, A. 1979. Ritualisation and the evolution of movement signals. *Behaviour* **72**, 77–81.

Zahavi, A. 1987. The theory of signal selection and some of its implications. In V.P. Delfino (ed.), *International Symposium of Biological Evolution.* Adriatica Editrice, Bari, Italy.

Zuckerman, S. 1932. *The Social Life of Monkeys and Apes.* Paul, Trench, Trubner & Co., London.

Zuk, M. 1984. A charming resistance to parasites. *Natural History* **4**(84), 28–34.

Author Index

Page numbers in *italic* type refer to tables and/or figures

Abele, L.G. and Gilchrist, S. 185–6
Adams, E.S. and Caldwell, R.L. 368
Alatalo, R.V. *et al.* 224, 236
Alcock, J. *350*
 et al. 248
Alexander, R.D. 324, 343, 380
 and Sherman, P.W. 335
Alexander, R.McN. 380
Altmann, S.A. 220
Andersson, M. 190, *191*, 206, 348, 374
 and Wicklund, C.G. 126
Aoki, S. 319, 331
Arak, A. 192, 247, *248*
Argyle, M. 350
Arnold, S.J. 11, 186
Austad, S. 167, *169*, *170*
Axelrod, R. 281, 290
 and Hamilton, W.D. 280, 282, *284*

Baker, R.R. and Parker, G.A. 101
Bakker, T.C.M. *see* Milinski, M. and Bakker, T.C.M.
Baldwin, J.E. and Krebs, H.A. 380
Balmford, A. 243
Barnard, C.J. 146
 and Sibly, R. 144
 and Thompson, D.B.A. *145*
Bart, J. and Tornes, A. 225
Bartz, S. *346*
Basolo, A. *369*
Bateson, P.P.G. 14, 22
Bednekoff, P. 65–6, *67*
Beehler, B.M. and Foster, M.S. 222
Beissinger, S.R. and Snyder, N.F.R. 239
Bell, G. *19*
Belovsky, G. 71, 72–3
Bensch, S. and Hasselquist, D. 232
Benton, T.G. and Foster, W.A. 331
Benzer, S. 10
Bercovitch, F. 289
Berthold, P. *et al.* 12
Bertram, B.C.R. 5, 121, *123*

Birkhead, T.R. *127*, *184*
 et al. 226
 and Møller, A.P. *229*, 243
 see also Møller, A.P. and Birkhead, T.R.
Blurton-Jones, N. 359, 361
Boag, D.A. *see* Rippin, A.B. and Boag, D.A.
Borgia, G. 106
Bowman, R.I. 374
Bradbury, J.W. 221
 et al. 221
 and Gibson, R.M. 222
 and Vehrencamp, S.L. 243
Bray, O.E. *et al.* *134*
Brockmann, H.J.
 and Dawkins, R. 264
 et al. 225, *256*, 264
 see also Dawkins, R. and Brockmann, H.J.
Brodin, A. 66
Brooke, M. de L. 209
 see also Davies, N.B. and Brooke, M. de L.
Brown, C.R. *136–7*, 138
 and Brown, M. 135, *136–7*, *139*, 230
Brown, J.H. *see* Kodric-Brown, A. and Brown, J.H.
Brown, J.L. *105*, 110–11, 117, 266, 291, 298, 316
 et al. 302
Bull, J.J. 259
Bulmer, M. and Taylor P. 335
Burke, T. 225
 et al. *228*
Burley, N. 206
 et al. 206
Busnel, R.G. and Klasse, A. 356
Bygott, J.D., *et al.* 5, 218
Byrne, R. and Whiten, A. 374

Cade, W.H. 11, 257, *258*
 and Wyatt, D.R. 257
Caldwell, R.L. 173
 see also Adams, E.S. and Caldwell, R.L.
Calvert, W.H. *et al.* 124
Capranica, R.R. *et al.* 205
Caraco, T. 138, 140
 et al. 64, 65, 140, *141*, 142
 and Wolf, L.L. 278
Carayon, J. 186
Carlisle, T.R. *see* Dawkins, R. and Carlisle, T.R.

Carpenter, F.L. 32
 et al. 113, 114
 and MacMillen, F.E. 113
Cartar, R.V. and Dill, L.M. 64
Caryl, P.G. 173, 374
 see also Paton, D. and Caryl, P.G.
Catchpole, C.K. 190, 374
 et al. 190
Catterall, C.P. *see* Elgar, M. and Catterall, C.P.
Chappuis, C. 352
Charnov, E.L. 53, 182, 206, 264, 292, *298*, 324
 et al. 130
 and Krebs, J.R. *18*
 and Skinner, S. 56
 see also Gross, M.R. and Charnov, E.L.
Cheney, D.L.
 and Seyfarth, R.M. 372
 see also Seyfarth, R.M., and Cheney, D.L.
Clark, A. 181
Clayton, D.H. 197, 206
Clutton-Brock, T.H. 37, *179*, 217, 221, 243
 and Albon, S.D. 161
 et al. *179*, 182, 220, 223
 and Harvey, P.H. 29, 34, *38*, *39*
Collier, G. *see* Jenni, D.A. and Collier, G.
Conover, D.A. 259
Corbet, P.S. *184*
Coss, R.G. *see* Owings, D.H. and Coss, R.G.; Poran, N.S. and Coss, R.G.
Cott, H.B. 88
Coulson, J.C. 209
Cowie, R.J. *54*
Cox, C.R. and Le Boeuf, B.J. 220
Craig, J.L. *see* Jamieson, I.G. and Craig, J.L.
Craig, R. 332, 343
Creel, S.
 et al. 304, 305
 and Waser, P.M. 304, 305
Crook, J. 25–6
 and Gartlan, S. 33
Crozier, R.H. and Luykx, P. 346
Cullen, J.M. *362*, *363*, 374
 see also Neill, S.R. St J. and Cullen, J.M.

Daly, M.   210
Darwin, C.   8–9, 38
Davies, N.B.   *234*, 238
  and Brooke, M. de L.   95, 96, 97
  et al.   309
  and Halliday, T.R.   *163*
  and Houston, A.   116
Dawkins, M.   *82*
  see also Guilford, T. and Dawkins, M.
Dawkins, R.   10, 13, 22, 30, 264, 279, 290, 324, 347, 386
  and Brockmann, H.J.   381
  and Carlisle, T.R.   211
  and Krebs, J.R.   92, 101, 374
  see also Brockmann, H.J. and Dawkins, R.
de Groot, P.   128
de Voogd, T.J. *et al.*   37
De Vore, I.   33
de Waal,   374
Diamond, J. *et al.*   114
Dill, L.M. *see* Cartar, R.V. and Dill, L.M.
Dominey, W.J.   252
Dudai, Y.   11
  and Quinn, W.G.   10
Dunbar, R.I.M.   243, 264
Duncan, P. and Vigne, N.   125
Dunford, C.   271
Dussourd, D.E. *et al.*   188
Dybas, H.S.
  and Lloyd, M.   125
  see also Lloyd, M. and Dybas, H.S.

Eberhard, W.G.   161, 326
Elgar, M.   132
  and Catterall, C.P.   132
Elner, R.W. and Hughes, R.N.   *59*
Emlen, S.T.   316
  et al.   181, 239
  and Oring, L.W.   216, 221
  and Wrege, P.H.   310, 311
Endler, J.A.   90–1, 100
  see also Houde, A. and Endler, J.A.
Enquist, M.   369
  et al.   *166*, 369
  and Leimar, O.   164
Erichsen, J.T. *et al.*   84, *86*
Erickson, C.J. and Zenone, P.J.   202*
Evans, H.E.   322
Evans, M.R. and Hatchwell, B.J.   173
Evans, S.M. *see* Jennings, T. and Evans, S.M.
Ewald, P.W. and Rohwer, S.

232
Ezaki, Y.   232

Fanshaw, J.H. *see* Fitzgibbon, C.D. and Fanshaw, J.H.
Feare, C.   280
Ferguson, M.W.J. and Joanen, T.   259
Fischer, E.A.   285
Fisher, R.A.   89, 178, 191–3, 266
Fitzgibbon, C.D.   122
  and Fanshaw, J.H.   368
Fitzpatrick, J.W. *see* Woolfenden, G.E. and Fitzpatrick, J.W.
Folstad, I. and Karter, A.J.   385
Foster, S.A.   *133*
Foster, W.A.   331
  and Treherne, J.E.   *124*
  see also Benton, T.G. and Foster, W.A.
Fretwell, S.   102–3
Fricke, H.W.   262, 306
  and Fricke, S.   263
Fujioka, M. *see* Yamagishi, S. and Fujioka, M.

Gadagkar, R.   344
Galef, G. and Wigmore, S.   129
Geist, V.   157
Ghiselin, M.T.   260
Gibbs, L. *et al.*   229, *230*
Gibson, R.M.
  and Höglund, J.   206
  see also Bradbury, J.W., and Gibson, R.M.
Gilbert, L.   186
Gilchrist, S. *see* Abele, L.G. and Gilchrist, S.
Gill, F.B. and Wolf, L.L.   *111–12*
Gilliam, J.   69
Gittleman, J.L. and Harvey, P.H.   86–7, *88*
Goldthwaite, R.O. *et al.*   101
Goss-Custard, J.D.   132, *135*
Gould, S.J.   30, 291, 386
  and Lewontin, R.C.   29
Grafen, A.   195, 270, 290, 332, 342, 343, 380
Greenberg, I.   327
Greenlaw, J.S. and Post, W.   225
Greenwood, P.J.   239, *240*
  et al.   240
Grey, R.   76
Grosberg, R.K. and Quinn, J.F.   290
Gross, M.R.   *249, 250*, 251, 252, 264
  and Charnov, E.L.   252

and Sargent, R.C.   210, 213
  and Shine, R.   211
Guilford, T.   88
  and Dawkins, M.   368
Gustafsson, L. and Sutherland, W.J.   21
Gwynne, D.T.   *203*

Haas, V.   126
Hack, M.A. *see* Rubenstein, D.I. and Hack, M.A.
Haldane, J.B.S.   266
Halliday, T.R. *see* Davies, N.B. and Halliday, T.R.
Hamilton, W.D.   125, 180, *253*, 254, 266, 268, 275, 328, 347
  and Zuk, M.   195
  see also Axelrod, R. and Hamilton, W.D.
Hammerstein, P. and Parker, G.A.   166
Hanken, J. and Sherman, P.W.   277
Harcourt, A.H. *et al.*   40
Hare, H. *see* Trivers, R.L. and Hare, H.
Harper, D.G.C.   118
Harvey, P.H.   377
  et al.   38, 40, 89
  and Pagel, M.D.   35, 47
  and Purvis, A.   35, 36
  see also Clutton-Brock, T.H. and Harvey, P.H.; Gittleman, J.L. and Harvey, P.H.
Hassell, M.P.   106
Hasselquist, D. *see* Bensch, S. and Hasselquist, D.
Hasson, O. *see* Nur, N. and Hasson, O.
Hatchwell, B.J. *see* Evans, M.R. and Hatchwell, B.J.
Heller, R. *see* Milinski, M. and Heller, R.
Hinde, R.A.   110, *360*, 371, 374
Hixon, M.A.   114
  et al.   118
Hobbs, N.T.   73
Hodges, C.M. and Wolf, L.L.   55
Hoffman, S.G., *see* Warner, R.R. and Hoffman, S.G.
Höglund, J.   36
  see also Gibson, R.M., and Höglund, J.
Högstedt, G.   20–1
Hölldobler, B.   *94*, 351
Holmes, W.G. and Sherman, P.W.   275, *276*
Hoogland, J.L.   *134*, 146, 271, *272*
Houde, A.E.   90, 196

and Endler, J.A. 196
and Torio, A.J. 197
Houston, A.I.
et al. 76
and McNamara, J. 64
see also Davies, N.B. and
Houston, A.I.; McNamara,
J.M. and Houston, A.I.
Houtman, A.M. *231*
Howard, R.D. *187*, 247
Hrdy, S.B. 218
Hughes, R.N. 76
Hunter, M.L. and Krebs, J.R.
353
Hurly, T.A. *67*
Huxley, J.S. 150

Ims, R.A. 215, 216
Inman, A.J. 118
Iwasa, Y. *et al.* 197

Jakobsson, S. *et al.* *165*
Jamieson, I.G. 308
and Craig, J.L. 308–9
Jarman, P. 28–9
Jarvis, J.U.M. 314
Jarvis, M.J.F. 379
Jeffreys, A.J. *et al.* *228*
Jenni, D.A. and Collier, G. 239
Jennings, T. and Evans, S.M.
144
Joanen, T. *see* Ferguson, M.W.J.
and Joanen, T.

Kacelnik. A. *50–1*, 52
et al. 104, 105, 118
see also Krebs, J.R. and
Kacelnik, A.
Kamil, A.C. *see* Pietrewicz, A.T.
and Kamil, A.C.
Karter, A.J. *see* Folstad, I. and
Karter A.J.
Kenward, R.E. ` *122*
Kirkpatrick, M. and Ryan, M.J.
368
Klasse, A. *see* Busnel, R.G. and
Klasse, A.
Kleiman, D.G. 218
Kodric-Brown, A. 197
and Brown, J.H. 194–5
Koenig, W.D. 309
et al. 312, 317
see also Stacey, P.B. and,
Koenig, W.D.
Komdeur, J. 303
Kramer, D.L. *see* Nakatsuru, K.
and Kramer, D.L.
Krebs, H.A. *see* Baldwin, J.E. and
Krebs, H.A.

Krebs, J.R. 106, 160
and Dawkins, R. 374
et al. *62*, 129
and Kacelnik, A. 76
see also Charnov, E.L. and
Krebs, J.R.; Dawkins, R. and
Krebs, J.R.; Hunter, M.L.
and Krebs, J.R.; Stephens,
D.W. and Krebs, J.R.;
Whitney, C.L. and
Krebs, J.R.
Kruuk, H. 126, *127*

Lack, D. 16, *21*, 28, 47, 225
Lacy, R.C. 346
Lande, R. *193–4*
Lank, D.B. *et al.* 239
Le Boeuf, B.J. 220
and Reiter J. *179*
see also Cox, C.R. and
Le Boeuf, B.J.
Leimar O. *see* Enquist, M. and
Leimar, O.
Leonard, M.L. and Picman, J.
232
Lessells, C.M. 22
Leuthold, W. 219
Lewis, M.J. *see* Pruett-Jones,
S.G. and Lewis, M.J.
Lewontin, R.C. *see* Gould, S.J.
and Lewontin, R.C.
Lightbody, J.P. and
Weatherhead, P.J. 231
Lill, A. 222
Lima, S. 63, 66
et al. 66, 68
Limberger, D. *see* Taborsky, M.
and Limberger, D.
Lloyd, M.
and Dybas, H.S. 125
see also Dybas, H.S., and
Lloyd, M.
Lorenz, K. 150
Lotem, A. *et al.* 99
Luykx, P. *see* Crozier, R.H. and
Luykx, P.; Syren, R.M. and
Luykx, P.
Lyon, B.E., *et al.* 225

McClintock, M.K. 5
Macdonald, D.W. and
Moehlman, P.D. 291
McKinney, F. *et al.* 207
MacMillen, F.E. *see* Carpenter,
F.L. and MacMillen, F.E.
MacNally, R. and Young, D.
366
McNamara, J.M.
and Houston, A.I. 66
see also Houston, A.I. and

McNamara, J.M.
McPhail, J.D. 245
Magurran, A.
et al. 145
and Seghers, B.H. 145
Major, P.F. *131*
Manning, A. 11
Marden, J.H. and Waage, J.K.
159
Marler, P. 356–7
see also Marten, K. and
Marler, P.
Marten, K.
et al. 355
and Marler, P. 355
May, R.M. and Robinson, S.K.
96
Maynard Smith, J. 16, 76 147,
148, 150, *153*, *155*, 173, 182,
213, *214*, 266, 379, 386
and Ridpath, M.G. 273
see also Slatkin, M. and
Maynard Smith, J.
Metcalf, R.A. 180, 182, 336
and Whitt, G.S. 326
Michener, C.D. 347
Milinski, M. 68, 103, *104*
and Bakker, T.C.M. 197
and Heller, R. 68, *69*
and Parker, G.A. *108*, 118
Mock, D.W. 22
Modell, W. 30–1
Moehlman, P.D. *299–301*
Moksnes, A. *et al.* 96, 97
Møller, A.P. 197, *198–201*,
207
and Birkhead, T.R. 36
see also Birkhead, T.R. and
Møller, A.P.; Soler, M. and
Møller, A.P.
Monroe, B.L. *see* Sibley, C.G.
and Monroe, B.L.
Montgomerie, R.D. and
Weatherhead, P.J. 23
Moodie, G.E.E. 245
Morris, D. 363
Morton, G. 352, *353*
Myers, J.P. *et al.* *121*

Nakamura, H. 97
Nakatsuru, K. and Kramer, D.L.
178
Neill, S.R.St J. and Cullen, J.M.
126
Nelson, J.B. 378
Nisbet, I.C.T. 188
Nottebohm, F. 352–3
Nur, N. 20
and Hasson, O. 195

Orians, G.H. 232
Oring, L.W. 221

see also Emlen, S.T., and
  Oring, L.W.
Oster, G.F. and Wilson, E.O.
  348
Owen, D.F. 65–6
Owen, R.E. and Plowright, R.C.
  335
Owen-Smith, N. 219
Owings, D.H. and Coss, R.G.
  101

Packer, C. 241, 289
  et al. 279
  and Pusey, A.E. 7, 8, 157
Page, R.E.Jnr. *et al.* 331
Pagel, M.D. *see* Harvey, P.H. and
  Pagel, M.D.
Parker, G.A. *55, 56,* 150, *151,*
  160, 173, 202
  et al. 176, 204
  and Knowlton, N. 118, 176
  and Rubenstein, D.I. 166
  and Sutherland, W.J. 107
  see also Baker, R.R. and
    Parker, G.A.; Hammerstein,
    P. and Parker, G.A.;
    Milinski, M. and Parker,
    G.A.; Sigurjonsdottir, H.
    and Parker, G.A.
Parker, W.S. *see* Pianka, E.R. and
  Parker, W.S.
Partridge, B.L. and Pitcher, T.J.
  133
Partridge, L. 22, 188
Paton, D. 371
  and Caryl, P.G. 371
Patterson, I. 146
Perrill, S.A. *et al.* 259
Perrins, C.M. 16, *17, 18*
Petrie, M. 202
  et al. 189
Pettifor, R.A. *et al.* 20, 21
Pianka, E.R. and Parker, W.S.
  19
Picman, J. *see* Leonard, M.L. and
  Picman, J.
Pietrewicz, A.T. and Kamil, A.C.
  79, 80, *81,* 82
Pitcher, T.J.
  et al. 129
  see also Partridge, B.L. and
    Pitcher, T.J.
Plowright, R.C.
  et al. 45
  see also Owen, R.E. and
    Plowright, R.C.
Pomiankowski, A., *et al.* *194*
Poran, N.S. and Coss, R.G. 101
Power, M. 105
Prins, H.H.Th. *et al.* 131
Pruett-Jones, S.G. and Lewis,

  M.J. 303
Pulliam, H.R. 138, *141*
  and Caraco, T. 146
  et al. 122
Purvis, A. *see* Harvey, P.H., and
  Purvis, A.
Pusey, A.E. *see* Packer, C. and
  Pusey, A.E.

Queller, D.C. 344
Quinn, J.F. *see* Grosberg, R.K.
  and Quinn, J.F.

Rabenold, K.N. 317
Rabenold, P.P. *et al.* 317
Ralls, K. *et al.* 240
Ratnieks, F.L.W. *339*
  and Visscher, P.W. 339
Redston, S. *see* Roper, T.J. and
  Redston, S.
Reed, T.R. 117–18
Reeve, H.K. 314
Reiter, J. *see* Le Boeuf, B.J. and
  Reiter, J.
Reyer, H.-U. 306, *307,* 308
  and Westerterp, K. 307
Reynolds, J.D. 239
Richards, D.G. *see* Wiley, R.H.
  and Richards, D.G.
Ridley, M. 47
Ridpath, M.G. *see* Maynard
  Smith, J. and Ridpath, M.G.
Rippin, A.B. and Boag, D.A.
  222
Robertson, R.J. *see*
  Weatherhead, P.J. and
  Robertson, R.J.
Robinson, S.K. *see* May, R.M.
  and Robinson, S.K.
Rohwer, S.
  and Rohwer, F.C. 171
  and Spaw, C.D. 98
  see also Ewald, P.W. and
    Rohwer, S.
Rood, J. 304
Roper, T.J. and Redston, S. 87
Rosenzweig, M.L. 110
Røskaft, E. *et al.* 172
Rothstein, S.I. 96, *97,* 98, *99,*
  100, 101, 118
Royama, T. *82*
Rubenstein, D.I.
  and Hack, M.A. 173
  see also Parker, G.A., and
    Rubenstein, D.I.
Rutberg, A.T. 218
Ryan, M.J. *223*
  et al. 191, 192, *223,* 368
  see also Kirkpatrick, M. and
    Ryan, M.J.

Sargent, T.D. 79
Schlenoff, D. *83*
Schmid-Hempel, P. *57,* 58
  et al. 54, 57
  and Wolf, T.J. 58
  see also Wolf, T.J. and
    Schmid-Hempel, P.
Schoener, T.W. 114, *115*
Searcy, W.A. and Yasukawa, K.
  231
Seger, J. 338, 340, 347
Seghers, B.H. *121*
  see also Magurran, A. and
    Seghers, B.H.
Selander, R.K. 37
Semler, D.E. 245
Seyfarth, R.M.
  and Cheney, D.L. 287
  see also Cheney, D.L. and
    Seyfarth, R.M.
Shepher, J. 240
Sherman, P.W. *134,* 270, 290
  et al. 314
  see also Alexander, R.D. and
    Sherman, P.W.; Hanken, J.
    and Sherman, P.W.;
    Holmes, W.G. and Sherman,
    P.W.
Shields, W.M. 271
Shine, R. *see* Gross, M.R. and
  Shine, R.
Shuster, S.M. and Wade, M.J.
  264
Sibley, C.G. and Monroe, B.L.
  37
Sibly, R. 143
  see also Barnard, C.J. and
    Sibly, R.
Sigurjonsdottir, H. and Parker,
  G.A. 171
Sillén-Tullberg, B. 90
Silverin, B. 172
Silverman, H.B. and Dunbar,
  M.J. 157
Simon, C. 125
Sinclair, A.R.E. 161
Slatkin, M. and Maynard Smith,
  J. 92, 101
Slobodchikoff, C.N. *see* Trune,
  D.R. and Slobodchikoff,
  C.N.
Smith, J.N.M. *et al.* 225
Smith, R.L. 206
Smith, S.M. 87
Snyder, N.F.R. *see* Beissinger,
  S.R. and Snyder, N.F.R.
Soler, M. and Møller, A.P. 96
Southwood, T.R.E. 22
Spaw, C.D. *see* Rohwer, S., and
  Spaw, C.D.
Stacey, P.B.
  and Koenig, W.D. 314

and Ligon, J.D.   303
Stander, P.E.   278
Stephens, D.W. and Krebs, J.R.
   66, 76
Strassmann, J.E. *et al.*   327
Sutherland, W.J. *see*
   Gustafsson, L. and
   Sutherland, W.J.
Syren, R.M. and Luykx, P.   *346*

Taborsky, M.   299
and Limberger, D.   299
Taylor, P. *see* Bulmer, M. and
   Taylor, P.
Thompson, D.B.A. *see* Barnard,
   C.J. and Thompson, D.B.A.
Thornhill, R.   *189*, 203, 264
and Alcock, J.   221, 243
Tinbergen, N.   4, 24, 42, 47, *82,
   84*, 110
et al.   24
Torio, A.J. *see* Houde, A. and
   Torio, A.J.
Tornes, A. *see* Bart, J. and
   Tornes, A.
Treherne, J.E. *see* Foster, W.A.
   and Treherne, J.E.
Trivers, R.L.   22, 176−7, 209,
   211, 280
and Hare, H.   334−5
Trune, D.R. and Slobodchikoff,
   C.N.   *133*

van Valen, L.   376
Vehrencamp, S.L.   144, *213*,
   311, 312
Verner, J.   118
and Willson, M.   232
Vigne, N. *see* Duncan, P. and
   Vigne, N.
Visscher, P.W. *see* Ratnieks,
   F.L.W. and Visscher, P.W.

Vogel, S. *et al.*   *383*, 384

Waage, J.K.   159, 185
see also Marden, J.H. and
   Waage, J.K.
Wade, M.J.   377
see also Schuster, S.M. and
   Wade, M.J.
Ward, P.   337
and Zahavi, A.   128
Warner, R.R.   216, 260, 263
et al.   260
and Hoffman, S.G.   261, 262
Waser, P.M.
and Waser, M.S.   356, *357*
see also Creel, S. and
   Waser, P.M.
Watson, A.   106, 172
Weatherhead, P.J. and
   Robertson, R.J.   236,
   see also Lightbody, J.P. and
   Weatherhead, P.J.,
   Montgomerie, R.D. and
   Weatherhead, P.J.
Weihs, D.   *133*
Wells, K.D.   217, 221
Werner, E.E.   69
Werren, J.   180
West Eberhard, M.J.   325, 326
Westerterp, K. *see* Reyer, H.-U.
   and Westerterp, K.
Whiten, A. *see* Byrne, R. and
   Whiten, A.
Whitfield, D.P.   173
Whitham, T.G.   107, *109, 110*
Whitney, C.L. and Krebs, J.R.
   205
Whitt, G.S. *see* Metcalf, R.A.
   and Whitt, G.S.
Wicklund, C.G. *see* Andersson,
   M. and Wicklund, C.G.
Wigmore, S. *see* Galef, G. and
   Wigmore, S.

Wiley, R.H.   374, *221*
and Richards, D.G.   355, 374
Wilkinson, G.S.   285, *286*
Williams, G.C.   16, 22, 182
Willson, M.F. *see* Verner, J. and
   Willson, M.F.
Wilson, D.S.   347, 376
Wilson, E.O.   265, *319*, 320,
   331, 347−8
see also Oster, G.F. and
   Wilson E.O.
Wolf, L., *et al.*   225
Wolf, L.L. *see* Caraco, T. and
   Wolf, L.L.; Gill, F.B. and
   Wolf, L.L.; Hodges, C.M.
   and Wolf, L.L.
Wolf, T.J.
and Schmid-Hempel, P.   58
see also Schmid-Hempel, P.
   and Wolf, T.J.
Woolfenden, G.E. and
   Fitzpatrick, J.W.   292, 293,
   294, 295, 297, 298, 302, 317
Wrege, P.H. *see* Emlen, S.T. and
   Wrege, P.H.
Wyatt, D.R. *see* Cade, W.H. and
   Wyatt, D.R.
Wynne-Edwards, V.C.   15

Yamagishi, S. and Fujioka, M.
   97
Yom-Tov, Y.   279
Young, D. *see* MacNally, R. and
   Young, D.

Zach, R.   44−5, *46*
Zahavi, A.   194, 195, 367
see also Ward, P. and
   Zahavi, A.
Zuckerman, S.   31
Zuk, M. *see* Hamilton, W.D. and
   Zuk, M.

Subject Index

Page numbers in *italic* refer to tables and/or figures

abortion, in lions 7
adaptation 8, 378−9, 380
  experimental studies 42−4
  predator-prey 77, 79−91
  testing hypotheses about
    25−9, 79−86
aggression 139−42, 335
  see also display; fighting;
    resource defence
alarm calling *see* calling,
    warning
alkaloids, protective, transferred
    during mating 188
alleles 9, 275, 319, 332, 375
alligators 259
altruism 2, 241, 265−90,
    291−3
  and co-operation, in social
    insects 318−48
  and trait groups 376−7
  see also helpers; helping
ambiguity, reduction of 363,
    364
ambivalent behaviour *360*
anemone fish 262−3
  helpers in 305−6
anis, groove-billed 345
  conflict in breeding groups
    311−12
anisogamy 175−6, 202
antelope 122, 219, 221, 222,
    368
  communication 349
antlers 30−1, 157, 181−2
ants 337, 366
  African driver (*Dorylus*) 320
  communication in 351−2
  fire (*Solenopsis*) 352
  leaf cutter 347−8, 352
  Myrmica rubra, life history of
    320−2
  worker−queen conflict
    334−8
aphids 319
  asexual reproduction 331
  gall 331
    competition 107, 109, *110*
aposematism 86−90
Arctic skua, threat signals 371
arms races, predator-prey
    77−101, 379
artifical selection 11

assessment 205−6
  behavioural 259
  and courtship 187−8,
    190−3, 194, 197
  and display 160−4
  sequential 164
association, and parental care
    212
attrition, war of 147−50
autonomic reponses 359, *360*

babbler, grey-crowned, helpers
    302
baboons
  anubis 33
  gelada 220
  hamadryas 220
  harems 220
  olive *32*, 241
    multi-male groups 220
    reciprocity *288*, 289
barnacles, resource defence
    110
bass, largemouth 70
bats 243
  hammer-headed 221
  pallid *133*
  vampire, blood sharing
    285−7
bee-eater, white-fronted 311,
    316
  helpers 310
bees 75, 327
  bumblebees 64
  Centris pallida, mating
    strategy 248, 257, *260*
  halictine, partially bivoltine
    cycle 340−2
  honeybees 331−2
    communication 320, 369
    foraging 53−4, 57−9, 73
    manipulation of 371
    social 347
    worker, stinging 269
  see also Hymenoptera
beetles 161
  rove, mimicry by *94*
birds 110, 350
  altruism in 241, 265−6,
    272−4
  biparental care 209−10
  body reserves and
    environmental
    variablility 65−6, *67*
  brain mass 30

calling *231*, 349
  alarm 356−7
  and habitat 352−6
clutch size 15, 16−21
conflict in breeding groups
    310−14
courtship feeding 188
detecting prey 80−6
display 190−93, 194−6
female dispersal 240
helping in 293−8
hunger and predation risk
    66, 68−9
living in flocks 120−2, 123,
    124, 125−6, 128−9,
    130−1, 132−8
natal dispersal 239−40
parental care 212
  male 188, 225−39
territoriality 105−6,
    111−18, 237−8
time budgets 138−43
  see also individual birds;
    plumage; tail length
birds of paradise 190, 221
blackbirds
  calling 349
  red-winged 229
    cuckoldry *134*, 230
    and polygyny 232
  yellow-headed 231
blackcaps (*Sylvia atricapilla*),
    migration studies 12−13
body reserves, and
    environmental
    variability 65−6, *67*
body size, and behaviour 246,
    247−8
body weight
  and foraging 58−9
  and home range size 35
  sexual dimorphism in 37−8
  and social organization
    28−9, 30
breeding *see* reproduction
brood parasites 93−100, 230,
    279−80
brood size 15−18, 20−1
Bruce Effect 7
budgerigars 349
buffalo 161
  multi-male groups 220
bullfrogs *260*
  breeding 188, 217
  female, mate choice 187
  resource defence 110

satellites 263
silent 244
bushbaby 32
butterflies
 anti-aphrodisiac smells 186
 monarch, communal roosting
  123−4
 speckled wood, strategies
  245, 260

calling
 aggressive, frogs and toads
  161, 163, 222, 368−9
 attracting enemies 270,
  357, 366
 birds 231, 39, 352−7
 and habitat 352−8
 mating
  bullfrogs 26−7
  crickets 257−8
  frogs 20−5, 221, 223
  natterjack toads 27
  sedge warbler 190−1
 recruitment 132
 territorial 116−17
 warning 270−2, 356−8,
  372−3
Camponotus truncatus 319
cannibalism 13, 20
castes 319, 321−2, 38
 morphological modifications
  320
 sterile 318, 319
  evolution of 322−8
  and haplodiploidy 328−31
catbird, grey 98−9
caterpillars
 cinnabar moth 87
 monarch butterfly 89
catfish, armoured 105
causal explanations 5, 6−7,
  31−2, 382−4
 lion behaviour 7−8, 8
Centris pallida 28, 257, 260
chaffinches, territoriality
  117−18
cheating, prevention of 172
chickadee, black-capped 66
chicks
 and distasteful prey 86−7,
  88
 seeing cryptic prey 82−3
chimpanzees 40, 218
chingolo see sparrow,
  rufous-collared
chirruping, in sparrows 132
choruses 221−2, 223
 see also calling
cicadas, life cycles 125
cichlid fish 299
 fighting in 164, 165, 166

Clark's nutcracker (Nucifraga
  columbiana) 66
clownfish see anemone fish
clustering 283
clutch size 15, 16−18, 20−1
co-operation see altruism;
  helping coefficient
  of relatedness 266−9,
  328, 329−30, 331
coloniality 26
 costs and benefits 135−8
 and predation 123−6
 see also groups
coloration
 bird see plumage
 communication through 351
 cryptic 79−86, 90−1
 female 239
 of guppies 90−1, 196−7
 and ritualization 361
 and strategy 245−6
 to startle 79, 83
 warning see aposematism
 see also display
communication 349−50,
  371−4
 co-operative 366
 and ecological constraints
  350−8
 honeybee dance language
  320, 369
 reactor response 358−67
 sensory channels 350−1
 signal variability 369−71
comparative approach,
  hypothesis testing 24,
  25−42, 45−6
competition 2, 9, 219, 384
 by exploitation and resource
  defence 102−18
 and display 205
 male dungflies 147, 149, 150,
  151, 152, 204
 and sex change 260−2
 within a group 130−2
 see also display; sperm
  competition; territories
competitor distribution
  102−10
confounding variables 30,
  38−9
contest behaviour 147−73
 see also display; territories
copulation
 bowl and doily spiders
  167−8, 169
 in dungflies 53, 55−6, 74,
  380
 forced 202, 244
 homosexual 186
 in lions 6−7, 157−9
 primate 40

copulatory plugs 185−6
costs and benefits 42−3, 385
 of alternative strategies
  256−9
 and behaviour 48−74
 of coloniality 135−8
 of communication 365−6
 determining group size
  138−43
 of territory sharing 116
 see also optimality models
cotingas 221
courtship 175, 187−8, 190−3,
  194, 197, 244, 358, 359
 significance of 205−6
covariance, genetic 193−4,
  196−7
cowbird, brown-headed 96−8
crabs
 fiddler 364
 shore, prey choice 59−60
crickets
 field 11
 male strategies 257−8
 katydid 203
croaking see calling
crows
 mobbing of 126
 smashing egg-shells 44−5
crypsis 78, 79−81, 100, 250
cuckoldry 134, 197, 211, 226,
  227, 229−30, 230, 252
cuckoos 93−6, 366
Curious Naturalists (Tinbergen)
  47

damsel fish 133
damselflies, contests 159
deception, in polygyny 236
deer
 fallow 221
  mate choice 223−4
 mule 157
 red 208
  competition 160−1, 162
  harem defence 219−20
  sex ratio 181−2
 roe, communication 350
demography, importance of
  344
desertion 209, 211
 female 238−9
 male 225−6
despotism 105−6
diet, and home range size 35
differences, non-adaptive 30−1
dilution effect 123−6
diminishing returns, curve of
  49, 50
dimorphism
 male 249−54

sexual, in primates 34, 37−40
disease *134*
selection for resistance to 194
dispersal, natal 239, *240*
displacement activities 359, *360*
display 150, 156, 221
aggressive 161−2, *162*
courtship 205, 358−9
ritualized 160−1
and sexual selection 190−97
synergistic 222
threatening 358
see also calling
DNA fingerprinting 226, *228, 278, 279,* 317
dogs, prairie *see* prairie dogs
dominance
female 182, 311−12
in flocks 139−40, 142−3
and plumage variability 171−3
doves, barbary, courtship behaviour 197, 201
dragonflies 245
sperm displacement mechanisms 185
Drosophila 204
mate choice 188
mutation and behaviour 10−11
segregation distortion 376
Drosophila melanogaster, selective breeding 11
ducks
courtship ritual 358, 359, *360,* 361
forced copulation 244
dungflies 75, 167, 204, 257, 384
competition for females 106, 147, 149, 150, *151, 152*
copulation 53, *55*−6, 74, 380
sperm competition 186
dunnocks
DNA fingerprinting *228*
helping in 309−10
variable mating systems 237−8

Ecclesiastes, quoted 3
ecology and dispersal 239−41
economic defendability 110−13, 118
ectoparasitism *134*
in cliff swallows 136−8
eggs
clutch size 15, 16−18
dumping of 230, 279−80
evicted from nests 311−12

mimicry 93−6
rejection of 95, 96−100
trading of 285
eggshell removal, from nest 42−3
elephant seals, harems 220
elephants 219
enzyme polymorphism 327, 336
enzymes, and behaviour 13−14
ESSs 149−150, 153−4, 155, 166, 249, 250, 257, 259, 263
mixed 255
and optimality models 378−82
and parental investment 213−14
sex ratio 333
tit for tat 283−4
eusocial insects *see* social insects
evolution
and gene frequency 9−10
see also evolutionary history; natural selection
evolutionarily stable strategies *see* ESSs
evolutionary history 4
of signals 358−67
exploitation 102−5
eyespots, and startle effect 83, *84*

feeding
associated costs 131−2
costs and benefits 48−54, 57−9
to foragers 59−65
and social organization 26, 27−9, 33, 35
see also food; herbivores; predators
feeding interference 132, 134
females
change to males 260−2
choosiness of 186−97, 206, 208, 221
conflict between 311−14
dispersal 215−16, 240
kin recognition 277
local resource competition 181
in mammalian mating systems 217−20
parental care 209−11
polygyny, costs of 231−4, 238
reproductive synchrony 5, 219−20

resource value 168
a scarce resource 176−7
see also mating systems; reproduction
fertilization
coho salmon 249−52
and parental care *211,* 211−12
sperm displacement mechanisms 184−5
see also copulation; reproduction; sneaking
fieldfares, group mobbing 126
fighting 102, 150−73
see also resource defence
figwasps
fights 157
male dimorphism 252−4, *260*
finches
Darwin's 374
zebra
mate choice 206
mating habits 22, 226, 229, *231*
fireflies 349
fish *133,* 259, 350
coloration 90−1
damsel fish *133*
display in sticklebacks 359
feeding strategies 68, 103−5
helpers in 299
muscle arrangement 380
nests 250, 252
parental care 210−11, 212, 213
reciprocity 285
shoals 129, 130
and sneaking 250−1, 252
tail length 369
see also cichlid fish; guppies
Fisher's hypothesis 190, 192−4, 196−7, 206, 236
fitness
assured 344
direct and indirect 266, 279, 293, 298, 304−5, 311
fitness curves 107, 109
flamingos 77, 120
flies
fruit *see Drosophila*
hanging *see* scorpionflies
mimicry 78
parasitic 257−8
see also dungflies
flocks
dominance within 139−40, 142−3
and ecology, weaver birds 27, 29−30
and food gathering 27, 126−32

mixed species *145*
optimal size 140, *141*
and predation prevention
  139, 140, *141*, 142
flycatchers 79
  great Kiskadee 87
  pied 243
    sexual conflict 235−7
food 34, 320
  and courtship 188
  and female dispersion
    215−16
  and flock size 132, 142
  foraging 48−74
  and home range size 35
  preferences 11
  regurgitation 285−7, 299,
    321, 351
  renewable, harvesting of
    130−1
  and social organization 26,
    27−9, 33, 35
  and tooth size 39−40
  whelk gathering by crows
    44−5
food storage, in the environment
  66
foraging, economics of 48−74
fowl
  domestic *362*
  jungle 361
frogs 243, *260*
  bullfrogs 263
    breeding 187, 188, 217
    resource defence 110
    silent 244
  competition 161
  cricket 205
  green tree 259
  leopard, song 191
  mating calls 205, 221, *223*
  Pacific tree 205
  tree *364*
  tungara, calling of 368−9
fruit fly *see Drosophila*
fulmars, variable threat signals
  369−70
functional explanations 4−5,
  382−4
  lion behaviour 7−8, *8*
future discounting 65

game theory 147, 147−73, 244
gametes 175−6, 328, 376
  order of release 211−12
gannets *360*
  adaptation of 378−9
geese
  Brent, foraging return times
    130−1
  imprinting in 275

genes 380
  and altruism 265, 266
  and behaviour 10−14
  and evolution 9−10
  and family smell 328
  and helping *298*, 324−5
  and mate choice 188, 194−5,
    196−7
  and nest sharing 326−8
  and phenotypes 290
  and salmon male strategies
    250
  selfish 291, 375−6
  see also alleles; coefficient of
    relatedness
genome, linked *346*
gibbons *32*
goldfinches *53*
gorillas 40−1
goshawks 120, *122*
  and prey alarm calls 358
grasshoppers 349
great bustards 221
green beard effect 275
grooming 287, 289
ground squirrels (*Spermophilus
  beldingi*) 134
  altruism 241
  co-operation and alarm calls
    269−71, 290
  sibling recognition 275−7
group extinction 92
group selection 14−15, 91,
  150, 376−8
  evidence against 15−18,
    20−1
groups
  avoiding predators 120−6,
    127, 144−6
  and food gathering
    126−32, *133*
  individuals in 144
  and resource defence
    116−17
  size of 34
    influences on 120
    optimal size 133−44
  see also social organization
grouse 221
  black 224−5
    male preference of 224
  red
    resource defence 106
    territorial defence 172
  sage, leks *221*
  sharp-tailed 222
guillemot colonies *127*
Guinness Book of Records *179*
gulls
  black-headed 126, *360*
    eggshell removal 42−3
    in mixed flocks *145*

herring *360*
guppies
  coloration of 90−1
  covariance in 196−7
  group living 144−6
  group size *121*

habitat
  and calling 352−8
  saturation of 295−6
  and strategy 245−6
Hamilton's rule 268−9, *274*,
  332, 338
hamlet fish, black, reciprocity
  285
handicap hypothesis 190,
  194−5, 197, 206
haplodiploidy
  and altruism 328−38
  and origin of eusociality
    338−43
harems 29, 161, 264
  permanent 218, 220
  seasonal 219−20
hawk-dove model 152−4, 249,
  255
hawk-dove-bourgeois model
  154−5, 159, 160, 166
hawks
  provoking alarm calling 357
  red-tailed 358
heat (oestrus), in lionesses 5,
  6, 7
helpers
  non-relative 304−8
  primary and secondary
    306−8
  specialized 314−15
helping 291−317, 345
  and haplodiploidy 340−3
herbivores, constraints on diet of
  70−3
herding, of horses 124−5
heredity 9−10
hermaphroditism 285
  protandrous 262−3
  protogynous 260−2
  sequential 263, 264
home range
  female 215−16, 218−20
  size of 35, 38
Homoptera *see* aphids
honesty, in communication
  367, 368
honeycreepers 111
hooknose salmon *see* salmon,
  coho
horns 30−1, 157
hosts, brood 93−100
hotshot model 222−4
hummingbirds 111, 221

rufous 113−14
hunger, vs. danger 66, 68−9
hyenas 127
Hymenoptera 319, 375
    female workers 321
    partially bivoltine 340−3
    sex ratios 182, 332−8
    see also ants; bees;
        haplodiploidy; wasps
hypothesis testing 24−47

ideal free distribution 106−10,
    234
imprinting 275
inbreeding 180
    avoidance of 240−1, 296
inbreeding/outbreeding cycles
    346
incest, in baboons 241
infanticide 239
    lion 7, 14, 204
information centres 128, 136
insects 221
intention movements 359, 360
interference 106, 221
isogamy 175

jacanas 239
jackals 366
    black-backed 299, 300−1
jacks (Caranx ignobilis), hunting
    anchovies 130, 131
jacks (salmon) see salmon, coho
jays
    blue 79, 83
        detecting prey 80−1, 82
    Florida scrub, helping among
        293−8, 345
    food storage 66
juncos
    dark-eyed 225
    yellow-eyed 64, 75
        time budgets 139−43

kakapo 221
katydid crickets, eating
    spermatophores 203
kin recognition 274−7
kin selection, and altruism
    265−78, 283, 289
kingfisher, pied, helpers
    306−8, 316
kites, Florida snail 238−9
kittiwakes
    mate fidelity 209−10
    reproductive rate 179
klipspringer 218
knot, feeding 134, 135
kob, Uganda 219, 221, 222

Krebs cycle 380

lactation 210, 263, 304
langurs 32, 218
learning 60−1, 63
learning deficiencies, mutant
    flies 10−11
lechwe 219, 222
leech predation 187
leks 206, 219, 220−5, 228, 242
Leptothorax ants 351
life-dinner principle, rabbits 92
lions 77
    co-operation 265
    infanticide 7, 14, 204
    mutualism 278−9
    reproductive behaviour 5−8,
        157−9
    suckling other cubs 5, 241
litter size, obligate monogamy
    218
lizards 364
    cryptic 78
load carrying, economics of
    48−59

magpies 229
    azure-winged 97
    clutch size 20−1
males
    benefits of helping 296−8
    change to females 262−3
    dispersal 240−1
    harem defence 219−20
    incubating clutches 239
    investment by 197, 201−2
    and lekking 220−5
    mate competition 183−6
    parental care 209−10,
        211−12, 225−39, 242
    polyandry, cost of 238
    potential reproductive rate
        179
    reproductive success limited
        176−7
    satellite 222, 246−7, 257,
        258, 263
    see also leks; reproduction
mammals 350
    altruism 241, 269−72,
        278−9, 287−9
    mating systems 217−20
    natal dispersal 239−41
    parental care 210
man
    communication 350, 356
    reproduction 179
manakins 221, 222
mandarin ducks see ducks
manipulation 279−80, 289,
        293
    in communication 365−6
    parental 324−5, 343−5
    of sex ratio 334
marginal value theorem 49, 50,
    51−3
    and reproductive decisions
        54−6
marmosets 40−1, 218
mate choice 206
mate competition, local 180−1
mate defence 240−1
mate guarding 183−4, 184,
    229, 384
mating see copulation;
    fertilization; reproduction
mating decisions 202
mating systems 178
    African ungulates 29
    and parental care 208−43
    variations in 216−17
    weaver birds 25−6
matings, multiple 187, 204−5,
    225, 277
menstrual cycle synchrony 5
Metapolybia aztecoides 326,
    345
midges, gall, life sacrifice 291
mimicry 78
    eyespots 83, 84
    of host chemical 93, 94
    host eggs 93−6
    male−female 264
mites, viviparous 180
mobbing 126
mole rats, naked, specialized
    helpers 314−15
mongoose, dwarf 304−5
monkeys
    colobus 37, 218
    mangabey, calling 356, 357
    rhesus, grinning 358, 359
    vervet
        communcation system
            372−3
        grooming 287, 289
monogamy 25, 38, 208, 210,
    237, 243
    in birds 225−6
    mammalian 218
    obligate 218, 225
    and sexual selection 178
moorhens, male investment
    201−2
moose 75, 218
    feeding strategy 70−3
moths
    cinnabar, distasteful
        caterpillars 87
    courtship gifts 188
    underwing, as prey 79−81
musk ox 157

mussels, chosen by crabs
    59–60
mutants, genetic, use of   10–11
mutation   89
mutualism   278–9, 289
Mymecocystus ants   *319*
Myrmica rubra, life cycle
    320–2

narwhals   157
Nasutitermes exitiosus   *319*
natal dispersal   239, *240*
native hen, Tasmanian, wife
    sharing   272–4, *275*
natural selection   8–10, 21, 150
    and colour pattern change
        90–1
    Hamilton's rule   268–9, *274*
    and helping   291–317
    see also genes
nest sharing   325–8, 345
nests
    adaptation of   11
    communal   311–12
    digger wasp strategies   254–6
    fish   250, 252
    parasitism in   136–8, *139*
    parasitoid wasps   323–4
    of weaver birds   26, *27*
newts, fertilization   186
nose length, sexual selection for
    193–4

odour *see* smell
oestrus
    in langurs   218
    in lionesses   5, 6, 7, 157–9
    in red deer   219–20
offspring
    and altruism   265–8, 289,
        291
    helping   324–5
    vs. sibling care   291, *292*,
        338–40
    see also siblings
On the Origin of Species
    (Darwin)   9
optimality models   44, *50*,
    51–3, 60–1, *61*, 73–4
    and ESSs   378–82
orang-utans   40, 218
ostriches, random scanning
    121, *123*
ovulation failure, lionesses
    6–7
owls, and prey alarm calls   358
ownership, and fighting   157,
    159, 165–6
oystercatchers, eggshell removal
    43

Paramecium reproduction   175
parasites
    and callers   257–8
    defence against   323
    and group living   128
    resistance to   200–1
    satellite males as   246–7
parental care
    and altruism   265–6, 291–2
    male   205
    and mating systems   208–43
    parasitoids   323–4
    and the provisioning rule
        308–9
parental effort   177, 183
parental investment   177 204
    see also parental care
partially bivoltine cycle   340–3
passerines
    European, clutch size   *21*, 21
    and song learning   37
paternity
    assessment of   277
    certainty of   201, 211, *298*
    measured by DNA   224–5,
        228
    protection of   229
    shared   237, *238*
    and subordinate males
        272–4
    see also DNA fingerprinting;
        matings, multiple
paternity markers   226
peacocks   190, 361, *362*
phalaropes   239
pheasants   190, *361*
    ring-necked   *362*
phenotype matching   277
pheromones   188, 335
philopatry model, benefits of
    303–4
phylogenetic trees   35, *36–7*
pigeon flocks   120, *122*
pigs, health of   195
plovers
    golden, in mixed flocks   *145*
    ringed, feeding   134, *135*
plumage
    and dominance   *171*, 171–3
    elaborate   190–91
    and habitat   26
    see also preening; tail length
Pogonomyrmex ants   352
Polistes apachus   337
Polistes metricus   180, 182,
    326–7, 336
Polistes variatus   180
*Politics and the English
    Language*, G. Orwell   3
polyandry   208, 237, 239
polygamy   25, 26, 38
    and sexual selection   *178*

polygyny   208, 226, 237, 242–3
    in birds   210, 230–4
    leks   206, 219, 220–5, 242
    resource defence   219, 231
    threshold model   232–3, *234*
polymorphism   80–1, 252, 255,
    258–9
    see also enzyme
        polymorphism
populations, genetic differences
    in   11–14
prairie dogs
    air-flow in tunnels   382–4
    black-tailed, alarm calling
        271–2
    ectoparasites   *134*
prawns   91
praying mantis   175
predation   34
    by leeches   187
    defence against   33, 38
    pressure of   312, 345
    prudent   92
    reduced by male aggregation
        222, *223*
    and selection   26–8
predation risk, and feeding rate
    68–9
predator–prey arms races
    77–101, 379
predators
    adaptations
        co-operation   265
        search image   80–1, *82–3*
    and prey calling   270, 357,
        366
    and prey social organization
        26–7, 29–30, 33
preening   359, *360*, 361
prey
    adaptations   79–91
    choice of   59–61, 63–4
    noxious   100
primates   243
    co-operation in   287–9
    communication   *360*, 371–3
    female dispersion   240
    female home range   218–20
    reproduction in man   *179*
    social organization of   31–41
Princess of Burundi cichlid fish
    299
Prisoner's dilemma model
    280–4, 290
promiscuity   *178*, 208
prosimians, male-biased
    investment   181
protective behaviour   358, *360*
protein   9
    provisioning rule   308–9
proximate factors   4–5
pseudopregnancy   305

puku 219

rabbits, life-dinner principle 92
rats, information flow 129
receiver psychology 368–9
reciprocity 280–9, 290, 293
recognition, kin 275–7
redirected attack 360
redshank, feeding habits 132
reed warbler 232–3, 243, 275
  as cuckoo host 95
regurgitation 285–7, 299, 321,
  351
rejection, of parasitic eggs 95,
  96–100
relatedness, coefficient of *see*
  coefficient of relatedness
reproduction 175–6
  asexual 331
  breeding constraints
    293–304
  and competition 107, 109
  conflict in breeding groups
    310–14
  cuckoldry *134*, 201, 211,
    226, 227, 229–30, *230*,
    252
  gametes 175–6, 211–12,
    328, 376
  in lions 5–8
  mating systems and parental
    care 208–43
  multiple matings 187,
    204–5, 225, 227
  sex change 260–3
  sex ratio 177–82, 183, 217,
    332–8, 340
  sex role reversal 239
  sexual competition and
    dimorphism 37–40
  sexual conflict 174–82,
    202–4, 234–8
  sexual selection *178*,
    183–202
  and testis size 40–1
  see also copulation; courtship;
    females; fertilization;
    males
reproductive decisions, in
    dungflies *55–6*
reproductive synchrony 5,
    219–20
resource defence 102, 105–6,
    110–18
  and mating systems 219, 231
resource depression *54–5*
resource value
  of females 168
  influence of 159–60
Rhytidoponera ants 337

ritualization 361–3
roaring, in deer 161, *162*,
    181–2
robin
  American 98–9
  European 115
Ropalidia marginata 344
Royal Jelly 322
ruff, multiple paternity 225
rumen, in moose 72

sacrifice, life 291
salamanders, fertilization 186
salmon, coho, male strategies
    249–52, 257, 259, *260*
sanderlings, territoriality *121*
sandpipers, spotted 239
satellite males 222, 246–7,
    257, *258*, 263
scent *see* smell
schools, guppies 144–6
scorpionflies (hanging flies)
  courtship feeding 188, *189*
  enforced copulation 202–3
sea horse 208
sea plantain 130–1
seals, elephant, reproductive
    rate *179*
search images 80–1, *82–3*
segregation distortion 375–6
self-sacrifice 291
selfish herd effect 90
sex allocation
  theory of 182
  see also sex change; sex ratio
sex change
  an alternative strategy
    260–3
  vs. sneaking 263
sex ratio 177–82, 183, 217,
    332–8, 340
sex role reversal 239
sexual conflict 2, 174–82,
    202–4
  dunnocks 237–8
  and polygamy 234–8
sexual dimorphism 34, *39*
  in body weight 37–8
  in tooth size 38–40
sexual selection 2, *178*,
    183–202
'sexy son' hypothesis 236
shearwater, manx, mate fidelity
    209–10
sheep, mountain 219
shell cracking 44–5, *46*
shoals 129, 130
shorebirds 221
  sex role reversal 239, 243
  social organization 132

shrimps 264
  mantis, bluffing 367–8
siamangs 218
siblings 266
  care of 266, 295, 324–5,
    329–30
  coefficient of relatedness
    267–8, 298, 346
  recognition of 275–7, 327–8
  see also offspring
signals *see* communication
singing *see* calling
size, body *see* body size; body
    weight
smell
  anti-aphrodisiac 186
  and communication 350,
    352, *360*
  family 328
  and information transfer 129
  in phenotype matching 277
snakes
  coral 87
  garter, food choice 11
sneaking 226, 227, 229–30,
    242, 244, 246, 250, 251,
    260, 261–2, 263
  see also cuckoldry; satellite
    males
social insects 269
  co-operation and altruism in
    318–48
  see also ants; aphids; bees;
    termites; wasps
social organization
  in African ungulates 28–9
  dominance in flocks
    139–40, 142–3
  living in groups 120–46
  primate 31–41
  weaver birds 25–8
  see also flocks; mating
    systems; social insects;
    territoriality
sodium, and moose 70–3
sound attenuation 353–5
sparrows 225
  chirruping 132
  Harris, plumage variability
    171, 171–2
  individual differences 144
  rufous-collared 352–3, *354*
species recognition 364
sperm competition 55, 56, 167,
    184–6, 204, 211, 229
  hypothesis 40–1
sperm displacement
    mechanisms 184–5
spermatophores 186, *203*
sphecid wasps *341*, 342–3
spiders, bowl and doily

167–71
squirrels 75
    feeding 66, 68
starlings 56–7, 75, 208, 230
    egg dumping 279–80
    flocking 125–6, 144
    load carrying 48–53, 74
    song 4
startle effect 79, 83
starvation, risk of 63–4
status signalling 171–3
sticklebacks 75
    colour morphs 245–6, 260
    feeding strategy 103–5
    and predation risk 68–9
    zig-zag display 359
stimulus pooling 222
stotting 368
strategies
    alternative 244–64
    fighting 102, 150–73, 380
suckling, communal, in lions
    5, 241
sunbirds
    golden-winged, resource
        defence 111–12
    scarlet-tufted malachite,
        plumage variability
        172–3
sunfish
    bluegill 75, 252, 257, 259,
        260
        feeding strategy 69–70
    cuckoldry 211
surgeon fish, blue tang 133
survival
    and genetic quality 194
    and reproductive effort 19
swallow bugs 136–8
swallows 197, 229, 230
    barn 198–201
    cliff, coloniality 135–8
    elongated tail feathers 367
    long-tails and female
        preference 198–201
sweat bee 327
synchrony
    in cicadas 125
    reproductive 5, 219–20

tactics 245
tail length 190, 191, 192–3,
    369
    swallows 198–201
tamarins 218
tandem running 351
taxonomic independence,
    problem of 34–5
temperature, and flock size
    139, 141

termites 319, 324, 330–1, 346
terns, courtship feeding 188
territories 217, 219
    inheritance of 296–8
    optimal size 113–14
    and parental care 212
    polyterritoriality 236
    quality of and breeding
        105–6, 303–4
    resource defence 110–18,
        130
testes, in primates 40–1
testosterone 172
Thompson's gazelle 122, 368
threat displays 361, 363, 368
threat signals, variability of
    369–71
time budgets 138–43
tit for tat co-operation 282–4
tits
    black-capped chickadee 66
    blue, brood manipulation 20
        extra-pair paternity 243
    great 75, 105–6
        body mass fluctuation 65
        brood size 16–18, 20
        and cryptic prey 84–6
        natal dispersal 240
        prey choice 60–1, 62
        song of 353
        territoriality 117–18, 160
        threat displays 361
    long-tailed 14
    marsh 75
    willow 66
toads 243
    assessment by 161, 163
    Brazilian 78
    common 161, 163
        explosive breeders 217
    croaking 161
    natterjack, alternative
        strategies 247, 248, 260
tooth size, sexual dimorphism in
    38–40
topi 219, 221
trait groups 376–7
Trigonopsis cameronii 325
turkeys, health of 195

ultimate factors 4–5
ungulates 221
    African, social organization in
        28–9
    lekking 219
utethesia ornatrix, courtship
    gifts 188

variance, and taxonomic levels
    34–5

variation 9
vegetation, harvesting of
    130–1
vertebrates, cf. social insects
    344–6
vigilance 68
    in groups 120–2
voles, grey-sided, female
    dispersion 215–16

wagtails, pied
    mutualism 278
    territory sharing 116–17
walrus 221
war of attrition 147–50
    asymmetric 165–7
warblers
    migratory behaviour 12–13
    reed 95, 232–3, 243, 275
    sedge, songs of 190–1, 192
    Seychelles 303
wasps
    digger 323
        female nesting strategies
            254–6, 260, 380–1
        great golden 326
    figwasps 157, 252–4, 260
    mate competition 180–1
    nest sharing 325, 326
    paper, nesting 326–7
    parasitoid 322–4
    partially bivoltine cycle 341,
        342–3
    sex ratio 180, 180–1
    social 344
    see also Hymenoptera
water bugs 206
waterbuck 219
waterfowl 206
weaver birds
    information flow 128–9
    social organization in 25–8
weight, body see body weight
whelk shells, cracking of
    44–5, 46
widow bird, long-tailed 190,
    191
wolves, territory urination 358
woodpeckers
    acorn
        conflict in breeding groups
            312–14
        helping in 309–10
    downy 75
        optimal sampling 63
    worker–queen conflict 332,
        334–8
    worker–worker conflict 339
worms, acanthocephalan,
    copulation

prevention 185−6
wrasse, blue-headed
  determination of spawning
    sites 216
  sex change 260−2
wrens
  fairy 303
  marsh 232

stripe-backed 316
Wytham Woods 240
  great tits 16−18, 105−6

Xiphophorus, tail lengths in
  369
Xylocoris maculipennis,

fertilization 186

young, stealing of 124

zebra
  Burchell's 220
  Grevy's 219